J. Durda

ONT

A Sierra Club Naturalist's Guide to

THE PIEDMONT

by Michael A. Godfrey
Foreword by Albert E. Radford

SIERRA CLUB BOOKS *San Francisco*

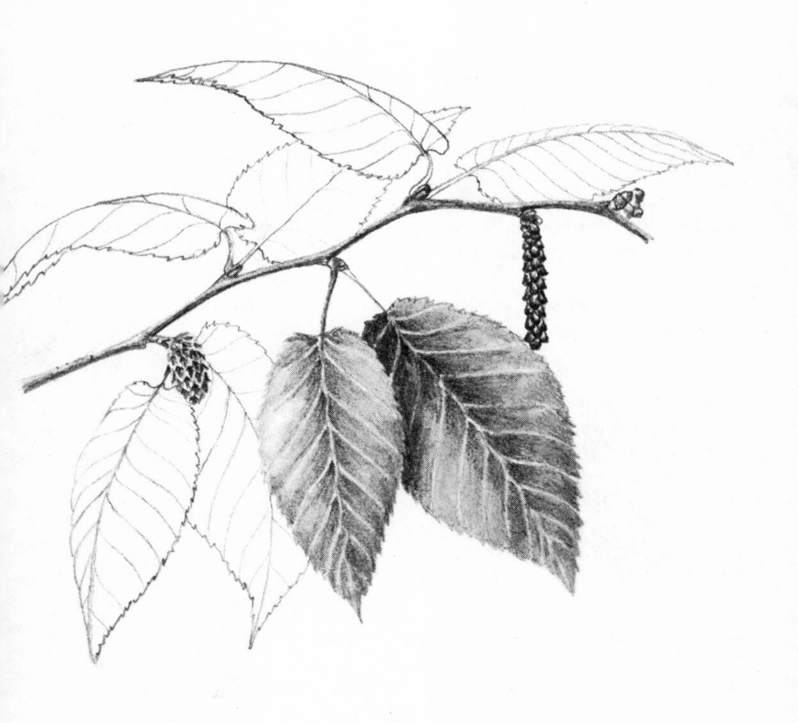

The Sierra Club, founded in 1892 by John Muir, has devoted itself to the study and protection of the earth's scenic and ecological resources—mountains, wetlands, woodlands, wild shores and rivers, deserts and plains. The publishing program of the Sierra Club offers books to the public as a nonprofit educational service in the hope that they may enlarge the public's understanding of the Club's basic concerns. The point of view expressed in each book, however, does not necessarily represent that of the Club. The Sierra Club has some fifty chapters coast to coast, in Canada, Hawaii, and Alaska. For information about how you may participate in its programs to preserve wilderness and the quality of life, please address inquiries to Sierra Club, 530 Bush Street, San Francisco, CA 94108.

Library of Congress Cataloging in Publication Data

Godfrey, Michael A 1940–
A Sierra Club naturalist's guide to the Piedmont.

Bibliography: p.
Includes index.
1. Natural history—Piedmont region—Guide-books.
2. Piedmont region—Description and travel—Guide-books. I. Title.
QH104.5.P54G52 574.974 79-22328
ISBN 0-87156-268-5
ISBN 0-87156-269-3 pbk.

Design by Drake Jordan based on a series design by Klaus Gemming

Printed in the United States of America

10 9 8 7 6 5 4 3 2 1

To my father
Arthur M. Godfrey
a Piedmont farmer

TABLE OF CONTENTS

FOREWORD

THIS NATURALIST'S GUIDE to the Piedmont of eastern North America exhibits the craftmanship of a keen observer and a perceptive and sensitive author richly versed in field knowledge of the area. His experience extends from early boyhood rambles in the fields and woods of northern Virginia to extensive travels in recent years throughout the province. Michael Godfrey has judiciously combined facts gleaned from the literature and from reliable consultants with years of careful observation to produce a readable, enjoyable, and accurate presentation of the natural history of the region. As a writer he has the knack of breathing life into dry facts and making vibrant the natural dynamics of the Piedmont. Many of his delightful word pictures are artistic gems, created without loss of scientific accuracy.

Mr. Godfrey has woven physiography, geology, topography, hydrology, biology, climate, and the successional story into a coherent and scientifically sound presentation of the ecological diversity of this highly disturbed region. His descriptions of the thistles, hayfields, pastures, dung piles, and woods will stimulate all those interested in the out-of-doors to have a look for themselves—to gain insight into some of the untold stories in the webs of life. The mesosere–xerosere–hydrosere approach provides a sound basis for comprehending the biological activities and diversity throughout the entire Piedmont. The state-by-state treatment of natural history at the end of each chapter makes the entire book more meaningful for the beginning and experienced field enthusiast and for individuals who may be restricted in travel. The author has been most successful in making us aware of every organism's dependence upon its environment, of the importance of the food chain and trophic levels, of predator–prey relationships, of plant–animal interactions, of community significance, and of the nature of succession and climax.

Mr. Godfrey describes the biologically unique natural areas that remain in the province and the endangered and threatened endemic species that occur in critical localized habitats. Discussion of the outstanding sites of national landmark significance completes his treatment of the Piedmont landscape. For the protection and conservation of our natural heritage, all readers are enjoined to comply with the author's advice in the introductory paragraphs of the last chapter.

This book should stimulate everyone—children to senior citizens—to observe the wondrous diversity and complexity of the flora and fauna of lawn, barnyard, garden, weed patch, woods, pools, and streams. It should provide fresh impetus for residents of both urban and rural Piedmont to really look at and appreciate their surroundings. Hopefully, it may even induce some to leave the TV tube and passive spectator sports for the active mental, physical, and spiritual rejuvenation of learning about their natural environment. Finally, I believe this book will bring to all of its readers an increased appreciation of our natural heritage and a sincere desire to help conserve and protect it for future generations to enjoy.

Albert E. Radford
Chapel Hill, North Carolina

ACKNOWLEDGEMENTS

I GRATEFULLY ACKNOWLEDGE my indebtedness to Dr. Albert E. Radford of the University of North Carolina Department of Botany whose writings and personal counsel shaped the content of this guide. Dr. Radford's work in the natural history of the Piedmont has materially advanced the scientific and popular understanding and has resulted in the preservation of biologically priceless tracts in the Piedmont and elsewhere.

My friend and field companion Stanley Alford provided primary guidance on the reptiles and amphibians and valuable advice on the birds and mammals. Dr. Elizabeth McMahan of the University of North Carolina Zoology Department was helpful with the insects and arachnids; I am thankful to her and to the noted lepidopterist and alpinist, Neil A. Stephens. It was my good fortune to have an extended dialogue with Dr. Robert K. Peet of the UNC Botany Department who is presently doing research on old-field succession in the Piedmont. For their patient assistance in identifying plants and animals, I am grateful to Drs. Jimmy Massey and Robert Wilbur, curators, respectively, of the UNC and Duke University herbariums; to Willie Koch of the UNC Botany Department; to Ray E. Ashton, Jr. and William M. Palmer of the North Carolina Museum of Natural History who helped with reptiles, amphibians and aquatic forms; to Angello Caparella whose versatility as a field naturalist yields a continuum of insights; to Dr. Anne Lindsey and to Dianne Wickland and Debbie Otte who fielded numerous inquiries on plant ecology. Dr. William A. Whyte and Dr. James R. Butler of the UNC Geology Department led me to a rudimentary understanding of Piedmont geology. Librarians throughout the Piedmont led me unerringly to the needed references no matter how obscure; some who were repeatedly helpful are Sue Bagnell of the UNC Zoology Library, William Burk and John Darling of the UNC Botany Library, and Kathleen Georg of the Gettysburg National Military Park Library. Dr. Helmut Mueller of the UNC Zoology Department has for some years been a principal advisor to me on raptors and his experience with birds of prey in the Piedmont was particularly helpful. I thank Janet Stephens for her patient help with the field work and typing, Dwight Stephens for giving me intimate looks at the xeric microcommunities on the Piedmont's vertical rock faces, and Jill Alznauer for feeding my assortment of injured birds and animals while I traveled in the Piedmont. I am particularly grateful to those who granted me welcome to their Piedmont lands.

Introduction

THIS IS A GUIDE to the geology, climate, and life systems of the Piedmont.

Chapter One defines the Piedmont geologically and outlines the geologic history from which the present contours result. A map of the Piedmont in three sections shows the major geologic features, roads, and cities.

Chapter Two describes the climate of the Piedmont and introduces the concepts of moisture and temperature effectiveness. Included is a discussion of the region's climatic history, with emphasis on the influence of climatic changes on the Piedmont's life forms.

Chapter Three examines the life systems on the cultivated lands of the Piedmont, primarily hayfields and pastures, two grassy habitats distinguished principally by the height of the vegetation. Microhabitats associated with hayfields and pastures are explored, with emphasis given to the woody, fence-line hedgerows.

Chapter Four introduces the concepts of primary and secondary plant succession. The latter is of particular importance in the Piedmont because of the vast acreage of abandoned farmland presently cloaked in successional vegetation. The social and economic causes of this abandonment are sketched. Also covered are the distinctions in drainage (the degree of moistness of the soil) so important in determining the types of vegetation that will flourish on a given piece of land.

Chapter Five describes the *mesosere*—plant succession (and attendant animal populations) on soils of moderate moisture content. Succession is traced from the appearance of herbs and grasses, through the establishment of woody plants, to the growth and—if succession is allowed to continue undisturbed—eventual climax stabilization of the forest.

Chapter Six describes the *xerosere*—succession on dry soils—with emphasis on the special stresses affecting plant selection and growth.

Chapter Seven describes the *hydrosere*—succession on wet soils and associated land and aquatic habitats. Alluvial soils, which are low-lying, and therefore hydric, are distinguished from the residual soils that support mesic and xeric regimes. The two distinct types of forest that grow on the Piedmont's alluvial soils—alluvial forest and swamp forest—are discussed. The description of aquatic habitats is divided between life systems in flowing (lotic) waters and those in still (lentic) waters. There are no natural lakes in the Piedmont, so the discussions of lentic waters are limited to man-made and beaver-built lakes and ponds. Also traced is the process of organic and mineral sedimentation which begins as soon as these impoundments are constructed and ultimately results in the building of hydric soils where once there was standing water.

Chapter Eight locates and describes some of the Piedmont's rare botanical communities. Many of these are National Natural Landmarks or have been nominated to the National Park Service for that status. Some were chosen for inclusion here because they are exceptional representations of the successional habitats described in previous chapters (for example, the Hutcheson Memorial Forest, at Millstone, New Jersey, which represents the climax of the mesosere in the northern Piedmont). Others are examples of unusual or specialized habitats (such as the serpentine barrens) not covered in the successional material. Still others are included for their uniqueness (among these are the Great Swamp, in Morris County, New Jersey, and the James River Arborvitae Bluff, in Buckingham County, Virginia). Because some of these sites represent extremes of dryness or wetness, it is important for the reader to be familiar with the general discussions of the xerosere and the hydrosere presented in earlier chapters.

Approaching the Piedmont

The Piedmont has been occupied by European man for nearly three centuries. Wilderness nestles in a few coves and bottomlands, but all but a few of our 85,000 square miles are or have been under human management.

Recognizing the impact of human occupancy on the character of the Piedmont, this guide is divided into segments detailing the life processes on active farmland and on lands previously farmed or otherwise vegetatively exploited but now abandoned to the forces of succession. I estimate that the two categories account for at least 95 percent of the Piedmont. The remaining segment is composed of diverse and widely scattered remote places and extremes of topography which have discouraged exploitation.

More than half the Piedmont is abandoned farmland in some stage of reforestation by the process of plant succession. Characteristic vegetation dominates at each phase of plant succession and determines the animal communities that are present. The vegetative mix is dictated by the nature of the soil, the most important property being the soil's ability to retain water. Moistness and successional advancement, then, are the two most important variables in defining the Piedmont's habitats. The material on succession in this guide is segmented into arbitrary ranges (wet, well-drained, and dry), and successional communities are described for each condition.

Use of This Guide

The Sierra Club Naturalist's Guide to the Piedmont can be used most effectively by first making some assessment of the moistness of a site being explored. Topography is usually the best key. Low, flat places are wet; moderate slopes are moist; and severe slopes and rocky hilltops with obviously thin soils are dry. Having estimated a site's moistness, the user is then in a position to turn to the material on the appropriate successional phase by estimating the maturity of the vegetation. (For example, a formerly cultivated hillside now dominated by briars, cedars, and broomsedge can be seen to be mesic; i.e., in the midrange of the moistness scale, and in the early woody phase of succession.)

Estimating moistness and successional maturity are important skills a field naturalist must devleop, particularly in a thoroughly manipulated landscape such as the Piedmont. With practice these skills, coupled with some appreciation of the properties of the underlying rock, will gradually yield a systematic means of dealing with the parquetry of contrasting communities in the Piedmont, and indeed in most of eastern North America. One can watch birds, brutalize butterflies, name plants, or follow tracks in the snow and still know nothing of the *community* in which he stands, of the interrelated lives of soil, plants, and animals. It is the intent of this book to help the reader to understand the workings of such communities, an approach that I hope will also be useful in refining his personal system of comprehending the land he lives on.

Plants

In the discussions of the habitat types, descriptions are given for the major plants characterizing the habitat. Generally these are the plants which dominate the vegetation at particular seral (successional) stages or which are prominent in the subdominant layers (such as the shrub and herbaceous strata of a mature forest). More than 3000 species of vascular plants grow in the Piedmont, and space permits us here to describe relatively few of them. In most cases those selected are the more important plants in their habitat from a life-systems standpoint; that is, they contribute most to the nutritive base of the habitat.

Animals

The principal forms of animal life associated with each habitat are described in sections consistent with normal taxonomic grouping (i.e., insects, arachnids, reptiles, amphibians, etc). The groups are arranged according to their approximate trophic rank (feeding order, or distance in the food web from the primary producers, green plants) rather than in the taxonomic sequence recognized by zoologists. Tables A-1 through A-5 in the appendix list the animals according to taxonomic rank and describe briefly the feeding and habitat preferences, behavior, and ranges in the Piedmont. Tables A-2, A-3, and A-5, the listings, respectively, of the reptiles, amphibians, and mammals, I believe are exhaustive of animals of these categories normally found in the Piedmont. Tables A-1 and A-4, which cover the lepidoptera and birds, list the most prominent species in the province from the standpoints of observability and trophic importance.

Names of Plants and Animals

The appearance on the printed page of an organism's Latin name can render a passage unintelligible to the nonbiologist and leave him or her wondering why the author doesn't stick to simple English. Conversely, the trained biologist soon suspects that any work making repeated reference to organisms by their folk or common names may be courting inaccuracy. In this guide the Latin scientific name is provided in conjunction with a common name (when available) at the first reference to the organism in each chapter. Thereafter the standardized common name is used without the scientific name.

It is helpful to distinguish between the types of nomenclature in the naturalist's lexicon. A common name is officially sanctioned for many (unfortunately, not for all) organisms by some group of interested scientists and laymen. The American Ornithologist's Union, for example, establishes the standard nomenclature for birds, and it recognizes the name "bobwhite" for the quail known so well throughout the Piedmont. "Partridge," one of several folk terms for the bird, has a poetic appeal but can lead to confusion because the same term is applied in New England to the ruffed grouse, while officially it is reserved for a highly specific taxonomic group of Old World sand grouse. The perils of using common names alone, without the scientific name, can be readily seen.

Scientific names follow standards established by Karl von Linne, the eighteenth-century Swedish botanist who, after latinizing his own name to Carolus Linnaeus, fathered the science of taxonomy (the classification of life forms) and gave Latin names to many of the life forms then known, including *Colinus virginianus*, the bobwhite beloved of the Piedmont. The Linnaean scientific names are composed of the name of the genus *(Colinus)*, which is always written with the initial letter capitalized, and an epithet to denote the species, not capitalized *(virginianus)*. When, in the opinion of taxonomists, further differentiation is necessary, a term for subspecies is added. We, for example, are *Homo sapiens sapiens*, a different animal from *Homo sapiens neanderthalis*, a now-extinct race of the same species. The terms "race" and "subspecies" signify equivalent taxa, although the term race, when applied to humans, has loose ethnic connotations different from the more rigid taxonomic usage. For plants the subspecific taxon is called *variety*, abbreviated *var*.

In the Field

A few cautions are in order for the user of this guide afield in the Piedmont.

First, ask before you enter private lands. Landowners are usually generous in granting entry when they meet their guests face-to-face and have the purpose of the visit explained. They can be notably disagreeable when welcome is presumed. Give the landowner a chance to gauge your intent and to explain about the presence of dangerous livestock and the like. When granted permission, treat the property as you would your own. In particular, don't climb wire fences. It bends them down. Instead, cross by going under the fence, through strands of barbed wire, or, better, go through a gate.

The problems facing the hiker in the Piedmont are few and are nearly negligible when approached with respect and caution. Comfort can be enhanced by appropriate dress, a matter best taught by experience. The table below lists the principal problems in the order of their likely severity and suggests a remedy for each.

Problem	*Remedy*
1. Thirst	Dehydration is not likely to become critical in the course of a day's outing, but it can be uncomfortable. In warm weather, particularly, carry a container of water.
2. Mosquitos	Around streams, lakes, alluvial and swamp forests, wear loose-fitting long pants and long-sleeved shirts. Insect repellants (such as Cutter's) with the active ingredient N,N-Diethyl-metatoluamide are effective.
3. Chiggers	Scrub with soap and warm water within one hour of exposure. Immediately upon returning from a hike is usually (not always) soon enough.

Problem	Remedy
4. Ticks	Where mosquitos are not a problem, wear short pants, the better to detect and remove ticks as they ascend. If a tick becomes embedded, a hot needle (stuck into the tick) may make it release. Remove upon discovery by whatever means necessary.
5. Hornets	Avoid nests. Run as fast as possible if attacked. Slap hard at hornets as soon as they alight. To end an attack, it is necessary to vacate the territory (200 yards or more) and kill the individuals which persist. An initial burst of speed is the best defense—alertness is the best preventative.
6. Lightning	Do not take shelter under tall trees and do not stand on exposed prominences during thunderstorms.
7. Saddleback caterpillars	Alertness avoids this attractive caterpillar's painful touch.
8. Snakebite	Rare in the Piedmont, the generally docile copperhead being the only venomous snake present in significant numbers. Rarely attacks unless provoked. Stay alert. Do not handle any snake until it is identified and remember that some nonpoisonous snakes are ill-tempered and can bite painfully. If bitten by a copperhead, get to a physician as quickly and calmly as possible—few encounters are fatal. Do not attempt amateur surgery.

PART I

The Piedmont Defined

CHAPTER ONE

Geography

THE EASTERN MARGIN OF THE PIEDMONT traces a gentle, open S from New York City southwestward to Montgomery, Alabama connecting an arc of urban centers which includes five state capitals,* America's largest city, and the national capital. The section of this arc between New York and Washington has grown into a single, unbroken sweep of urbia. Southward, major metropolises alternate with farm towns along this recurvate crescent that could be called the cradle of American culture.

The alignment is, of course, no accident. The colonists, after setting up seaports at the mouths of the major rivers, sailed upstream as far as they could and founded commercial centers at which to gather the surpluses of what was then known of the continent's interior. These outposts were built where the sluggish rivers of the Coastal Plain reach the steps of crystalline rocks that mark the rivers' navigable limits. On each Atlantic-bound stream, rapids and, in some cases, falls abruptly halted all but the most adventurous traveler at a front which early boatmen called the "fall line."

At the fall line the flatlands of the Coastal Plain give way to gentle, rolling hills. "The west," as this new hilly land was then known, was in places rocky. Its reddish, clayey soils contrasted with the dark sandy loams of the Coastal Plain. The privileged were granted the preferred alluvial lowlands which followed the new region's rivers (the Delaware, the Susquehanna, the Potomac, the James, the Roanoke, and the Savannah). Along these watercourses they built their stately manors and trotted their fine mounts. The less well-to-do had to follow their oxen in the uplands, where by the middle of the eighteenth century their settlements reached westward to a landscape where towering domes and spires spiked the rolling uplands, halt-

*Trenton, Richmond, Raleigh, Columbia, and Montgomery.

ing ultimately at a solid front of mountains—hostile then, untillable still. The isolated hills (monadnocks) preceding the wall of ridges in the west of the region reminded settlers of the Piedmont or foothill sections of southern Europe (the Italian *Piemonte* means 'foot of the mountain'). In time, the entire upland province bounded by the fall line on the east and the mountains on the west became known as the Piedmont. The reference proved durable.

In the 1930s the U.S. Coast and Geodetic Survey commissioned studies which yielded precise geologic definitions of the nation's physiographic provinces. These studies substantiated what the first inland travelers had observed about the soil and topography: that landforms in the region typically progress inward from beach to tidal marsh to coastal plain to the rolling, clayey Piedmont and the steep, rock-strewn Blue Ridge Mountains.

Nevin M. Fenneman was the USCGS's man on the scene. After making his descriptions to the government, Fenneman wrote geologic texts which remain definitive on the physiography (the study of the land's physical features above and below the surface) of the United States. His descriptions are lucid and precise. They offer us a deeper identity, a suddenly-discovered old friendship with the land on which we live.

Fenneman viewed the Piedmont as the easternmost of four physiographic provinces comprising a region collectively called the Appalachian Highlands. To the west of the Piedmont (of its southern two-thirds, at least) lie the Blue Ridge Province, the Valley and Ridge Province (also called the folded Appalachians), and the Appalachian Plateau Province.

The Piedmont is underlaid by a bed of crystalline rocks (largely granite) of Precambrian and Paleozoic age (400 to 600 million years). Immediately below the surface of the adjacent provinces are rocks of very different origin. This helps to explain the Piedmont's distinctive features: the presence of rapids along the rivers at the fall line, its soil types, the monadnocks (isolated hills of resistant rock rising above a peneplain, an extensive area that owes its overall low relief to erosion), and the mountainous scarp that looms above the western edge of the Piedmont.

Geologists emphasize that it is the rocks beneath the surface and not the lay of the land which demarcate the provinces, though in many places these subcutaneous formations offer striking contrasts in topographic relief. Witness the abrupt change in the landscape at the fall line at Havre de Grace, Maryland, where the Susquehanna has cut through the rocks of McGees Fault. Today the Susquehanna's rapids are three miles (5 kilometers) upstream of the fault at Port Deposit. Below the rapids' original location at the fault is the flattened Coastal Plain; upstream, precipitous hills.

In places the mountain boundaries of the Piedmont are indistinct to observers on the surface. Traveling west from Washington, D.C., on Route 7, for example, we reach the Catoctin Ridge at Leesburg, Virginia. This ridge rises abruptly several hundred feet above the peneplain and gives the impression that one is leaving the Piedmont and entering the Blue Ridge. South of the Potomac River, however, the Catoctin Ridges are separated from the Blue Ridge by ten to fifteen miles of peneplain, so it is not until we reach Bluemont on Route 7 or Hillsboro on Route 9 that we confront the Blue Ridge scarp. North of the Potomac the Catoctins lie hard against the scarp, and to avoid confusion Fenneman gerrymandered them out of the Piedmont and into the Blue Ridge. It is a geologic inconsistency for which the modern naturalist is grateful because there is no readily apparent difference in the life schemes of the Catoctins and the eastern expressions of the Blue Ridge north of the Potomac. Driving Route 15 in the vicinity of Frederick and Thurmont in Maryland, we see the inner limits of the Piedmont—the Catoctins—rising as an obvious boundary immediately west of the road.

The Piedmont's present surface is a vast plain of rolling knolls and hillocks, dissected gently by minor streams, more boldly by the creeks and rivers. The geologic term for such a landscape is peneplain, a plain not yet worn entirely smooth by the ineluctable agents of erosion. The province stretches southwest from the Palisades along the Hudson River for 1000 miles (1600 kilometers) to the Black Belt of central Alabama. The peneplain widens across Delaware and southeastern Pennsylvania, constricts to 25 miles (40 kilometers) at Washington, D.C., then expands again to a maximum breadth of 125 miles (200 kilometers) in North Carolina. Elevations tend to be lower in the northern

Piedmont for two reasons: the plane of the bedrock dips toward the northeast, and the typical Piedmont crystalline foundation is replaced there by Triassic sediments of lesser resistance to erosion.

Overall, the slope of the province is quite gentle. As a consequence, erosion, at least in recent geologic times, has not been rapid. Under a mantle of lush vegetation the rock beneath the Piedmont peneplain has had time to weather into saprolites (decomposed rock lying in place) which are readily available for incorporation into the soils by the life processes at the surface. Hence, most of the soils in the province are termed residual, meaning they were formed in place by the mixing of underlying saprolites with decayed organic matter, rather than formed elsewhere and transported by the enterprises of water and wind. Due to the composition of the parent rocks and saprolites, Piedmont soils are clayey. Their nutritive value varies locally from splendid to squalid and is typically reflected in the vigor of the plant societies dwelling at the surface.

The Piedmont soils and substrates differ fundamentally from those of adjacent provinces. East of the fall line the crystalline bedrock, dipping seaward, bears a wedge of sediments which have washed off the Piedmont, and perhaps provinces to the west as well, mainly during the Cretaceous era, 70 to 90 million years ago. These sediments have accumulated to a depth of 11,000 feet (3350 meters) at Cape Hatteras, forming a clastic wedge ("clastic" means composed of the transported fragments of older rock) which feathers onto the crystalline bedrock along a mild gradient called the fall zone. At intervals, the Atlantic has advanced and retreated across the Coastal Plain, with the result that most of the land east of the fall line has once, or perhaps repeatedly, been what geologists call "high energy beach." The sledge-hammer surf pulverized, and the sea dissolved, all but the most obdurate materials, mostly quartzite—in a word, sand. Consequently, the Coastal Plain is covered by sandy soils over relatively young sedimentary rocks, which in turn cover the same crystalline basement that underlies the Piedmont. The sediments deepen seaward.

It is often difficult to separate the Piedmont from the Coastal Plain. Viewing the Piedmont's eastern margin from, say, Roanoke Rapids, a fall-line town named for the rocky limits to upstream travel on the Roanoke River, we find the fall zone hills undulating into the Piedmont and the river becoming gradually impassable. Fenneman's advice on distinguishing between the provinces is to look at the soil. Feel it. The tops of some fall zone hillocks are sandy—Coastal Plain by definition—while the slopes below might well be clayey, and therefore Piedmont. This apparent paradox is the result of erosion cutting into the feathered margin of the Coastal Plain.

The Piedmont's eastern boundary shows chaotic irregularity. It changes perceptibly, perhaps, in so brief a time as a human life span. The two provinces extend fingers, whole hands into one another along the fall zone hills. Islands of Coastal Plain lie deep within the Piedmont and vice versa. Only occasionally while traveling a southern fall-line road do we come upon a prominence affording a definitive prospect of the Coastal Plain falling smoothly away to the sea. Such perspectives are more readily gained north of the Potomac.

Fenneman placed the western boundary of the Piedmont "approximately where the metamorphic rocks of the Piedmont end against upthrust Precambrian formations," again a geologic definition. The western frontier is in places more complex than the fall line, and we must find means of making the personal interpretations to separate the Watchungs from the Ramapos (described below), the Catoctin from the Blue Ridge, the monadnocks which spike the inner southern Piedmont from the foothills of the mighty Smokies. It is helpful to fix in mind the major features of the western margin, especially those of the sections we live near or plan to explore.

The northern limit commonly proposed for the Piedmont is the Palisades, which form the west bank of the Hudson River. Some Piedmontologists (it's always a treat to coin a word) urge that Manhattan Island and some of adjacent Connecticut be included. But the northern boundaries are complex enough where they are not controversial, the Wisconsin Glacier having trifled with the courses of the Passaic and Raritan rivers, scooping out the Great Swamp and dumping an irregular terminal moraine into the Piedmont's northern border zones. So let us use the magnificent trap rock scarp of the Palisades ("Trap rock" is rock trapped between two layers of different rock, in this case intrusive material trapped between layers of Triassic sediments.) as the Piedmont's northern limit. Additional trap rock layers appear as the Watchung Mountains just west of Paterson and Montclair and Plainfield, New Jersey. We get into the mountains, and out of the Piedmont, west of Suffern and Boonton and Morristown, New Jersey, where the Reading Prong of the New England (physiographic) Province, called the Ramapo Mountains, rises above the Triassic lowlands of which the Watchungs are geologic constituents. The Reading Prong serves as the Piedmont's western boundary as far south as its terminus at Reading, Pennsylvania. A simplified sketch of these features is provided in Figure 1.

South of the Reading Prong and east of the Susquehanna River there is a section where no mountains separate the Piedmont from the Valley and Ridge Province. To further obscure the boundary here, the Great Valley lies to the west of the Piedmont in this gap so that we see lowlands to the west of such Piedmont upland prominences as South Mountain, Pennsylvania and Welsh Mountain. The city of Harrisburg lies approximately at the Piedmont's inner limit in this intermountain hiatus.

In southern Pennsylvania the Blue Ridge Mountains serve as the western provincial border for their entire length, yielding that function in northern Georgia to the Cartersville Fault. At some locations the Blue Ridge rise abruptly above the peneplain, presenting a formidable and unmistakable scarp known as the Blue Ridge Front. Elsewhere, high hills in the western Piedmont rise to heights approximating those of the Blue Ridge, obscuring the boundary. The key to locating the boundary of the Piedmont here is the peneplain. The Blue Ridge Mountains are continuous. That is, once they begin, the relief between the ridges never returns to that of the peneplain. So from a vantage in the mountains overlooking the Piedmont, say at Peaks of Otter, west of Lynchburg, Virginia, we look eastward for the first (westernmost) expressions of the low relief of the Piedmont's rolling peneplain. Any prominences rising from the peneplain are not Blue Ridge peaks but Piedmont monadnocks.

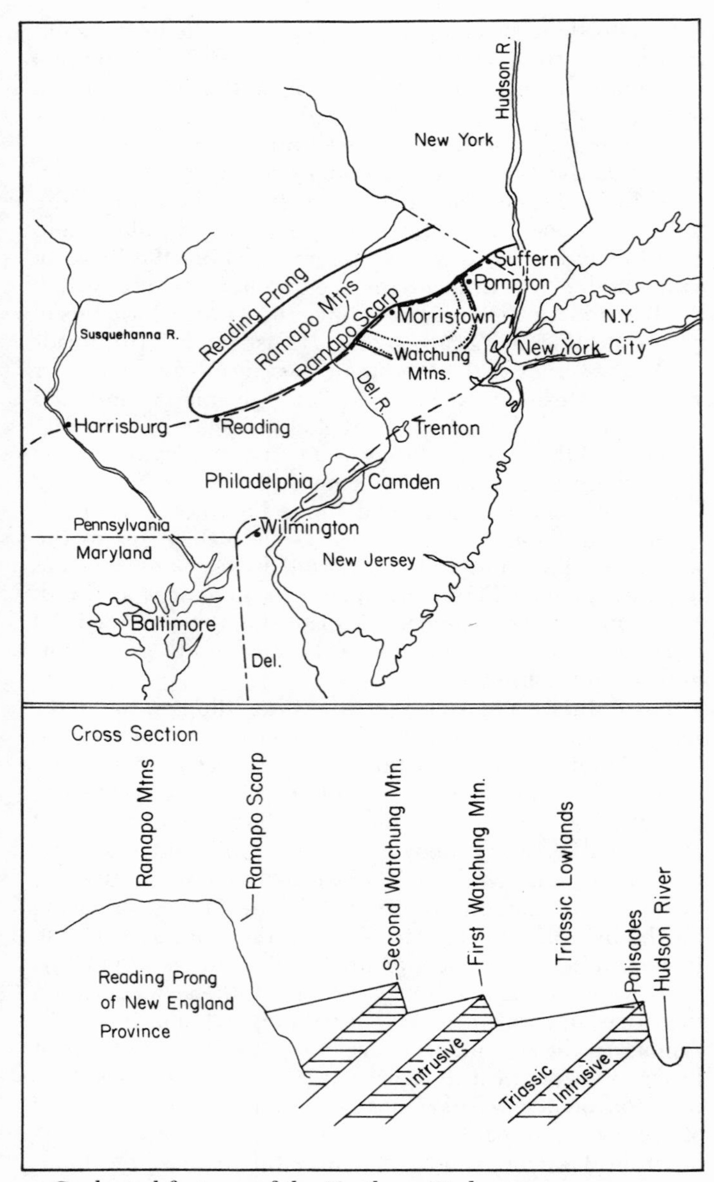

1. Geological features of the Northern Piedmont

Beneath the Surface

The slope at the fall line is steeper than that of the Piedmont to the west or of the Coastal Plain to the east. A cross-sectional sketch (Figure 2) shows the relationship of this slope to neighboring structures and reveals how geologists interpret this gradient in terms of the Piedmont's history and present topography.

Figure 2, a stylized and generalized representation most applicable to the southern two-thirds of the Province, depicts the ancient granites of the Blue Ridge thrust abruptly up through the somewhat younger rocks of the Piedmont. The rocks of the Blue Ridge are clearly the more resistant to wear. Along much of the boundary a conspicuous scarp separates the Blue Ridge and the Piedmont rock, suggesting that faulting as well as differential erosion may account for some of the difference in the elevations of the two provinces.

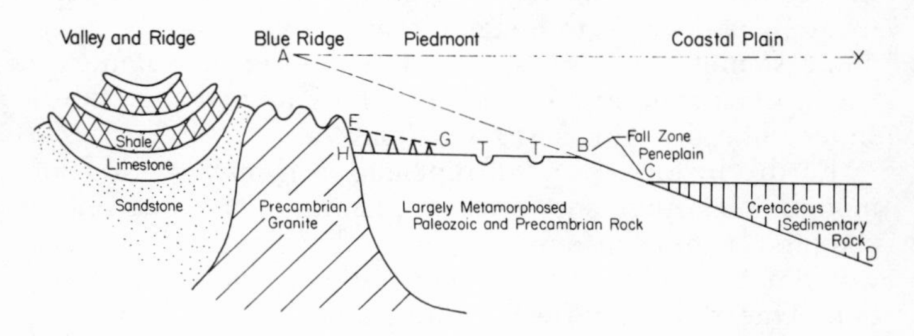

2. The peneplain and the erosion cycle

The fall zone peneplain is exposed at BC. Eastward (CD) it is buried beneath sediments carried off the Piedmont and the Blue Ridge. Geologists reason that this peneplain once extended westward (perhaps as far as A) and upward above the highest present elevations in either province. This ancient peneplain probably had a gentler slope, say AX, until deformation of the earth's crust tilted it to its present slope of ABCD.

More recent erosion cycles (uplifts of the earth's crust resulting in rapid erosion) are represented by the monadnocks in the western Piedmont, the tops of which are the remains, or perhaps roots, of what geologists call the Schooley Peneplain (FG), and by the Piedmont's present surface, which bears the geologic reference of the Harrisburg Peneplain (HB). Peneplains more recent still are the Triassic basins (T) etched into the softer materials laid down on the Piedmont's crystalline bedrock in Triassic times but eroded more recently, perhaps as late as the Pleistocene, to form the Piedmont Lowlands. These depressions include the Newark, Gettysburg, Culpepper, Richmond, Danville, and Deep River basins.

What is clear from the present profile of the Piedmont and its adjacent provinces is that the Piedmont and the Blue Ridge were once of much grander stature. The vast clastic wedge which we call the Coastal Plain and continental shelf once rested on the present remnants of these older uplands. And the fall zone peneplain? What does it imply about mountain ranges older still? Peneplains are graded from mountain ranges over hundreds of millions of years; the Himalayas will one day be ground to Piedmont-like contours. It is possible to imagine that the same gentle Piedmont, so meticulously manicured by the Amish in Pennsylvania, once sent spires of ice and rock toward the stratosphere like the Alps and the Himalayas.

Earth's history is one of repeated orogenies (periods of mountain raising) and erosion planings. The crags and couloirs looming in our fancy over the present Piedmont would not even have been the first of their kind at this location. One billion, three hundred million years ago, in a Herculean convulsion known as the Grenville orogeny, there was born a belt of mountains as grand as any on earth today. They towered over the lifeless land (a billion years would pass before the first plants would root ashore) from Nova Scotia westward to Texas and Arizona. Remnants of these masses are alleged to be visible at some locations as far south as Alabama. Because there was no vegetation to hold the rock in place as the chemistry of time and weather worked its decomposition, no saprolites or soils were formed. Erosion sluiced the mighty Grenvilles to the sea in a scant 150 million years.

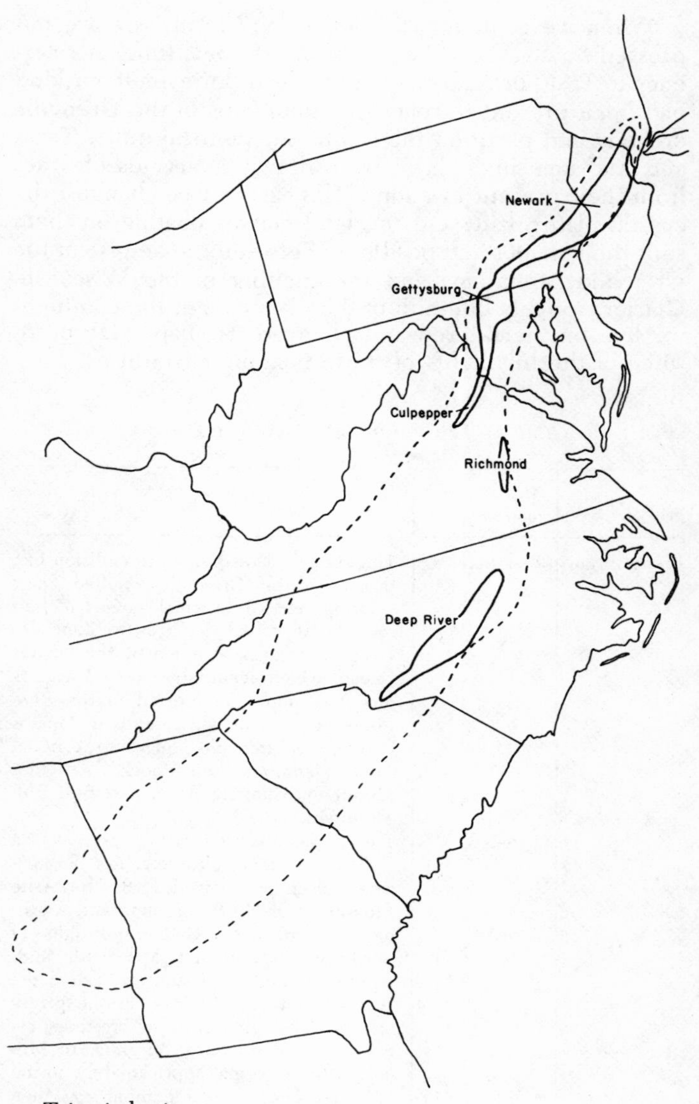

3. Triassic basins

When we contemplate ancient land forms, we are impressed how vastly they change in geologic time. A reference to Texas or Alabama may mislead, for even if our kind had been present to roam the mountains of the Grenville orogeny and partition them into our political units, Texas and Alabama might have been drawn thousands of miles from their present locations. The earth's face changes; the crustal plates drift, and the land masses floating on them shift their shapes and positions. Between the demise of the Grenville Mountains and the melting of the Wisconsin Glacier, roughly one-fifth of the planet's life, the continent we live on meandered and changed its shape. Here are some of the highlights of those tectonic adventures:

Table 1: The Piedmont's Geologic History

Period	*Millions of years ago*	
Precambrian	850–800	The crustal plates, which in collision had thrust up the Grenvilles, pulled apart creating cracks into which flowed molten rock. A rift called the Brevard Zone developed and was occupied by the Iapatos Ocean, which eventually covered what is now Piedmont and crested in the Blue Ridge. The clean white sands of Iapatos beaches became the quartzite caps of such Piedmont monadnocks as Pilot Mountain, Hanging Rock, and Bull Run Mountain.
	750–410	The restless crustal plates produced a period of violent volcanic action. Submarine volcanoes spewed forth the Ashe Mountains of North Carolina and southern Virginia, the Catoctin Ridges of northern Virginia and Maryland, and, most important, an island arc, called the Raleigh belt of volcanoes, in the protoatlantic. The remains of the arc today recline from Petersburg, Virginia, to Milledgeville, Georgia, approximately on the fall line. The entire Piedmont was then submerged, and the region occupied by the worndown Grenvilles might even have been under water. Volcanic ash settled into the sea to form slate, as in the

Table 1: The Piedmont's Geologic History [continued]

Period	*Millions of years ago*	
		Carolina Slate Belt. Mud mixed with limestone developing in the sea bottom, and the resulting rock later metamorphosed (metamorphic rock has been altered chemically and physically by heat and pressures within the earth) into the imperfect marble quarried today near Charlotte, North Carolina.
	620	Some worms were fossilized in the muds of turbidity currents (currents caused by the slow movements of silt-laden waters) and were recovered recently near Durham, North Carolina. These are among the oldest life forms yet recovered and are capital among the very few found in the Piedmont from whose rocks nearly all signs of life have been cooked and twisted by metamorphic forces.
Mid-Cambrian	475	The southern sweep of the island arc was well below the equator. New York City, had it existed then, would have been an equatorial resort.
Late Cambrian to Ordovician	450 400	The crustal plates of the ancient continents of Africa and North America began to grind together; where, in terms of our present notions of coordinates relative to Greenwich and the equator, is subject to obfuscation. Over tens of millions of years the island arc was pushed inshore, and the rocks of the Piedmont and Blue Ridge were thrust upward as the continental land masses piled into one another.* The two continents joined into one gigantic land mass called Pangaea.
Mississipian and Pennsylvanian	300 to 230	To the west of the new mountains there were lowlands and marshes festooned with lush growths of the primitive plant life newly arrived from the sea. Giant ferns and club mosses the size of oak trees fell into the bogs and failed to rot completely. Coal resulted from their fossilization.

*This is the Appalachian Orogeny.

The forces involved in continental collisions are difficult to imagine. Before the end of the Appalachian orogeny the lands we now stand on had been thrust into the clouds while being squeezed northwestward 100 miles (160 kilometers). The Folded Appalachians are the remnants of sedimentary layers toppling off the uplifted Blue Ridge and Piedmont. The analogy of a loaf of sliced bread tilted westward until it fell is often used as an aid to envisioning the origins of the Valley and Ridge Province. The analogy might well be extended backward to include the kneading and baking in the great earthen ovens of frictional heat and mechanical deformation which destroyed all but the most protected Piedmont fossils and washed their granular remains to the sea. The stresses of this great leavening effectively obliterated the past.

The tectonics which shaped our continent ended with the close of the Paleozoic era. The Piedmont's history in the Mesozoic and Cenozoic eras was largely one of erosion. The Anapurna-like peaks born of the fires and upheavals of the Appalachian orogeny were eventually worn down to form the surface of the fall zone peneplain. As the continental crustal plates slid apart, the ancient peneplain tilted seaward; the western surface rose several thousand feet, and was worn down by erosion. Then the land rose again, and again was worn to a peneplain, the Schooley, to which the Piedmont monadnocks are testament. A final uplift ushered in the present erosion cycle, giving us the Piedmont (Harrisburg) peneplain of today. Its gentle, sensuous swells do not even whisper of the peaks of the past.

The Primal Piedmont

Portaging the fall line rapids in the late 1600s, European man looked up through a vaulted canopy of deciduous giants; a different vegetation from that of the Coastal Plain, particularly different from the pine stands of the fall zone hills. We can only imagine this experience, for we have no analogue in the province now to give us a taste of the sensation. The primal Piedmont is gone utterly.

The act of imagination, though, is worth the effort. This is what I see in the Piedmont of 300 years ago: the canopy claims the direct sunlight completely by filling every gap with a broad-leafed parasol, cantilevered aloft. No part of the jade awning is within one hundred feet of the ground. The trees holding it aloft are chestnuts, white oaks, mockernut hickories and tulip trees, immense and widely spaced. Only around the younger dominants, say those under 200 years old, could two lovers link hands. Many are more than four feet (120 centimeters) in diameter.

Chestnut predominates, occupying half the canopy. It peppers the ground with mast, a fundamental food source for deer, turkeys, passenger pigeons, jays, squirrels, opossums. The oaks add their acorns to the mix of energy and proteins flowing through the same birds and animals. But the hickory yields its treasure only to the chattering gray squirrels and the silent, lumbering fox squirrels—they alone can gnaw through the bone-hard hulls.

There are smaller trees, too, in the forest of my fancy; their crowns stop respectful fathoms beneath the first branches of the dominants. Sourwoods make hexing gestures downward with floral sprays arrayed like fingers on skeletal hands. Burly, gnarled dogwoods redden the subcanopy in autumn with leaf and fruit. Lower on the slope squat massive beeches and red mulberries. The beeches occasionally hold a position of canopy rank. Redbud, woody legume of the deciduous forest, dangles its bean pods directly on the bark of its older branches. A healthy shrub layer of viburnums and vacciniums contributes, respectively, arrow shafts and blueberries to those who desire them. The shrubs stand guard over a stunning profusion of woodland flowers.

Checking fantasy carefully against botanical record I envision a stand of mixed yellow ladies' slippers and showy orchis, orchids found occasionally today on rich wooded inclines in the Piedmont. Nor is it unrealistic to imagine riotous throngs of trout lily, spring beauty, and other spring ephemeral wildflowers, even false rue anemone, reduced now to rarity. And watercress in the stream—it is there still.

Succession

The term climax defines that cloak of vegetation which ultimately covers the land in the absence of disturbance over a sustained period. Stability is the salient characteristic of the climax growth, and the stability results from diversity.

In some instances a disturbance, such as fire, occurs at regular intervals and the climax becomes arrested at some vegetative mix short of an undisturbed conclusion. The prairie grasslands of the Great Plains and the Sandhills Longleaf Pine Forests of the Coastal Plain are examples.

If the climax vegetation is disturbed, a process called secondary plant succession acts to restore it. Secondary succession may be observed on nearly all the nonurban land in the Piedmont that is not under immediate cultivation. After a forest is destroyed or after a farm is abandoned, a series of generally predictable plant associations succeed one another until the combination of plants stabilizes in a self-perpetuating community. That is the climax.

Another process, primary plant succession, inducts previously barren surfaces into the society of life. Road cuts, erosion scars, natural outcrops of rock are surfaces subject to primary succession's influence. Like secondary succession, the primary process is broadly predictable. It typically involves an initial infection by lichens advancing onto a lifeless substrate, producing acids which etch into and help decompose the rocks, and sifting the wind for chaff which gathers about the lichen to form organic buildups. Mosses often germinate in the lichen islands, hastening the organic accumulation. Then pioneer flowering plants arrive to continue the soil-building process by prying with ever-stronger roots at the underlying rock or rubble, ingesting the previously rockbound minerals and returning them in their own decomposing detritus along with compounds of carbon captured from the air through photosynthesis. It was through an initial labor of primary succession that the plants, emerging from the sea in the early Devonian period, made the planet inhabitable for the rest of us by creating the soil and putting free oxygen in the air.

I think of primary succession as a process of soil building, of secondary succession as one of soil retention.

The Piedmont is either plowed, paved, or in succession. Habitats controlled by primary succession are small and isolated. Secondary succession is the principal life trend over most of the province, especially in the south. Together these processes are of such importance that they dictate the outline of our approach to studying the habitats of the province.

Land use patterns are not uniform so let's divide the land mass into three sections for a focused discussion of each.

North of the Potomac, abandoned farmland is not so profuse as elsewhere. The Pennsylvania Piedmont in particular is meticulously cultivated, in places almost to the point of sterility from a natural systems view. There are woodlots composed mostly of deciduous hardwoods used for building material and firewood. Weedy places and fields of sedges and saplings are uncommon. The New Jersy Piedmont is a little wilder, abandoned farmland slightly less rare. Maryland's part of the province shows a healthy patchwork of hardwood stands, abandoned fields, active farmland, and pine woods (though north of the Potomac, especially north of Maryland, the pine stage is inconspicuous or absent in the successional march). Red cedar is more common than pine, and certain deciduous pioneers (locust, cherry and dogwood) are in places more common than either.

The central Piedmont between the Potomac and the Savannah rivers supports considerable active cropland and a wealth of herding enterprises, but is principally a forested landscape. Many of the abandoned fields have been left to grow whatever arrives by wind or bird. Such lands, where man has ceased, for the moment, to contest with nature, offer splendid laboratories in which to watch the advance of plant succession. The majority of these abandoned central Piedmont fields are not easily recognizable as such, for most of them were abandoned decades ago and have already passed through the weedy and shrubby stages into stands of pine, mixed pine–deciduous, or maturing deciduous woodlots.

The Piedmont south of the Savannah River has experienced a similar decline in tilled acreage over the past half century and its natural succession stages are similar to those in the central Piedmont. If there is a difference in these

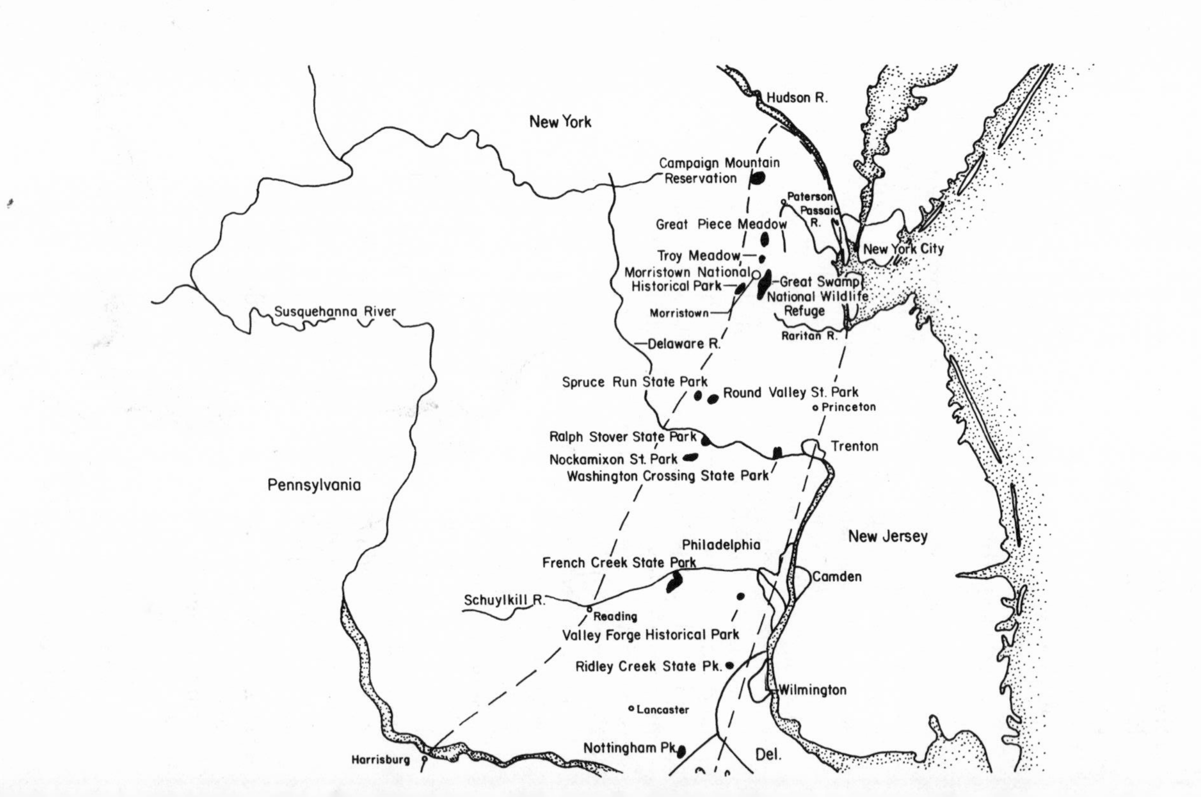
Hudson R.
New York
Campaign Mountain Reservation
Paterson
Passaic R.
Great Piece Meadow
New York City
Troy Meadow
Morristown National Historical Park
Great Swamp National Wildlife Refuge
Morristown
Susquehanna River
Raritan R.
Delaware R.
Spruce Run State Park
Round Valley St. Park
Princeton
Ralph Stover State Park
Trenton
Nockamixon St. Park
Washington Crossing State Park
Pennsylvania
New Jersey
Philadelphia
French Creek State Park
Camden
Schuylkill R.
Reading
Valley Forge Historical Park
Ridley Creek State Pk.
Wilmington
Lancaster
Nottingham Pk.
Harrisburg
Del.

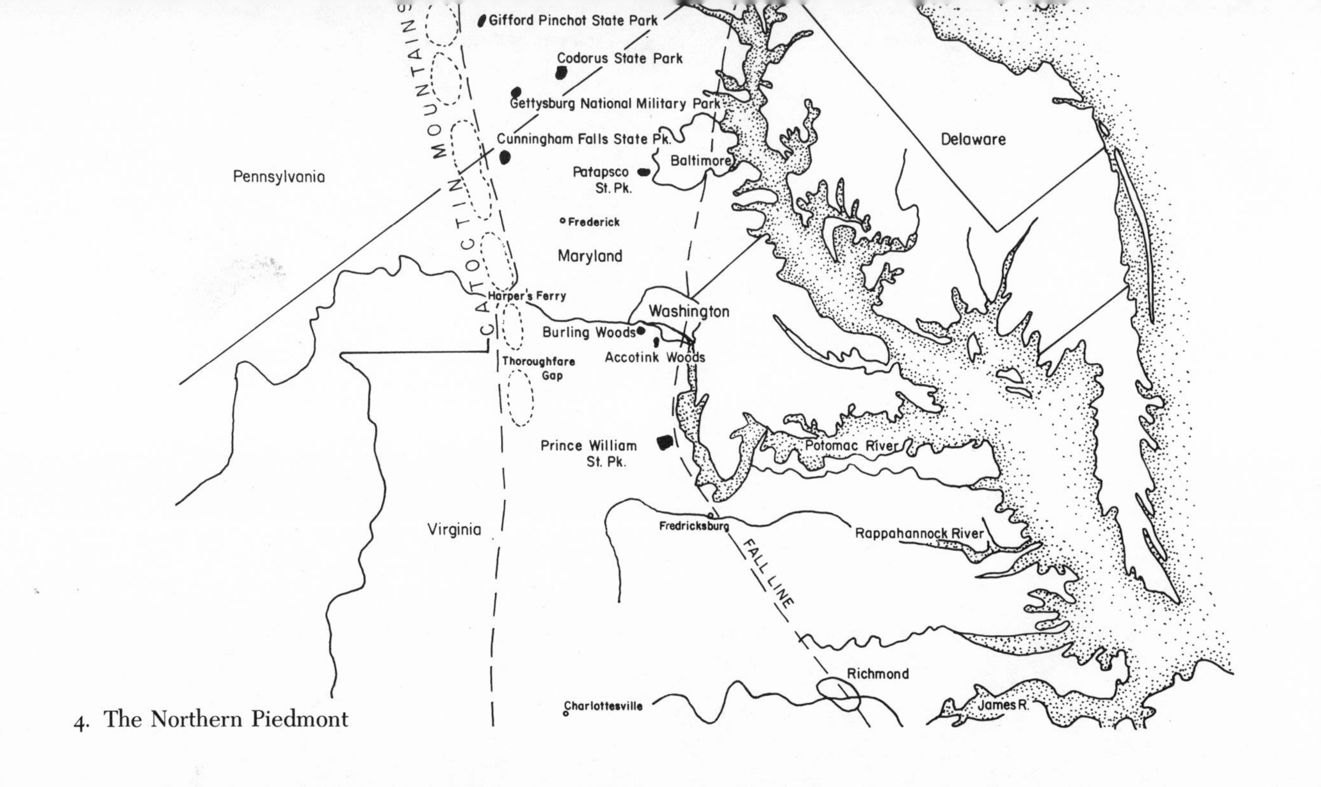

4. The Northern Piedmont

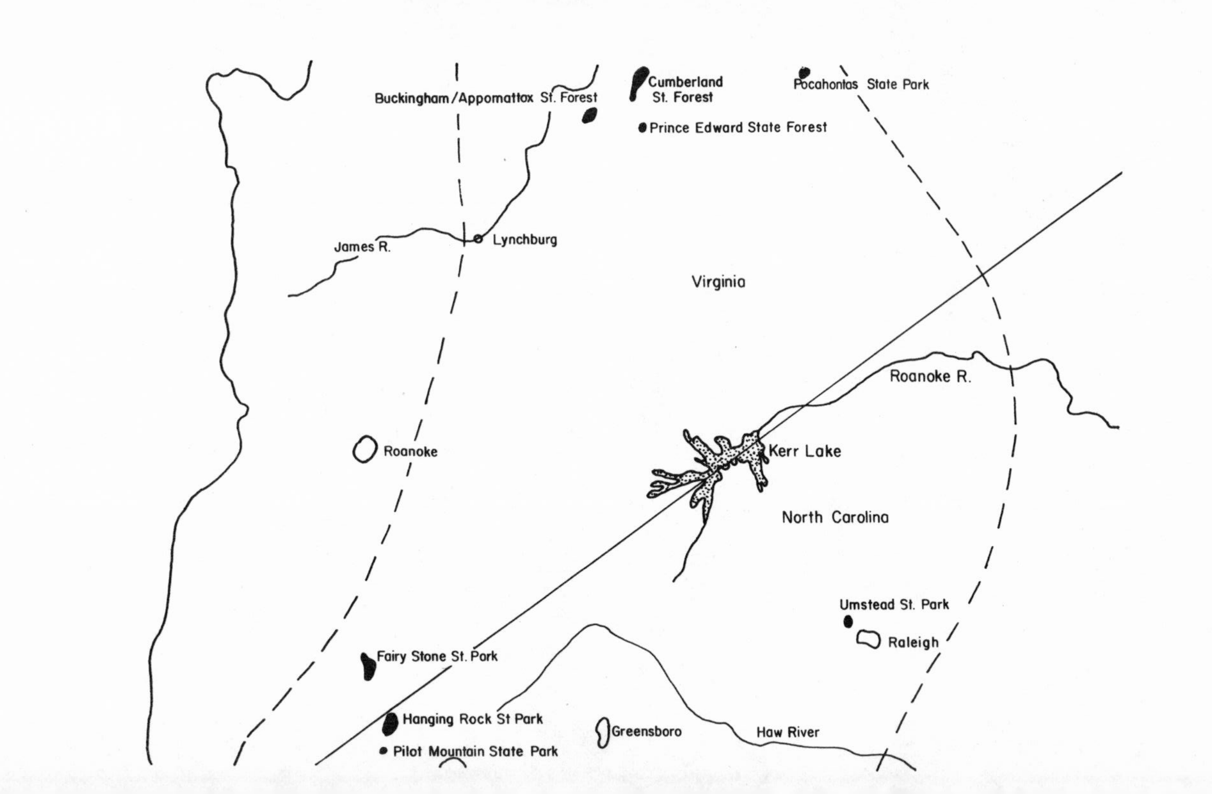
Buckingham/Appomattox St. Forest
Cumberland St. Forest
Pocahontas State Park
Prince Edward State Forest
James R.
Lynchburg
Virginia
Roanoke R.
Roanoke
Kerr Lake
North Carolina
Umstead St. Park
Raleigh
Fairy Stone St. Park
Hanging Rock St Park
Greensboro
Haw River
Pilot Mountain State Park

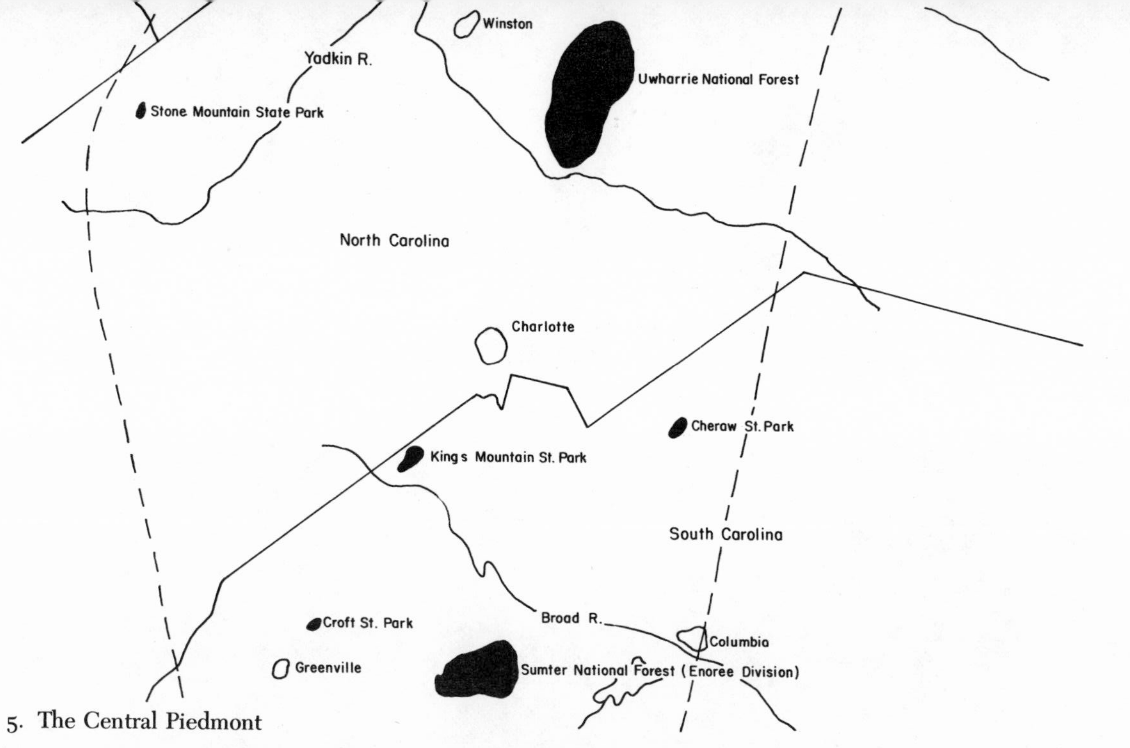

5. The Central Piedmont

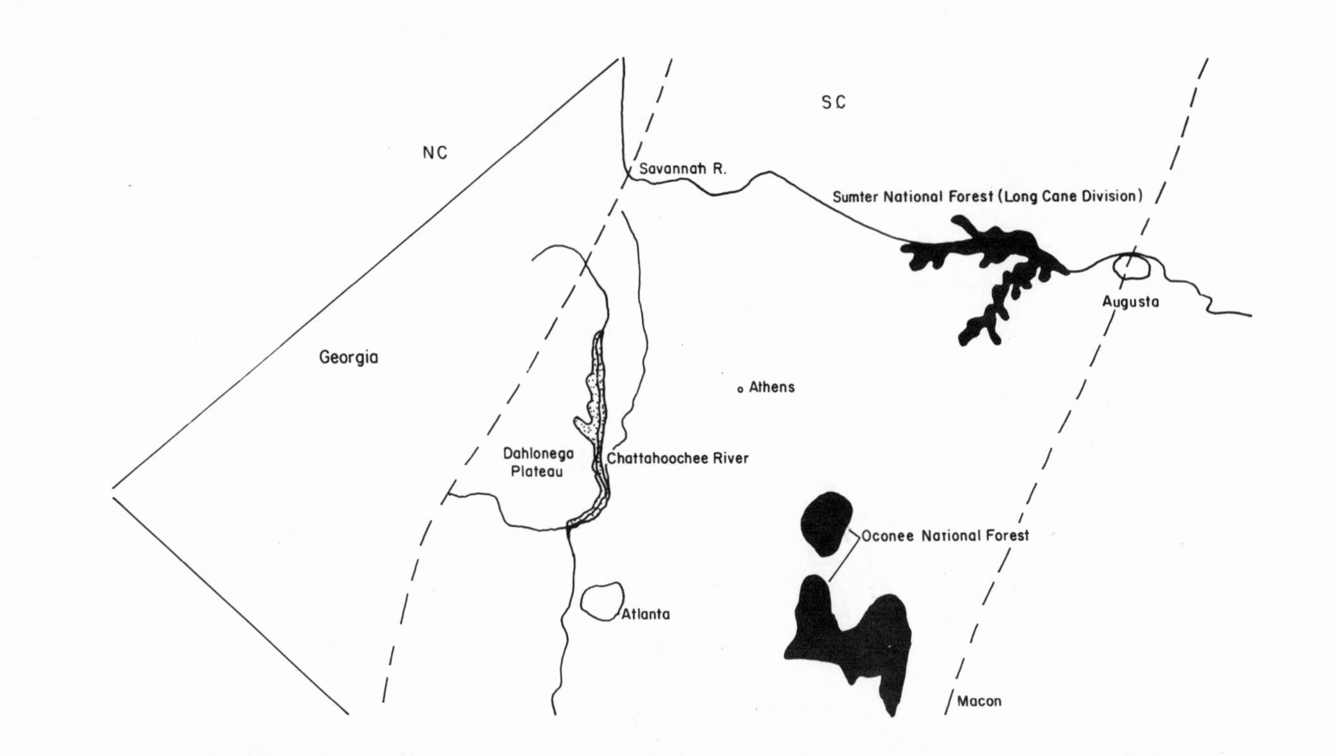
NC
SC
Savannah R.
Sumter National Forest (Long Cane Division)
Augusta
Georgia
Athens
Dahlonega
Plateau
Chattahoochee River
Oconee National Forest
Atlanta
Macon

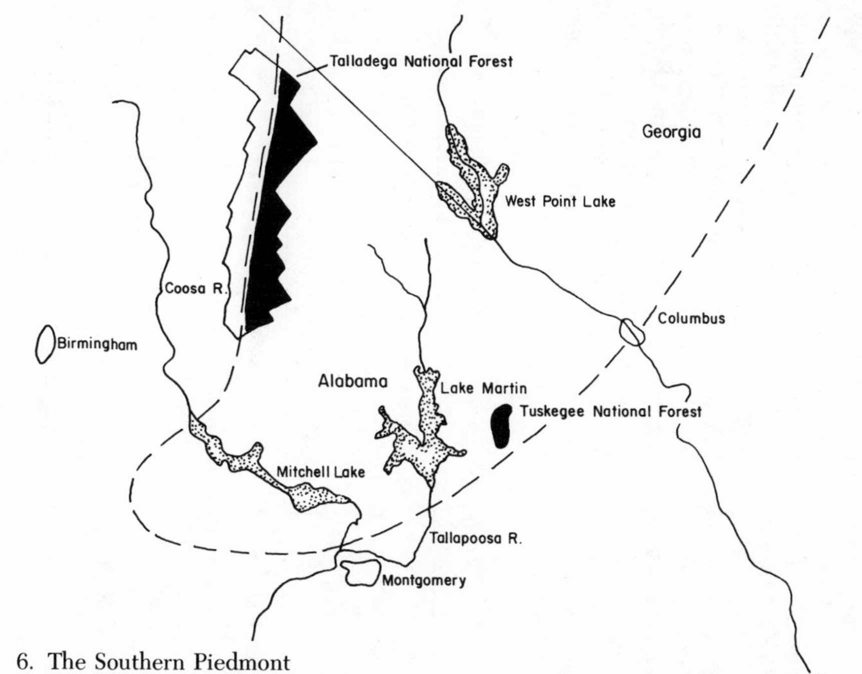

6. The Southern Piedmont

divisions' landscapes it probably results from a kind of short-circuited succession at work in the south. Many of the old cotton fields were planted in rows of pines, usually loblolly, simplifying the succession process. Traveling Piedmont Georgia and Alabama I am struck by the vast acreages devoted to growing grist for the paper mills.

My estimate of the vegetative status of the Piedmont is shown in Table 2.

More than half the Piedmont is now dominated by some phase of plant succession; these phases constitute important habitats. They will receive premier attention in this guide. Active farmlands occupy the next larger portion and, except for the chemically sterilized fields devoted to row crops, these lands are vibrant habitats. We will also examine the aquatic habitats: streams, ponds, rivers and marshy places. Finally, we'll look at the Piedmont's unique and exotic habitats such as the Great Swamp of New Jersey; the serpentine barrens of Pennsylvania, Maryland and Georgia; the Granite outcrops of the central and southern sections; and the monadnocks overlooking the peneplain from Sugarloaf Mountain, Maryland, southward through Georgia. Ours is an ancient life system, and it has assumed a quiet grandeur. Some landscapes tower and impose; ours deepens.

Table 2: Vegetative Status of the Piedmont

	Areas controlled by man		*Areas controlled by plant succession*	
	urban	*active farmland*	*abandoned farmland*	*woodlots never farmed*
North of the Potomac	20%	50%	25%	5%
Potomac to Savannah	5%	30%	60%	5%
South of the Savannah	5%	25%	65%	5%

CHAPTER TWO

Weather and Climate

IT IS REMARKABLE that a strip of land one thousand miles long and oriented generally from north to south should have an approximately homogeneous climate. It is even more startling to note that in comparing the outlines of Fenneman's physiographic provinces with those of the Piedmont's plant growth regions, one quickly sees that only the Piedmont is defined similarly by the geologist and the climatologist. Most other physiographic provinces are overlain by a quilting of the climatic boundaries that sometimes bear no apparent relation to the landforms.

Weather is defined as the condition of the atmosphere with respect to wind, temperature, and moisture at a particular place and time. Climate is weather over an extended period. It is the climate that dictates which plants will grow in the Piedmont and thereby molds the character of its life systems. Temporary, freaky perturbations of weather do little to change the composition of plant communities. Given certain average conditions of temperature and moisture, the plants will adjust, sort themselves into the unique chronologies and spacial relationships which shape the sere at each location.

Temperature and moisture are the cornerstones of climate. They are varied and tempered by other considerations and they do not act independently on living organisms, but we separate them here in order to quantify their measurements and their effects on the life of our province.

Temperature

It is not uncommon for the temperature range at many locations in the Piedmont to span one hundred degrees or more in a single year, particularly so in the northern part of the province where the high temperatures in summer approximate those in the south and where the lows of winter approach or fall below zero. North of the James River, minimum yearly temperatures are normally around 0°F (−18°C); lows of less than 10°F (−13°C) are uncommon south of the James. Probably the most significant measure of temperature in a biological sense is the average temperature during the growing season—in the Piedmont, 70° to 80°F (21° to 27°C) throughout. The 75° line runs northeastward across the province from the Dahlonega Plateau, in Georgia, to Washington, D.C.

The growing season is an important selecting factor in the vegetative mix and is of great significance to agriculture. From approximately the Yadkin River, in North Carolina, northward, the last frost occurs between May 1 and May 31, and the first frost of autumn halts the growing season between October 1 and October 31. South of the Yadkin, the last frost of spring occurs on the average in March and the first frost in November. Consequently, no part of the Piedmont south of New Jersey enjoys less than 180 frost-free days per year. New Jersey averages 160 frost-free days. The outer Piedmont in Alabama and Georgia is free of frost for up to 230 days per year. The Georgia Piedmont shows the most pronounced variation in its growing season, with 180 days on the Dahlonega Plateau and 230 days at the fall line.

The Dahlonega Plateau averages about 1400 feet (425 meters) in elevation and experiences the same 180-day growing season of the inner Piedmont in Pennsylvania, 600 miles (1000 kilometers) to the north, but 1000 feet (350 meters) closer to sea level. This gives us a direct look at the climatologist's principle that altitude has the same effect on climate as does latitude; specifically, that a thousand-foot change in altitude effects the climate to about the same extent as a six-hundred-mile change in latitude.

Moisture

The entire Piedmont, excepting a 40-mile (65-kilometer) corridor along the Potomac River, receives a bountiful 40 to

50 inches (100 to 125 centimeters) of rainfall per year. The Potomac corridor averages just under 40 inches. The Piedmont's precipitation is reliable enough over the long term so that neither prolonged drought nor excessive flooding are significant as limiting factors in selecting the natural vegetation. Moreover, the Piedmont has no wet or dry season; precipitation is distributed amenably over the annual cycle. Slightly more than half—55 to 60 percent—falls during the growing season, a critical time, of course, if the moisture is to benefit the life system. The southern Piedmont gets slightly more rain during the growing season than does the northern part.

Snow and ice are part of the precipitation mix and their equivalent in rainfall is included in the preceding figures. From time to time, local weather produces freezing rain which accumulates as laminations of ice on all exposed surfaces, sometimes causing considerable damage to trees. The trees thus damaged are more often coniferous than deciduous, so the effect of the Piedmont's periodic ice storms is to hasten the progress of the sere through the stage of pine dominance. Wet snow may have the same effect. The Piedmont of Pennsylvania and New Jersey receive totals of 30 to 60 inches (75 to 150 centimeters) of snow annually, Maryland and Virginia get an average of 10 to 30 inches (25 to 75 centimeters), and the Piedmont from North Carolina southward receives less than 10 inches. Generally, from the James River northward, snows accumulate sufficiently in most winters to become an adaptive consideration in the lives of plants and animals.

Effectiveness

The raw statistics on warmth and moisture can do little, by themselves, to describe our climate, for the rainfall might quickly evaporate and the warmth could be lost to cold ocean currents passing offshore or could be radiated back into space. Because the plants and their dependent higher life forms are able to utilize the temperate features of the Piedmont's climate, the warmth and moisture are said to be *effective*. The moisture is *effectively* distributed over the annual cycle and it is retained in rather deep, fine-grained soils. The Piedmont's winds blow at a moderate 6 to 8 mph on the average in comparison with, say, the Midwest, where the unobstructed winds blow at a year-round average of 15 mph. Wind velocity is an important factor in determining the rate of evaporation.

From the Potomac north, the Piedmont receives sunshine for approximately 240 days each year; to the south, the sun shines on about 260 days annually. More significantly, 60 to 70 percent of the days during the growing season are sunny, a factor favorable for plant growth when there is adequate moisture. Cloudy days, occurring roughly one in three, have the effect of reducing moisture loss and stabilizing temperature—the amount of radiation inbound decreases, but so does the amount lost. Temperature extremes, in all seasons, are less on cloudy days than on sunny days.

Temperature and moisture are considered together in a measure known as relative humidity, the amount of moisture present in the air relative to the maximum possible at that temperature. The capacity of air to hold moisture is a direct function of its temperature; a rise in temperature enables the air to absorb more water, and cooling reduces that capacity. Air carrying a given quantity of water per unit of volume may be cooled sufficiently to become saturated; the temperature at which saturation occurs is called the dew point. Hence, if the temperature is 60° and the dew point is 56°, we know that the relative humidity is high and that if there is further cooling, precipitation is likely.

Relative humidity has a controlling effect on the rate of evaporation. Air of high relative humidity, for example, discourages moisture loss at the land surface. In the Piedmont, the relative humidity averages about 50% in all seasons.

Winds and Landforms

The planet's winds integrate the provinces' weather systems, moving air masses from place to place and making local climate less parochial than they would otherwise be. In a global sense, winds are powered by the differential in solar radiation received at the equator and at the poles. Air is warmed at the equator, rises, and flows northward and southward. In the northern hemisphere, it cools and returns to earth at the Tropic of Cancer (latitude 23½ degrees). Similar loops circulate poleward from the temperate zone. Meanwhile, the earth's eastward rotation adds a longitudinal component to the general flow of the air masses. The overall effect of these forces in North America is to cause a system of prevailing westerly winds. These westerlies would perhaps blow steadily and consistently were it not for the phenomenon of pressure centers—air masses or weather systems whose winds revolve around the centers of pressure. The winds move clockwise around a high pressure system; counterclockwise around a low. Normal atmospheric pressure is that which will support a column of mercury 29.92 inches tall; an air mass whose pressure departs from the normal by one inch of mercury would constitute a good, strong "high" or "low."

Because the winds circulate clockwise around a high, an approaching high is usually accompanied by, or borne upon, northwest winds. Winds from the polar and subpolar regions tend to be cold and dry, so highs in the Piedmont are associated with fair weather. Lows, whose counterclockwise rotation brings in moisture-laden air from over the Atlantic, generally bespeak rain. In the zone where two or more pressure systems meet, there is usually active weather, and this zone of clashing air masses is called a "front." Low-pressure air, coming from over the ocean, is warmer than the cool masses comprising a high-pressure system, and when a high meets a low, the dense cool air wedges beneath the warm air of the low, forcing it upward and cooling it. The result is precipitation. In summer, towering cumulus clouds, energized by the temperature differentials of clashing air masses, rise sometimes to 60 thousand feet (18 thousand meters), boiling with vertical currents of perhaps 500 mph. Thunderstorms and quick, heavy rains develop in the anvil-topped cumulus. In the extreme cold of such heights (air cools at the rate of 3 degrees Fahrenheit/1.7 degrees Celsius per thousand feet) raindrops freeze. Falling into strong updrafts, they may rise to receive another lamination of water, and another, eventually falling as hailstones. In winter there is not enough solar energy absorbed into the air masses to foment such atmospheric violence, so the meeting of a high and a low in cool weather usually results in more prolonged and gentle rain or snow.

Meteorologists say that all airborne moisture is absorbed as air masses pass over the oceans. At landfall, the air is forced upward by mountains, grows cool, often below the dew point, and loses some of its moisture. This accounts for the rainy climate, for example, of our Pacific Northwest. In eastern North America, air masses moving eastward on the prevailing westerlies similarly rise and lose moisture when they reach the Appalachians. The Piedmont, immediately to the east, receives some of this rain. The gap in the Blue Ridge Mountains north of the Potomac, and the reduced height of the ridges which exist in the area, may explain the slightly drier climate in the Potomac corridor (see page 12). Very broadly, we are concerned in the Piedmont with the effects of three involvements of the air masses with landforms: the dry, cool, "continental" nature of the prevailing westerlies, from which much of the moisture has been wrung by the Rockies; the further precipitation which results when the Appalachians again force the westerly air upward; and the arrival of moisture from the Atlantic on easterly winds, usually those associated with a low. Rarely does a sustained wind blow from the east that does not bring rain. In any event, we must recognize that any discussion of weather involves, necessarily, enough simplifications and omissions to hint at the staggering complexities which confront the meteorologist.

Climatic History

The great glaciations of the past are certainly the most exciting events in the earth's climatic history. There have been, apparently, not only the Pleistocene glaciers which have just passed or which may not yet be finished, but at least two, and possibly as many as five, others. All that is known is that from time to time the earth cools below a critical point, the great snows begin and much of the planet's water becomes trapped in circumpolar sheets of ice, miles thick.

The increasing area under the ice cap raises the proportion of the solar radiation reflected back out into space (the albedo), leaving less to warm the earth. General cooling is thereby accelerated, and it continues until so much of the earth's water is imprisoned that the cycle of evaporation and precipitation is impeded.

Libraries bulge with theories explaining and interpreting the glaciations, but none have inspired a consensus. Among the explanations advanced are theories involving the rise of mountain ranges, the wandering of the poles, and the passing of cosmic dust between the earth and sun. The recent confirmation of the theory of continental drift (plate tectonics) strengthens the case of the wandering poles. Still, there are paradoxes aplenty, and while much is speculated, nothing, in terms of a causal explanation, is known.

Geologists are certain, however, that there began about 35 million years ago, after 100 million years or more of warm global climate, a cooling trend which resulted in the most recent series of glaciations. The ice advanced and retreated several times during the Pleistocene epoch and worked a profound effect on the earth's biological communities and on the evolution of man. It was probably the last glacial advance (the Wisconsin) that brought *Homo sapiens* to North America across the land bridge that connected Siberia with Alaska. The Wisconsin Glacier began its retreat 10 to 15 thousand years ago. The paleobotanical record shows that interglacial periods prior to the Wisconsin Glacier were as warm or warmer than the present climate, so it is considered likely that the ice will advance yet again, probably 10 to 20 thousand years hence.

Before then there will be numerous minor shifts in the climate, as there have been in the recent past. For example, an era of worldwide frigidity was in progress at the time of Christ. The period from A.D. 1600 to 1850 was characterized by such cold that it is called by climatologists the Little Ice Age. The effects linger in the minds of some of our older kinsmen who recall skating on waters which now freeze indifferently, if at all. Skating downstream on the Potomac from Leesburg, Virginia, to Great Falls (20 miles/32 kilometers) was a popular exploit at the turn of the century.

A significant warming effect may have resulted from the

explosive increase in the release of carbon dioxide (CO_2) into the atmosphere by man's burning of fossil fuels; approximately 13 percent more of the gas was at liberty in the atmosphere in 1950 than a century earlier. Carbon dioxide absorbs radiation in the infrared part of the electromagnetic spectrum where the earth's outbound radiation is most intense. The effect is to trap this energy within the atmosphere, warming the climate. At present there is a consensus that this "greenhouse effect" is being overridden by a cooling trend of other origins, but eventually the increased CO_2 is bound to have its impact on the climate. Ironically, this may hasten another coming of the ice, for precipitation—the concomitant of a warming climate—must precede glaciation.

By isotope dating, by boring into the bottoms of oceans and old lake beds and into glacial ice, by examining the pollen and vegetative debris in peat bogs and by persistently stalking the other corridors of the geophysical archive, scientists long ago concluded that the climate is dynamic, ever-changing. The Gulf Stream shifts, the poles wander, the continents migrate and collide, climates are permanently in flux, and life forms are constantly adapting.

Microclimate

For the individual organism, the conditions of wind, temperature, and moisture immediately surrounding it are of greater importance than the overall climate of an area, for local conditions can vary markedly from the overall average. As far as the individual is concerned, particularly if the individual is a plant or an animal whose mobility is limited, the microclimate it occupies *is* the climate.

Irregularities of terrain help shape some microclimates, offering shelter from wind, shedding or concentrating water, or blocking direct sunlight. Rocky, north-facing seeps, such as nestle on certain wooded slopes in the Piedmont, provide the coolness and moisture necessary in the microclimate of the craneflies and salamanders. The misted, nearly vertical banks of streams flowing through woodlands are climatically suited as substrate for the gametophytes of ferns. Only a few yards above, on the south-facing boulders through which the stream cut thousands of years earlier, crisp lichens crinkle in the sunlight, parched and baked, but home nonetheless to tiny homopterans and the minute salticid spiders that prey upon them. The groundhog hibernates in an excavated microclimate of his own design that is far warmer than the surface. His engineering hints at the talent shown by many creatures in manipulating their microclimates to advantage, though only the very cleverest—such as caddis flies and humans—fashion protective garments.

Plants, by far, exert the most powerful influence over the concept of microclimate. As the sere progresses at a given site, the plants—beyond producing the food and oxygen that sustain the animal kingdom—temper and ameliorate the conditions of many habitats. They break the winds, which can desiccate in summer's heat or intensify the hiemal cold. They shade the sun. Most significantly, the plants take moisture from the soil and transpire it through the stoma in their leaves as vapor. The microclimate in the forest is thereby humidified; but the effect of transpiration on temperature is even more dramatic. Water absorbs heat during the liquid-to-gas conversion and in doing so cools its surroundings. In a deciduous forest the transpiration rate may be great enough to lower the temperature on a summer afternoon by ten degrees Fahrenheit below the temperature in the shade outside the forest.

Effects of Climate on Life Forms

Each plant tolerates a range of temperature and moisture conditions. When the limits are exceeded the plant dies, for it cannot relocate or manipulate its microclimate. Each Piedmont plant is present because part, at least, of the province is climatically amenable. Those which do not occupy the entire province are absent from the forbidden part almost certainly by reasons of climate. The loblolly pine, to cite one of hundreds of possible examples, stays to the south of the Potomac—a frontier it observes with almost political fidelity—because the winters to the north are too cold. The northern Piedmont is occupied instead by the Virginia and pitch pines.

Where there are notable disjunct communities of a given species in the Piedmont, climate, again, is probably the cause. The *Ribes* and cypress, for example, at Stevens Creek, S.C., and the white pine slopes along the Deep River in Chatham County, N.C., are disjunct populations, glacial relicts. In the case of the subtropical plants at Stevens Creek, the advancing glacier(s) drove the species from all but this peculiarly favored habitat. The glacier-inspired boreal climate brought the white pine to the Deep River. The retreating ice (which, of course, halted far to the north) was followed by a warmer climate which eliminated all white pines from the southern Piedmont except those along the Deep.

The effects of climate on animals have drawn some interesting generalities. Diversity, for example, is said to decrease with coldness of climate. An example commonly cited is that the circumpolar regions in the northern hemisphere are attended by only two grazing ungulates—the musk ox and the caribou (or, in the palearctic realm, the reindeer)—as compared with the numerous forms of deer, sheep, goats and bovine ungulates of warmer climes. The validity of this argument is perhaps greater on a continental scale than on a local basis, more valid in winter than in summer and less valid with regard to birds than to mammals. In fact, examples to the contrary are abundant. The number of terrestrial rodents in the high Rocky Mountains and other western ranges considerably exceeds the rodent diversity of the much warmer Piedmont. The nesting birdlife in the far north is anything but scant when one considers the legions of passerine birds, waterfowl and shorebirds, which overfly the temperate zone to nest on the tundra. There are ten species of owls which breed in the Canadian forests, five in the Piedmont.

Another generalization is that within any one species, the size of individuals increases with coldness of climate. It is said also that the limbs and protruding members are smaller in cold regions than in warm and that coloration tends to be lighter in cold climates, darker in warm. It is often necessary to venture outside a given species and make comparisons of, say, the kit fox with the arctic fox in order to present a suitable example, for species which occupy the cold regions are most often limited to the habitat and do not have southern races available for ready comparison. Man is one who does, and the long-limbed, dark-skinned equatorial African is often compared with the Eskimo for this purpose. Proponents note also that the world's smallest people, the forest pygmies of Africa, occupy a warm climate, but some of the pygmies' neighboring tribesmen are among the most massive and stately of humans.

We apply these generalities to the Piedmont, aware that there is validity and error in each. We may use them to help fix the Piedmont in the biological continuum which accompanies the range of the planet's climates. We know that climate limits and defines our habitats and all others; hence, we divide our study of the Piedmont into moisture regimes, for, temperature being mild, moisture is the limiting factor in our life systems. The foregoing maxims may be safely applied to the Piedmont's life forms as sensible first approximations, the refinements and applications being the responsibility of the Piedmont naturalist.

PART II

Cultivated Lands

CHAPTER THREE

Farmlands

THE LATIFUNDIA of earlier eras have undergone patrimonial and economic division for enough generations to reach an equilibrium of size and productivity manageable for a single family, perhaps extended. The working farms of the Piedmont are rarely larger than 400 acres (160 hectares). Large estates exist, though they are perhaps as often fashioned from the tax laws as from the surpluses of the land.

Grand or humble, the working farms of the Piedmont have in common a nurturing feature that makes them fecund homes for life systems: diversity. Piedmont farms are not single-crop monocultures; they are polycultures of mixed husbandry, agronomy, and sylvaculture. The more fundamental units grow their own meats and poultry; orchards and grapevines are often cultivated; keeping a milk cow or two is common practice; and home-use apiaries are not unheard of. The vegetable garden, active summer and winter, is universal to Piedmont working farms. The main enterprise tends to be a mixture of cash crops (corn, soybeans, tobacco, small grain) and herding. Most of the cropland is devoted to growing food for the beef and dairy herds common in the province. Normally, a cow (or a horse) can be sustained year-round in the Piedmont on an acre of cropland and an acre of pasture, so typically we find that cleared land on working farms is about equally divided between the two. The hillier fields are often devoted to permanent pasture.

If a single feature could be said to distinguish a working farm, it would be the corridors of vegetation adorning the fences—hedgerows. The hedgerows are more than simply windbreaks preventing erosion or serving as shelters for wildlife. They are ecotones, zones where different communities meet, where life is triply enriched by the overlap of organisms from each adjacent community and by those who live only in the ecotone. Hedgerows, as we shall see, are of critical importance to the agricultural life systems of the Piedmont—so much so that we will find it most meaningful to view the cultivated vegetation in the field and its bordering hedgerows as a unit. In these descriptions of the agricultural habitats, a hedgerow border is assumed, and its contributions to the life system are discussed integrally with those of the adjacent field. A field without hedgerows will not support so rich a community as one with hedgerows.

Row Crops

There is no question that a life system—an interplay of plants, pollinators, pests, predators and scavengers—exists in a field of row crops in spite of the short (perhaps four months) life of the plants, the chemical toxins applied, and the fact that any monoculture is weak for lack of diversity. These communities are, however, so varied that a reliable description could be composed best by an interested naturalist on the spot in a given Piedmont cornfield or wheat field. There are numerous "pests," native and alien insects which are present in unnatural numbers because of the unnatural plantlife upon which they feed. Natural controls tend to be weak because the predators have longer life cycles and fewer offspring than the "pests" and therefore have less opportunity to escape the applied poisons through mutation. This genetic expedient has enabled such insects as the Japanese beetle to develop nearly pesticide-proof strains (ecotypes), while their natural controls are killed off.

I would emphatically not discourage the personal investigation of cropland life systems; I think they can be especially interesting for what they lack. Cornfields in particular display simplified interrelationships which are easily observable. But so much depends (inversely) on the pesticides that a generalized description of Piedmont row crop life systems is hardly possible. With an inexpensive handbook on insect pests, you can pursue the issue locally; a discussion with the farmer regarding the chemistry he has applied might lend perspective.

Hayfields

Hayfields, now that's another matter. Land planted in forage grasses (grasses such as timothy or orchard grass) and legumes (clover, *Lespedeza* spp., and other cultivated members of the botanic family Fabaceae) tends to stay in that configuration for several years, often semipermanently if the tract is too hilly for plowing. Periodic mowing and removal of the hay halts the succession process and prolongs the tenure of the planted grasses, which provides time for the development of relationships between the plants and their attendant fauna.

A hayfield is a field planted in forage grasses and small-seed legumes for the purpose of cutting, baling, and storing the aboveground portion of the plants for use after the growing season. Unlike many plants which grow in rapid spurts in early spring, the hayfield grasses and legumes have been genetically selected to grow more or less evenly throughout the warm months, provided there is adequate rain. Two to five cuttings are expected over the growing season; three is about average. Expecting to cut and dry the grasses and legumes for hay or to chop them green for storage in a silo, the farmer does not let his livestock graze the hayfield. If the field were grazed it would become a pasture.

Occasionally a Piedmont farmer will grow a pure stand of clover or orchard grass, let it mature, then cut, dry and thresh it for the seed to be sold as a cash item. But as a rule, good, all-around hay is a mixture of grasses and legumes. Horse hay usually contains more legumes, cattle hay more grasses. Hayfields typically have several varieties growing, flowering, and producing seed by different schedules. And hayfields are treated with considerably less chemistry and inorganic fertilization than are fields devoted to row crops, so the hazard of poison is reduced. Except on soils with specific mineral deficiencies, the need for fertilization of hayfields is diminished by the presence of the legumes, whose wonderful property it is—and among higher plants this is a function believed unique to the family—to capture free nitrogen from the air and fix it in the plant tissue in combination with oxygen. This natural fixation of nitrogen by the bean family is the principal global means by which nitrogen is made available in the food web for growth. All growth—plant and animal—requires protein synthesis within the organism, and nitrogen is the pivotal element in protein. No nitrogen, no growth. Hayfields containing legumes are rich in nitrogen—organically (not synthetically) fixed nitrogen in readily metabolized compounds free of chemical contamination.

Hayfields, rich in nutrients, thickset with tall cover, periodically mowed for harvest and renewed growth, free of heavy-footed livestock who squash nests and pack topsoil, enriched by populations of stray grasses (crabgrass, Dallis grass) and "weeds" like vetch, chicory and other asters, bordered by the heavier cover and complementary food sources in hedgerows, are a wonderfully diverse habitat.

Animals of Hayfields

Insects

Grasses are principally pollinated by the wind. I have, however, pictures that show syrphid flies on the flowers of Dallis grass and tiny bees dangling from the sexual apparatus of panic grass. Other pollinators of the grasses surely include ants, who forage over every living surface of every habitat, and members of the "hunting" families of spiders (such as Lycosidae and Salticidae), who in their travels through the grassland forests could not help but transport pollen.

The flowers of legumes, always recognizable by the "keel and standard" arrangement of their sepals and petals, are showy organs that attract insect pollinators. Flies (order Diptera), bees (order Hymenoptera), and butterflies (order Lepidoptera) buzz and flicker among the heads of red clover and bicolor lespedeza. Among the butterflies, hairstreaks (family Lycaenidae), and crescents (family Nymphalidae), in particular the ubiquitous pearl crescent, *Phyciodes tharos*, dance from flower to flower like chaff upon the wind. Sulphurs (family Pieridae) feed upon clover and alfalfa as larvae but return the favor as adult pollinators. Whites and orange tips (also pierids) are nondestructive hayfield pollinators.

Bumblebees, honeybees, and an assortment of much smaller, iridescent-bodied bees, from the family Halictidae, attend the legume and weed flowers and occasionally the grass flowers. A characteristic of nectar and pollen-gathering insects is their fidelity to a single plant species, at least temporarily. An individual bumblebee, for example, does not divide its attentions promiscuously among competing plants, but maximizes its efficiency as a pollinator by visiting only a selected species. When all else is quiet, the hum of pollinators in a hayfield where legumes are in flower sets the undertone and rhythm for an ensemble of birdsong and insect stridulation unique to the habitat.

Those who stridulate, or "sing," by scraping parts of their bodies together, are in the order Orthoptera, a group embracing the grasshoppers, katydids, crickets, mantids, walkingsticks, and cockroaches. Not all orthoptera stridulate, but those who do use the sounds as a means of bringing the sexes together; the calls are unique to each species and are given only by the males, though in a few species the females make a weak response. The presence of orthoptera is not universally applauded by Piedmont farmers, for the group feeds mainly on the growing shoots of forage crops. The short-horned grasshoppers (family Acrididae), some of which are recognizable by the colorful hindwings and snapping mechanical sounds produced by their ungraceful efforts at flight, can be notably destructive. Most populations are controlled, however, by a variety of arachnids, birds, reptiles, mammals, and predatory insects, so that the damage done by the grasshoppers and other orthoptera lags well behind the growth rate of the plants. But grasshoppers are important to the matrix of life in a hayfield, a coalition of interrelated ecological niches the collective effect of which is a net benefit to the forage crop. The grasshoppers are among the lower heterotrophs; that is, they are among the first to take the food manufactured by the plants. They figure prominently in the food chain at a fundamental level. Because of their numbers and potentially explosive populations, the interplay of natural controls bearing upon the acridids is of immediate economic significance to the Piedmont herdsman. In the absence of these natural controls, forage cropping might become infeasible, for the costs of chemical supports for the production of hay could wither the operating margin of many Piedmont farms.

Who, then, controls the orthoptera? Other orthoptera—the mantids—to a small extent. Arachnids are active eaters of grasshoppers. The orb-weaving spiders (family Araneidae), particularly those of the genus *Argiope*, called "writing spiders" after the zigzig decorations they weave into their capturing webs, are notable grasshopper gourmets. In a Piedmont hayfield it is common to see a banded argiope, *A. trifasciata*, pounce on a grasshopper that has hopped into the web while escaping your approach. The argiope twirls the grasshopper with its rear pairs of legs, playing out sheets of webbing that quickly enshroud the victim, who is then bitten, injected with digestive juices, and stored for later feasting. The daddy longlegs, also an arachnid and more specifically called a harvestman, often preys on grasshoppers. They are in the order Opiliones, of which the family *Phalangida* are well represented in the Piedmont and well attuned to the abundance of grasshoppers.

Reptiles

Reptiles figure prominently in the control of hayfield rodents. The black rat snake, *Elaphe obsoleta obsoleta*, and the black racer, *Coluber constrictor constrictor*, are commonly found in tall grasses, among other habitats, throughout the province. They move silently along the runways of the rodents, grabbing the victims with lunges too quick for the human (and probably for the rodent) eye to follow. Both kill by throwing several loops around the victim to hold it in place while it suffocates in the snake's mouth. The eastern garter snake, *Thamnophis sirtalis sirtalis*, visits the hayfields, especially the more poorly drained parts, to hunt insects and amphibians. Poisonous snakes are rarely found in grasslands of the Piedmont, with the occasional exception of the copperhead, *Agkistrodon contortrix*, a relatively mild-mannered viper—I have inadvertently stepped on one without being attacked—whose principal food is mice. Snakes are usually careful, when visiting grasslands, to remain well-concealed for they are frequent prey of the red-tailed hawk and the horned owl.

Birds

The major burden for the control of insects in hayfields, as in most habitats, rests with the birds. Many competent field naturalists believe that the common crow, *Corvus brachyrhynchos*, much maligned and persecuted, is the principal avian controller of grasshoppers. Leaping and dancing ("crowhopping," to be colloquially precise) in the high grasses, feeding in family groups, crows take numerous orthoptera and, as we shall see, rodents. The big corvid may be the staunchest wildlife ally of the type of farming which has developed in the Piedmont.

The eastern meadowlark, *Sturnella magna*, denizen of the tall grasses throughout the province in all seasons, is another important orthoptera controller. The meadowlark and the grasshopper sparrow, *Ammodramus savannarum*, strongly favor the hayfield habitat, and Piedmont farming practice has doubtlessly increased their numbers substantially. The grasshopper sparrow not only eats grasshoppers—witness Eliot Porter's eloquent portrait of the bird, grasshoppers in mouth—but its territorial song imitates the stridulation of the meadow grasshopper, *Orchelium vulgare*, a high-pitched, buzzing *tsick-tsick-zhhhhh*. The call is faint and ventriloquistic, easily blending with the general droning of summer orthoptera. It is nearly useless in locating the bird. We are left to speculate whether this strange vocal similarity is a case of parallel evolution, where the selective forces of the grassland habitat have somehow fashioned a tonal convergence in two vastly different organisms, or whether one of these meadow-dwellers imitates the other. And, if so, why? An additional service of the grasshopper sparrow is the control of weevils that infest hayfield legumes.

A host of additional bird species gain a major part of their livelihood from hayfield insects. Many have an interest in the lepidoptera, especially the larvae, as well as the orthoptera, and some divide their attention between the hayfields and other habitats. It is therefore appropriate to recognize the role of the overgrown hedgerow: these strips of woody vegetation bordering the hayfields provide part of the hunting territory and most of the protective cover for certain bird and mammal species. These ecotone dwellers scan the adjacent hayfields from hunting perches in the hedgerows, gaining the best advantages of both habitats. The blue grosbeak, *Guiraca caerulea*, brown thrasher, *Toxostoma rufum*, chipping sparrow, *Spizella passerina*, field sparrow, *Spizella pusilla*, indigo bunting, *Passerina cyanea*, and mockingbird, *Mimus polyglottus*—a list by no means exhaustive—feed heavily in the fields but nest and find singing and hunting perches in the hedgerows.

Twice yearly, during the spring and fall migrations, hordes of bobolinks glean the tall-grass hayfields for seeds and insects. The flocks of spring bobolinks gleam with the gold and silver backs of the males contrasting with their ebony underparts, an unusual violation of the principle of protective countershading, by which the lightness of the underparts relative to the rest of the body gives an even-shaded appearance. Fall bobolinks returning from nesting grounds in the taiga resemble outsize grasshopper sparrows. The two can occasionally be compared in Piedmont hayfields.

Chief among the open-country raptors in the eastern United States is the red-tailed hawk, *Buteo jamaicensis*. It is the largest eastern *Buteo* (the genus of the soaring hawks), with broad, thermal-riding wings spanning four feet (120 centimeters). The red-tail selects a hunting perch on an exposed limb at the edge of the grasslands and scans the habitat for hundreds of yards. The larger rodents discussed earlier, including young groundhogs, as well as cottontails, skunks, weasels, and muskrats provide food for *B. jamaicensis*. Toads and snakes are frequently taken. Prey which has been recorded in the red-tail's diet but which is not ordinarily associated with grasslands, the gray squirrel, for example, may often be scavenged from along roadsides. Photographing at the nest of a pair of red-tailed hawks in North Carolina in June 1977, I watched an adult arrive with a well-flattened squirrel not unlike one I had seen on the road on the way to the nest. Passing the spot on my way home four hours later, I noted the squirrel was gone.

At night the horned owl and, to a lesser extent, the smaller barred owl and the much smaller screech owl, emerge from the woods to hunt rodents in the open spaces. The horned owl, *Bubo virginianus*, is a massive raptor—

the largest North American bird of prey ouside of the eagles. Hayfield rodents and rabbits, being active at night, are likely to come under the all-seeing eye of a horned owl perched on a fence post or a tree at woods' edge. So strong is this owl (European members of the genus are referred to as the eagle owls) that it occasionally takes predatory mammals ranging in size up to the gray fox. Skunks and stray house cats are not uncommon prey. The snakes and amphibians living in hayfields provide a significant portion of the horned owl's diet.

In winter the diurnal raptors of the Piedmont hayfields include the American kestrel, *Falco sparverius*, and the marsh hawk, *Circus cyaneus*. The kestrel also nests in the northern Piedmont, being most in evidence as a breeding bird north of the Potomac. The kestrel, formerly called the sparrow hawk, takes a varied diet from the open grasslands; roughly half of its food is large insects, and the remainder is divided about equally between small birds and rodents. The availability of mice and insects makes Piedmont hayfields prime kestrel hunting territory. The little falcon sits on the bare limbs in fencerows or on utility wires scanning the grasses. It often pauses in flight to hover for long moments over prospective prey, facing the wind, motionless except for the shallow, quivering wing strokes that keep it aloft until it plunges down to strike. The marsh hawk flies tirelessly on easy wing strokes just above the grass tops, quartering the fields, especially the lower meadows, in search of rodents, reptiles, small birds, and insects. In all provinces where the marsh hawk winters, it shows a riparian (wetland) preference, but is frequently seen hunting stands of tall grass on the slopes in the Piedmont.

Crows, in all seasons, are among the principal hayfield mousers. For some reason, I was surprised to learn this, but the relationship of corvid to rodent was fixed in my mind by an injured crow who was in my care. The crow picked indifferently at his grain and table scraps. "Give him a mouse and watch what happens," said an ornithologist friend. The crow severed the mouse with a single blow of his beak, then tossed up and swallowed the halves with alacrity. Other large birds not commonly associated with the hayfield indulge an occasional taste for its rodents and insects. The great blue heron, *Ardea herodias*, ventures from the shallows of farm ponds into the nearby hayfields to whirl and dance, wings outstretched, in pursuit of mice, voles, and grasshoppers. The American bittern, *Botaurus lentiginosus*, well concealed in tall grass, tours the hayfields at twilight.

Mammals

Of the mammals of Piedmont hayfields, the shrews (order Insectivora, family Soricidae) are the premier insect-eaters. Orthoptera, particularly grasshoppers and crickets, are the dietary mainstays of shrews living in hayfields. These mouse-size hunters haunt a variety of habitats, and the three species found in Piedmont meadows show no habitat preference beyond being terrestrial. They are the southeastern shrew, *Sorex longirostris*, least shrew, *Cryptotis parva*, and short-tailed shrew, *Blarina brevicauda*. They are active wherever insect prey and protective cover are available, and hayfields meet these requirements admirably. The masked shrew, *Sorex cinereus*, occupies a vast boreal range that includes moist habitats, wooded and open, in the northern Piedmont. This species exemplifies the exceptionally high metabolic rates characteristic of shrews; heart rates of 1200 beats per minute have been recorded. To sustain such a high metabolism, the shrews eat voraciously, usually consuming more than their own weight each day. With populations of up to 25 of each shrew species per acre in good pasture habitats, these minute hunters, active day and night, account for a lot of orthoptera. Of course, they do not confine their attentions to orthoptera, nor even to insects. Shrews devour practically any invertebrate encountered—snails, worms, slugs—as well as other small mammals such as immature mice and voles.

The mammalian predators of Piedmont hayfields are shy and secretive, operating at night or under cover of tall grass. One group of particularly ardent mouse hunters which rarely penetrates the human awareness is the weasel. I have seen only one, a long-tailed weasel, *Mustella frenata*, which had been killed by a mowing machine in a hayfield in the Piedmont of northern Virginia. Though rarely seen, this species is common throughout the Piedmont, as it is in most of the United States. In diet it concentrates on small mammals up to the size of the cottontail; the kill is made by biting through the skull. With its long slender body, the long-tailed weasel is well adapted to pursuing mice through hayfield grasses, though it does not confine its hunting to grasslands. The literature lists one other weasel, the short-tailed, *Mustella erminea*, as present in the northern Piedmont, but its habitat preference is wooded, so we are probably safe in assuming the long-tailed to be the main weasel at work in Piedmont hayfields. The species does not confine its interests to rodents; the long-tailed weasel frequently embellishes its diet with birds. There was the regrettable instance of a half-dozen guinea poults given me by a neighbor in Piedmont, North Carolina. After the birds were half grown, a weasel took a fancy to them and for some days running there was a single dead guinea in the yard, unmarked except for the telltale fang punctures at the base of the skull and the neatly opened throat.

Another mustellid common in the Piedmont north of the Roanoke River is the striped skunk, *Mephitis mephitis*, who hunts a variety of habitats and is attracted by the insects and rodents of the hayfields. Omnivorous, the striped skunk in a single night's foray through the grasslands might dine on mice, insects, grubs, worms, birds' eggs, berries, and carrion. It does not hibernate, but takes advantage of the groundhog's winter torpor to share the shelter of its dens. The smaller spotted skunk, *Spilogale putoris*, shares with its larger relative the southern Piedmont. In any region occupied by skunks, the animals' presence is betrayed by the powerful scent, produced in glands common to all mustellids but so developed in the skunks as to be an almost totally effective defense. To the skunks' disadvantage, the great horned owl has no sense of smell.

Of the gray fox, *Urocyon cinereoargenteus*, a popular field guide to the mammals says, "A wonderful mouser; rarely invades poultry yards; probably wholly beneficial." This small canid, shy and nocturnal, feeds on a variety of fruits, insects, and vegetables, as well as small mammals on which it pounces, catlike, with its forepaws. Habitat, like the diet, is varied, resulting in a versatility of life style that is well matched to the Piedmont landscape, the mouse-rich hayfields and hedgerows in particular. Unlike the red fox, *Vulpes fulva* (believed by some to be conspecific with the European red fox, *Vulpes vulpes*), the gray fox climbs with alacrity, often to forage or to escape danger. A gray fox might well climb to take grapes, while the red fox, also common in the Piedmont, must wait till the fruit falls. Hedgerows are of particular importance to foxes as sheltered runways and as arbors of persimmons, blackberries, grapes, pokeberries, and wild cherries—sources of fructose to sweeten the high-protein cuisine of the adjacent hayfields.

The red fox is less nocturnal than the gray and can be seen in the fields by day. Fox hunters attest that the red fox is a tireless runner, leading a more merry chase than the gray, who might soon tire of the sport and take to a tree or to earth. The red fox, it is said, enjoys trifling with the dogs, sitting down to scratch within sight of the pack, then widening his lead as necessary. Both foxes are versatile predators of great importance in all habitats; they are particularly valuable in the hayfields of the Piedmont, where lagomorphs (rabbits, hares, and the like), rodents, and insects, if unattended, can change profits to penury. Foxes den in groundhog burrows usurped and enlarged to accommodate a half-dozen or more playful young. Alternate dens are maintained to which the family can be moved when danger threatens.

Rodents are sustained directly by the forage plants in Piedmont hayfields. Their populations, if unchecked by predation, can do considerable damage to the crop; all are characterized by unbelievably rapid reproduction. The deer mouse, *Percomyscus maniculatus*, lives in the hayfields of the northern Piedmont, although this rodent's catholic acceptance of habitats cautions us against associating the species with hayfields more than with other quarters. In hayfields the deer mouse feeds on roots, green shoots, seeds, and some insects. It nests in underground burrows within its feeding territory. The white-footed mouse, *Percomyscus leucopus*, includes the entire Piedmont in its extensive range. It prefers somewhat more overgrown habitats than the deer mouse but frequently occupies hayfields, especially those with lush fencerows or wooded borders. Its diet is effectively the same as that of the deer mouse. The meadow jumping mouse, *Zapus hudsonius*, a hibernating species, also accepts a variety of habitats within the Piedmont as far south as the Carolinas. Its preference is for moist meadows and its diet is typical of the small hayfield rodents, mostly new growth and seeds of the forage crops.

Probably the most numerous rodent of the Piedmont hayfields is the meadow vole, *Microtus pennsylvanius*, which occupies rank growths of legumes and grasses on both damp and well-drained soils. The meadow vole's range includes most of northern North America, dipping southeastward to encompass the majority of the Piedmont. Its fare is typical of other small rodents, but the impact of the species, due to its size (twice that of mice) and numbers, is especially heavy.

The hispid cotton rat, *Sigmodon hispidus*, five to eight inches (13 to 20 centimeters) long without its six-inch (15-centimeter) tail, includes the southern half of the Piedmont in its range. A grassland rodent of dynamic populations, this animal gives birth to as many as a dozen young at a time, produces up to nine litters yearly and reaches sexual maturity in 40 days. Its breeding season is the entire year. *S. hispidus* is a moderate-size rodent with a big appetite. Needless to say, the loss of a few of its principal predators could result in significant depredation to the hay crop. Its method is insidious; it works at ground level, and considerable thinning could result from this rodent's selective cropping of whole stems before the collective effect would be noticed. In addition to its fondness for forage grasses and legumes, the hispid cotton rat eats the eggs of ground-nesting birds.

The rice rat, *Oryzomys palustris*, slightly smaller than the hispid cotton rat but with a longer tail, occupies the province south of the Potomac. It shows a preference for moist, grassy conditions and is often a force in the natural community at work in the poorly drained portions of hayfields.

In the Piedmont grasslands north of the Roanoke River the groundhog, *Marmota monax*, is the rodent to be reckoned with. Woodchuck is the officially sanctioned common name, but groundhog or whistle pig are the terms to use in the Piedmont if you wish to be understood. My field guide to the North American mammals records the groundhog's weight at five to ten pounds (2 to 5 kilograms); I recall weighing one at 18 pounds (8 kilograms). In operating a cattle farm in the northern Virginia Piedmont, my family's initial experience with this marmot was one of constant warfare characterized by poor morale on our side. The groundhogs quickly replaced any losses of theirs so that the populations gave the appearance of being stabilized at saturation density. Our losses included several horses and cattle which had to be destroyed due to broken legs or irreparable sprains suffered stepping into concealed dens.

On balance, however, my experience with the groundhog helped me gain a sense of patience indispensable in observing nature. I have waited for hours behind stumps and stone fences to observe these wary, ever-watchful animals and to learn something of their life. Burrowing into the clayey soils in hayfields, more often into rock piles and hedgerows with access to hayfields and pastures, the groundhog is a fact of life in the northern half of the Piedmont. Hayfields and pastures have been a boon to the groundhog, who may have all but abandoned its traditional foods in favor of cultivated grasses and legumes where they are available. Because these plants do not grow much in winter, the groundhog passes the colder months in continuous hibernation, the males arising in February or March to trundle to the den of a prospective mate. The groundhog's is a true hibernation, not simply an energy-conserving, intermittent doze. It is characterized by very low heart and respiration rates and extremely frugal consumption of the carbohydrates taken from last summer's grasses and stored as body fat. During hibernation an individual may lose over half its body weight.

A healthy population of rodents is one appropriate to the carrying capacity of the habitat; it is not necessarily a large population. In the absence of predators, rodents afield in a habitat of very high carrying capacity, such as a protected (from grazing animals) hayfield, could multiply explosively. A single pair of hispid cotton rats, for example, is capable of producing between spring and winter enough offspring to overrun and destroy a hayfield. Such a population could not be considered healthy. Fortunately, the predators, being opportunists no less than their prey, are quick to seize upon a lucrative feeding circumstance and they will be found, barring some unnatural force such as environmental contamination or persecution by humans, wherever there is prey. Stated differently, if a hawk or a fox is present, it is needed to control the prey.

Pastures

The vegetative base which Piedmont farmers use in their pastures is nutritively similar to that of hayfields. But plant species in pastures have an additional burden to bear; they must be able to withstand the abuse of grazing. Varieties of fescue grass, *Festuca* spp., are popular pasturage because the foliage is guarded by a hardened outer layer which resists crushing. Among pastureland legumes, clovers of the smaller-leafed varieties are hardy. Grazed legumes and grasses must be capable of continued growth with all but a little of the foliage eaten away. They must be able to grow in a microclimate unprotected from cold, sun, and wind because they lack the yard-high cover of the rank growths of hayfields.

Among the creatures who prefer to live in grassy open spaces, the length of the grasses, and therefore the depth of the cover, is of critical importance. The fauna occupying a well-grazed pasture will differ significantly from that of the same field if left ungrazed to be cut for hay or silage. As we will see, this is especially true, or at least especially noticeable, among grassland birdlife. Piedmont pastures tend, over the long term, to carry approximately their maximum load of grazing ungulates, and the length of the grass cover, therefore, tends to be at a minimum year-round. In the main, our discussion pertains to well-grazed pastures with grass of three inches (7 centimeters) or less. Under conditions of heavy grazing (but not overgrazing), the plants conduct their biological functions very close to the ground. They flower and spread their principal photosynthesizing leaves within a fraction of an inch of the soil, the better to avoid having these critical organs eaten. The shoots they send up vertically provide a means of taking advantage of any opportunity to enlarge the plant's scale of operations, but can be sacrificed without significant detriment. This contrasts with the growth habits of hayfield plants, which after a few cycles of tall growth and moderate cutting (say, to within six inches [15 centimeters] of the ground) accomplish their reproduction and food-making well up on the stalk. For fauna, the tall growth of the hayfield provides good protective cover but perhaps reduced food near the ground. We must consider each of the grassland animals separately to explain their preferences for long or short grasses. For example, a yard-high tangle of legumes might be just fine if you are a meadow vole and can gnaw down a stem to eat the leaves and flowers in the safety of your tunnel, but the same cover might make foraging and quick escape difficult if you are a killdeer. A bobwhite quail, on the other hand, could eat its fill of lespedeza seeds from high on the plant, then leap to flight on strong legs and short, fast-beating wings, almost regardless of the cover.

In a well-grazed pasture largely shorn of tall cover, the bordering hedgerows take on added importance. Without hedgerow cover many of the perching birds (order Passeriformes) and other ecotone life forms are eliminated from a situation where they are sorely needed to protect the sometimes scanty grasses from insect and rodent depredation. As in the previous chapter, we will assume for the following discussion of animal life in pastures that hedgerows are present. However, because the biological composition of fencerows reflects local and regional trends in plant succession, our investigation of fencerow life systems themselves will be deferred to a later discussion of plant succession.

Weeds, too, are significant in pastureland life systems. The cutting edges of hooves sometimes denude the soil along well-traveled routes, and here pioneer plants germinate in the disturbed soils where conditions were too harsh for the cultivated plants. Poke, *Phytolacca americana*, and thistles, *Carduus* spp., are prominent pastureland weeds serving the very important function of preventing erosion of the exposed soil by holding the soil in place with their roots and by keeping the livestock from the spot while it recovers. Poke is handsomely adapted to this pioneer role; its seeds can lie dormant for long periods (50 years has been confirmed) to germinate only when the soil is disturbed. The plant is totally dependent on birds and animals for dispersal; the seeds can germinate only after being chemically scarified in the digestive tract of a vertebrate. And pokeberries are important wildlife food. The winter annual thistles send their seeds onto the wind to touch down on bare spots, perhaps, and to germinate as opportunities arise. The thistles have a unique and highly developed relationship with a particular fringillid (the family Fringillidae are the seed-eating birds) that releases the pappus-borne thistle seeds. We will soon explore this example of pastureland mutualism which Audubon found so picturesque.

The thistle is an aster, a member of the botanical family Asteraceae. The asters comprise a large and important group of plants in many parts of the eastern deciduous biome. They are especially important in the open spaces of the Piedmont. The thistles and some other asters (like fireweed, *Erechtites hieracifolia*, groundsel, *Senecio vulgaris*, and ragweed, *Ambrosia* spp.) are particularly successful in pastures because they are not eaten by the livestock, while their competitors are. They are often able to grow robustly while the forage crops around them are being shaved. As we contemplate a pastureland life system our attention is drawn to the savannah-like aspect lent by the poke plants and asters to what might otherwise be a landscape without cover or diversity. The weeds stand above the bald spots, holding the soil in place where the grass roots have failed and offering haven for a variety of microcommunities.

There is a major distinction between the vegetation, and therefore the higher life forms, on a hillside and that in an adjacent bottomland. The term *mesic* applies to soils of moderate wetness such as those on well-drained slopes. *Hydric* soils are characteristic of the flat and poorly drained bottomlands. The difference in drainage is one to which plants are very sensitive. Frequently in a Piedmont pasture the rolling slope of a hillside flattens onto the floodplain of some small stream, resulting in a marked increase in average year-round soil moisture. The pastureland grasses and legumes that thrive on the mesic slopes have difficulty competing with the rushes, *Juncus* spp., and sedges (family Cyperaceae), grasslike plants of hydric inclination which thrive as weeds in the open wet places. (Weeds, it is said, are plants for which man has found no use.) The mesic weeds as well yield to their hydric counterparts in the bottomland pastures; in late summer the magenta floral crowns of ironweed, *Vernonia noveboracensis*, dominate the wetland meadows, and several species in the asteraceaeous genera *Eupatorium* and *Helianthus* are common in the low open places. This is our first encounter with the mesic–hydric distinction, the major realignment in the life scheme that results from the presence or absence of a slope. We will find that the drainage factor produces wet and dry vegetative associations at all stages of plant succession, and we will meet the concept repeatedly in this guide. This factor can complicate the happy labor of getting to know the Piedmont's life systems, but remember this simple rule: hydric and mesic vegetations rarely mix—a given plant species typically grows either on the slopes or in the bottoms, not in both.

Animals of Pastures

Insects

As in hayfields, the orthoptera are prominent pastureland insects. Naturally, those adapted to life on the ground (crickets) and in the short grasses (short-horned grasshoppers) are most numerous. Ants are significant for their role of aerating the soil, a chore of particular importance where heavy-footed ungulates pack the upper layer.

Because there are relatively few grasses and legumes in flower in a well-grazed pasture, the pollinating insects are reduced in number. Skippers and small butterflies such as hairstreaks are occasionally seen darting through the foliage at shoelace height, but the larger, more spectacular butterflies are rare. Hunting wasps are common in pastures, coursing low over the grass tops in search of spiders and lepidopteran larvae. Cicada killers, *Sphecius speciosus*, are spectacular hunters, arriving in a controlled crash at their den entrances bearing paralysed prey twice their size. Cicada killers often build their nests in the bare spots in pastures, and in a small way no doubt contribute to the aeration and conditioning of the soil.

A word here about wasps. They are broadly divisible into species which are social and solitary. The social wasps are dangerous. They are colonial nesters, and when they perceive a threat to the nest they attack en masse. Their stingers are used for protection as well as for hunting, sometimes only for protection, since not all social wasps are hunters. The yellow jackets and hornets (family Vespidae) are social creatures who nest in inconspicuous tunnels in the ground. To inadvertently tread on the entrance of a nest is, at best, to spoil an outing. The paper wasps, builders of the dolorous gray globes sometimes seen in shade trees in Piedmont pastures or fencerows, are similarly to be avoided.

The solitary wasps, however, are harmless and offer fascinating life history studies. Their mode of operating is to dig a nesting tunnel or, in the case of mud-daubers (subfamily Trypoxyloninae), to construct chimneys resembling organ pipes and partitioned internally into individual cells, capture a spider or insect, sting it to induce paralysis but not death, and seal it in the nesting cell with an egg. The egg hatches into a wasp larva, eats the prey which was left for provisioning, pupates, then leaves to begin its adult life. Watching a solitary wasp on its hunting errands, you might see a small fly (probably from either family Sarcophagidae or family Tachinidae) or another, smaller wasp called a cuckoo wasp (family Chrysididae), follow the hunting wasp's every move or somehow anticipate the movements as a pilot fish foreshadows a shark. These are cleptoparasites who follow the prey-laden wasp into her chamber and lay their own eggs on the provisioning victim. The cleptoparasites' eggs hatch first, eat the provisions and usually the host's larvae, pupate, and depart. Pasturelands and other principally open-country habitats in the Piedmont are excellent sites for observing the solitary wasps. Howard Ensign Evans' *Wasp Farm* is a delightfully readable celebration of the ways of these falcons of the insect realm.

Reptiles and Amphibians

The principal rodent-eating snakes of the province will be present in proportion to the prey. Look for the black rat snake, *Elaphe obsoleta obsoleta*, the black racer, *Coluber constrictor constrictor*, the copperhead, *Agkistrodon contortrix mokeson*, and occasionally, because of the presence of toads, the eastern hog-nosed snake, *Heterodon platyrhinos*. The latter is a harmless mimic of the copperhead. Its defense is to flatten its head until it assumes the menacing triangular shape of the pit vipers, a contortion accompanied by hisses and lashing movements. When this fails, the hognose rolls onto its back, mouth agape, and plays dead. If righted, it reinverts itself to emphasize that it has died and should no longer be molested. The copperhead is the only poisonous snake in the Piedmont which is not a rarity; having encountered it only twice, however, I would consider it uncommon. On one occasion I actually stepped on the viper; on the other, I placed my hand on the ground (I was on my hands and knees picking grapes in my overgrown Piedmont yard) within six inches of a copulating pair. In neither instance was I attacked. I have a friend, however, a Piedmont farmer, who has been bitten twice, resulting in messy, necrotic wounds taking weeks to heal. Fatalities from snakebite, incidently, are much rarer than fatalities from hacking, amateurish attempts to treat snakebite.

The copperhead has shown me the more tolerant side of its personality, but remember to approach it cautiously. The red-tailed hawk and the horned owl are more suited than we are to keeping it in check.

From North Carolina southward in the eastern Piedmont, where grass is tall enough, the upland chorus frog, *Pseudacris triseriata feriarum*, may be seen clinging and jumping, monkeylike, to the grass stems. Where grasses are short enough, the American and Fowler's toads, *Bufo americanus* and *B. woodhousei fowleri*, pursue insects, mostly at night.

Birds

During the nesting season the chipping sparrow, *Spizella passerina*, is the prominent fringillid foraging for seeds and insects in the closely cropped grasses, but this diminutive sparrow retires in winter to the weedier, overgrown fields, roadsides, and hedgerows. In winter the savannah sparrow, *Passerculus sandwichensis*, replaces the chipping sparrow in the pastures, eating mostly beetles. In all seasons the ringing cry of the killdeer, *Charadrius vociferous*, echoes over the Piedmont pastures, hydric and mesic. On the slopes the killdeer subsidizes the farmer by eating grasshoppers, beetles, and weevils from the forage crops and ticks, flies, and mosquitos from around livestock. The killdeer nests on bare, gravel-strewn spots in pastures, limping away from the nest with a heart-rending distraction display when a human approaches. The eastern meadowlark, *Sturnella magna*, is a common pastureland bird, conspicuous in song and color. The meadowlark likes its grasses somewhat longer than lawn length but shorter than the hip-high hayfield growths. On the slopes its summer diet focuses on the grasseating orthoptera; in winter the emphasis shifts to weed seeds and grubs (beetle larvae). Another icterine, the common grackle, *Quiscalus quiscala*, forages in winter pastures in great numbers, also taking seeds and grubs. The family Turdidae is represented in Piedmont pastures by the robin, *Turdus migratorius*, and the eastern bluebird, *Sialia sialis*, whose principal interests in all seasons are short-grass insects and other invertebrates. Bluebirds are especially fond of pastureland crickets. Robins collect in large flocks to forage in the sod for earthworms and beetle larvae, and, occasionally, to sit on a south-facing hillside and stare, as if entranced, at the winter sun. I have walked into the midst of a scattered flock of robins deeply engaged in this strange seance.

Where pastures adjoin pine forests the pine warbler, *Dendroica pinus*, frequently joins the family groups of chipping sparrows, specifically seeking out their company and occasionally pirating insects from the small fringillids. Crows, of course, frequent short-grass fields for essentially the same attractions that occasion their visits to the hayfields—mice and grasshoppers. In winter, flocks of water pipits, *Anthus spinoletta*, though preferring the plowed fields, frequently wheel above the pastures in wide arcs, then make their stair-step descent en masse into the more closely cropped pastures in search of seeds and insects.

Wetland pastures are the only habitats utilized in the Piedmont by the common snipe, *Capella gallinago*, a diurnal sandpiper (member of the family Scolopacidae) who probes the mud of open bottomlands for worms and other invertebrates. The song sparrow, *Melospiza melodia*, is at home in the pastured bottomlands, though not exclusively so, preferring accommodations with some weed cover or streamside vegetation (such as willows) but feeding principally in the wet pastureland. The elaborate ringing territorial call, heard in the Piedmont as far south as South Carolina (the bird winters throughout the province but does not sing on the wintering grounds), is one of the sounds we specifically associate with pastured bottomlands, as is the snipe's rasping *kzzzrt*, given when flushed.

The predatory birds of mesic pastures are those of the hayfields (the red-tailed hawk by day, the great horned owl by night) plus the American kestrel, *Falco sparverius*, where grasses are of medium height, and the loggerhead shrike, *Lanius ludovicianus*, in greatest numbers where grasses are short. The shrike nests and winters throughout the Piedmont. Its favored nesting circumstance is a small, isolated tree in a pasture, ideally a locust or hawthorn with spikes on which prey (in summer mostly grasshoppers) can be impaled. The winter habitat is the same, with a dietary switch to mice and small birds spied from a lookout perch on a roadside utility wire, a fence, or a hedgerow tree. The kestrel hunts the Piedmont's pastures in winter, taking the same or larger prey as the shrike, swooping to kill from a high perch or from a hover. Popular field guides show the entire province to be within the kestrel's nesting territory, but I have never seen evidence of a nest south of the James River.

Lowland pastures are a favored hunting habitat of the marsh hawk, *Circus cyaneus*, wintering in the Piedmont. Meadow voles, mice, and, in the southern Piedmont, rice rats, are principals in the diet; reptiles, insects, amphibians, and small birds complement the fare. The short-eared owl, *Asio flammeus*, hunts the same habitat in the Piedmont by night, and because it is sometimes active by day, can occasionally be spotted quartering a pastured bottom, like the marsh hawk, alternately flapping and sailing, holding its wingtips upward to keep them out of the grass tops.

Mammals

A pasture's attractiveness to rodents, especially the smaller species, relates directly to the length of the grasses. The populations in a sparingly grazed pasture might approximate those in a hayfield, while rodents could be scarce in a closely cropped pasture. The author would be hard pressed to cite rodents who prefer short cover over tall. The meadow vole, *Microtus pennsylvanicus*, and the deer mouse, *Percomyscus maniculatus*, frequent pastures throughout the province. In the southern Piedmont there is also the hispid cotton rat, *Sigmodon hispidus*, on the pastured slopes and the rice rat, *Oryzomys palustris* (*Oryza* is the genus of rice), in the grassy bottomlands. Groundhogs, *Marmota monax*, are common in pastures and are a real peril to livestock; mysterious lameness and even broken legs are often attributed to a well-concealed entrance to labyrinthine burrows of this eastern marmot. There is a long-term self-correcting factor here, though. The groundhog, wary and jealous of its privacy, responds to heavy grazing by abandoning the more exposed dens in open pastures in favor of protected sites in the fencerows and rock piles.

The least and short-tailed shrews, *Cryptotis parva* and *Blaria brevicauda*, are present and actively in pursuit of the orthoptera where grass cover is adequate. The eastern cottontail, *Sylvilagus floridanus*, may actually prefer the shorter, sparcer grasses of pastures over the rank tangle of hayfields, provided the protection of fencerows is hard by. In the northern Piedmont, the New England cottontail, *S. transitionalis*, inhabits the brushier, hillier sections, often feeding in pastures. The two rabbits are not easily differentiated by sight in the field, and probably with even greater difficulty by taste at the table. The striped skunk, *Mephitis mephitis*, forages for the fruits of weeds, mostly poke, and for mice and insects, as do the red and gray foxes, *Vulpes fulva* and *Urocyon cinereoargenteus*. The latter trio of predators have all been known to make use of groundhog dens, sometimes ad hoc, often by semipermanent appropriation. The white-tailed deer, *Odocoileus virginianus* is frequently seen sharing Piedmont pastures with domestic livestock, especially where adjacent cover promises quick escape.

Microcommunities

Livestock Droppings

Domestic grazing animals are the principal consumers of pastureland forage crops. The even-toed ungulates, especially cattle, have long, complex digestive tracts designed to process vast volumes of vegetative material. These animals spend the majority of their waking hours eating, chewing cuds of regurgitated food, or otherwise processing what they have swallowed in a digestive apparatus which begins with a four-chambered stomach and continues through a lengthy intestine. The process is efficient but not totally efficient, and a considerable amount of undigested material passes through, bearing notable caloric and nutritive value.

The vigor of the microcommunities which develop in and on pastureland droppings attests to this. No sooner, literally, does the dung hit the ground than it is aswarm with flies and beetles. Skippers and butterflies sometimes dab for moisture and carbohydrates, but the diptera (flies) and coleoptera (beetles) dominate. Arriving by air, the quarter-inch beetles of the family Hydrophilidae plow into the steaming riches. The initial flies and beetles lay their eggs in the soft dung, which provides food for their larvae. The young live in the galleries channelled in the crust which forms on the outside surface but feed in the soft interior. The initial coleoptera are joined by certain scarab beetles (family Scarabacidae) who specialize in the coprophagous life style; these are the dung beetles and tumble bugs. The latter, deified by the ancient Egyptians, construct a ball of dung, roll it some distance away, oviposit on the ball, and bury it as food for the hatchlings. Ground beetles (family Carabidae) arrive as the dung heap matures to scavenge the detritus and fungi from beneath. I have seen the larvae of at least one lepidopteran, probably a noctuid moth, dwelling beneath cow dung, presumably having been there since hatching.

After the dung pile hardens, it begins to soften to a powder on the inside. The lower organisms play an important role in decomposing (composting) the droppings at this stage, not unlike they do in the rotting of wood, processing materials so that they become palatable to higher organisms.

A series of fungi reveal their fruiting bodies on or under the dung at intervals over the decomposition process. Most have been in the dung as it passed through the ungulate, probably as ungerminated spores. The fruiting bodies, of course, are not the chemically active parts of the fungi; food gathering and assimilation are accomplished in the minute, threadlike hyphae which together form the fungal mycelium laced through the dung and on which the mushrooms appear. The fungi associated with ungulate dung in many cases accept no other substrate or food source, and their entire life cycles depend upon the spores being eaten with the forage crops to which they adhere and being chemically treated in the digestive tracts.

After an extended period of being attacked by coordinated teams of bacteria, fungi, and insects, each continuing the process beyond the chemical simplification worked by its predecessors, the remaining fiber and nutrients are returned to the soil by earthworms and by the ubiquitous colonies of ants.

The churning colonies of secondary heterotrophs (the contributing ungulate having eaten the plants and thereby become the primary heterotroph) naturally attracts predatory attention. Salticid and lycosid spiders lurk in the crevices and craters of dung waiting to pounce on flies. Opossums, foxes, and raccoons are known to vandalize the cow pies, breaking into the crust to get the larger beetles and their larvae. Birds are notably in attendance at these banquets, gleaning wasted grains and gobbling insects. The larger icterines (cowbirds, grackles, and meadowlarks) gain a large portion of their diet in this manner. Crows, great blue herons, and turkey vultures, to name a few larger birds, can be seen at these pastureland foci of life, too.

We of the animal kingdom pass through us a great deal more of what we eat than we fix in our tissues; body wastes are central to the cycles of the more important life-associated elements, notably carbon and nitrogen. The liberation of the essential materials in animal droppings is certainly no less important than the recycling of the tissues themselves after death. Those who scavenge upon (thereby retaining the materials at a fairly high trophic level) or decompose (and release the elements into the terrestrial and atmospheric pantries for later use by autotrophs—plants) the droppings in a Piedmont pasture telescope the global process of recycling.

Thistles and Other Asters

Standing tall above a closely cropped pasture, individual plants of *Carduus*, the principal thistle genus, and other aster genera often shelter vibrant microcommunities. Those on thistles are especially active because the plant's leaves are tipped with stiff, sharp spikes the collective effect of which is to shield all within the foliage against intrusion from outside. A careless gesture of the human hand or foot will readily demonstrate the effectiveness of the fortifications. The thistle provides food, shelter, and breeding places for scores of organisms whose lives interrelate because of the plant.

Feeding directly on the thistle are numerous varieties of aphids and treehoppers who are, more often than not, "farmed" by colonies of ants. These homopterans pierce the vascular system of the plant with mouth parts specialized for the purpose and suck out carbohydrate-laden sap from which they produce and excrete upon demand by ants a sugar-rich "honeydew." When stroked by the antennae of ants, the homopterans produce minute jewel-like spheres of it; in the absence of stroking the excesses are sprayed at random into the air. The ants defiantly protect their charges; the larger members of the subfamily Formicinae (carpenter ants), five-eighths inch (16 millimeters) long, effectively guard against all but the more determined predators. Lepidopteran (moth and butterfly) larvae, snails, click beetles (family Elateridae), and grasshoppers also eat the thistle foliage.

The flowers of thistles are large pink inflorescences composed of many florets packed tightly in an urn-shaped cluster. These inflorescences attract pollinators from all corners of the dipteran (flies, mosquitos, etc.), hymenopteran (bees, wasps, ants, etc.), and lepidopteran realms as well as a single representative from the class Aves, the ruby-throated hummingbird, *Archilocus colubris*. It is not unusual to see bumblebees and their lepidopteran mimics, the bumblebee moths (members of the family Sphingidae, whose wings are largely devoid of scales) on the flowers of the same thistle. The hummingbird's purpose in visiting the thistle flowers is probably to gather with their darting tongues small insects from between the florets rather than to tap the minute nectaries, but there is no question that the birds assist in the transfer of pollen.

Multiple levels of predation are readily observable on the thistles of a Piedmont pasture. Small salticid spiders pounce on flies and skippers who might be visiting the thistle to scavenge fallen droplets of honeydew. These spiders might become the prey of larger spiders or perhaps of the five-lined skink, *Eumeces fasciatus*, who might expose himself to a passing crow. Or the same salticid, after snaring a colorful bush cricket (subfamily Trigonidiinae), might leap into the web of an orb-weaving spider (the big, black-and-gold argiopes are common on thistles) who could later be taken by the Carolina mantis. This mantid shows a well-established relationship with the thistles, synchronizing the changing of its body colors with the browning of spent thistle foliage toward the end of the growing season. Thistle and mantid are both a fresh pea-green in spring's new growth, mottled brown on green by late summer, and dusky brown by the onset of frost. Several groups of predators station themselves at the narrowed waist of the floral urn, just below the exposed corollas, to nab unwary pollinators. Crab spiders (family Thomisidae) and ambush bugs (family Phymatidae) are notably adapted to this hunting mode and are numerically strong on Piedmont thistles. Another group of the true bugs (members of the order Hemiptera, those insects identifiable in the field by the rearward-pointing triangular plate, or scutellum, just aft of the thorax) which perform an active predatory role on the thistles is the assassin bug (family Reduviidae). Moths, butterflies, and grasshoppers are principals in the diets of assassin bugs. Many can inflict a painful bite.

Perhaps the thistle's most striking functional kinship is

with the American goldfinch, *Spinus tristes*. As Audubon recorded in his depiction of the goldfinch on a thistle, the life modes of these two organisms are closely interlocked. The goldfinch delays nesting until midsummer, when the first thistle flowers have gone to seed. The downy pappus, an airfoil which expands upon being released to catch the wind and carry the seed, is used to line the "wild canary's" nest, and the male feeds regurgitated thistle seeds to his mate while she broods and to the young when they hatch. By late summer, the feeding families of goldfinches are visiting the thistles together, tearing into the urns to feed on the ripening seeds, which they inadvertently share with the wind, thus helping disperse the next generation of thistles.

Some of the pastureland asters other than the thistles also sustain interesting communities worthy of investigation. Note, for example, how the crab spiders (particularly *Misumena* spp.) change color to match the white flowers of boneset, *Eupatorium perfoliatum*, in the bottomlands or the yellows of the goldenrod or helianthus on the slopes. And the assortments of soldier beetles (family Cantharidae), butterflies, and bees on the flowers of ironweed, purple-flowered regent (in anticipation of the plant species that will dominate later stages of succession) of the grassy wetlands, rival the insect communities at the thistles on the slopes.

PART III

Lands in Succession

CHAPTER FOUR

Plant Succession

ONE OF THE QUALITIES separating the organic and inorganic realms is the power of living things to heal themselves. Machines rust and wear without maintenance; rocks crack, crumble, and are carried to the sea. Only living things come into being with the ability to repair their damaged substance.

The concept applies to systems of living things perhaps even more strongly than to individual organisms. The cut finger heals but leaves a scar devoid of nerve endings, hair, glands, and pigment. The forest, though, can restore itself completely, leaving hardly a vestige of the original lesion. It has only two requirements—seeds and time.

Imagine a gap in the primordial Piedmont canopy of oak, hickory, and chestnut, perhaps a large gap resulting from a roaring holocaust which burned thousands of acres. The seeds to restore the forest are constantly adrift on a sea of wind, water, migrating birds, cache-minded rodents, roving predators. The germ cells arrive on every moving vector, and some are already present in the soil, perhaps even activated by the fire. They germinate—not all at once or at random, but by a carefully orchestrated sequence which results in one dominant plant species (and its associated community of life) succeeding another until, if enough time passes without additional disturbance, the original forest community is restored. This restorative sequence, aimed at regenerating the climax plant community after a disturbance, is called secondary plant succession.

More fundamental disturbances can expose surfaces which have not previously sustained life: glaciations, crustal upheavals, marine invasions and retreats, landslides, volcanoes, erosion, bulldozers. Any lifeless surface carries a certain affront to the plant kingdom. The response is a relentless pressure to invade and colonize. Botanists call this process primary plant succession. It is, as with secondary succession, mostly a question of seeds and time. The time involved is usually much greater than in secondary succession, and the "seeds" tend, at least in the very early stages of primary succession, to be the reproductive bodies of primitive plants such as lichens, mosses, and ferns, and they are more commonly called spores. We will look closely into the primary succession process in our exploration of the granite outcrops of the southern Piedmont.

In the 30 thousand years man has been a guest upon this continent, he has made fundamental changes in the nature of the forest. It is no longer relevant to speak of an isolated rent in an otherwise untorn blanket of climax deciduous forest in the Piedmont, and it is legitimate to wonder about the relevance of placing major emphasis on forest regeneration (another term for secondary succession as it applies to the Piedmont) where there are few if any primal forests to regenerate.

The relevance of our interest in succession is established by noting that although the forest is gone, its restorative force is still very much at large. It is precisely *because* of the destruction of the primal deciduous forest that succession is a power so much in evidence in the Piedmont. The tendency of nature to reclaim her own is so persistent that we are obliged to recognize that now, as always, the land is in one of two conditions: climax forest or some stage of forest regeneration. The negligible fraction now occupied by climax communities only emphasizes the importance of the successional vegetative stages cloaking the majority of the province.

How, one asks, can it be seriously contended that all the land, Piedmont and beyond, is occupied either by a climax community or some phase headed toward the climax? Vast cities bituminate the land with lifeless asphalt; farms bare unnumbered acres by the plow; playing fields, golf courses, airports, racetracks, subdivisions, and pastures widen the countryside with "open space" where there is no climax community nor anything that appears to be headed that way.

But let some time pass, and not so very much, and we see the open spaces begin to fill with weeds, the weeds yield to trees. Asphalt, when not under constant maintenance, crumbles as plants germinate in organic collections on its surface or, in a not-infrequent show of force, punch up through the pavement. Open fields, unmowed, are overtaken by weeds in a single summer, by woody plants within ten years. Scrubby tangles of young broadleaf trees occupy the yard of a house where friends lived a decade ago. Pines, come to think of it, crowd together in the field one played in as a child.

Succession is an important biological force. It is unsuppressable over the long term. It poses a constant, ubiquitous threat to man's ability to maintain the open space needed for his vital activities. With so much said, we will simplify this inquiry if we exempt from the immediate control of successional forces the lands we actively farm and on which we live. It is a deceptive exemption at best, for although we perceive ourselves to be in control of these lands now, and might reasonably expect to retain control for centuries more, the potential for successional usurpation will always be present. The subject of succession is presently a frontier of biological inquiry into which plant ecologists have advanced far enough to pause over new complexities; for our explorations we will use the simplest models consistent with what contemporary science tells us and with what we can observe around us.

The Terminology of Succession

Succession is a process of vegetational change in which plant communities replace one another until a climax equilibrium is reached. All the vegetational phases taken together make up the *sere* at a given site. The sere, of course, varies from place to place and even at a particular site may change from, say, one clear-cutting of a forest stand to the next. The sere also shifts gradually as climates change and as plant species migrate.

The last seral stage is termed the climax, a condition in which vegetative change ceases because an association of plants has developed which perpetuates itself at the site. On poorly drained soils the sere is called a hydrosere, implying that there is an expected sequence of plant communities for each site where soils are habitually moist. There is in addition a school of thought, perhaps archaic, which contends that as the hydrosere progresses it will become less hydric and more mesic; that is, more toward the midrange of moistness as a result of elevation gained due to organic accumulation over time. A xerosere is the vegetative series on a xeric, or habitually dry, site, and likewise implies to some a trend toward more mesic conditions. The term mesosere embraces all successional stages on sites which are well-drained but adequately watered to avoid drought stress. Postclimax suggests a condition more mesic or more mesophytic (the Greek *phytos* means 'plant') than the expected climax.

Effects of Drainage on Succession

The U.S. Department of Agriculture recognizes seven moisture classes applying to soils. Ranging from dry to wet, they are:

1. Very dry–xeric (excessively drained)
2. Dry–xeric (somewhat excessively drained)
3. Dry–mesic (well drained)
4. Mesic (moderately well drained)
5. Wet–mesic (somewhat poorly drained)
6. Wet–hydric (poorly drained)
7. Hydric (very poorly drained)

The extremes of this range do not apply to the Piedmont except perhaps where there are localized anomalies. We are mainly concerned with the middle half, or so, of the soil-moisture continuum, and to avoid being confused by terminology we can simplify the terms describing Piedmont soil moisture to:

1. Xeric (dry)
2. Mesic (well drained)
3. Hydric (wet)

Because virtually all parts of the Piedmont receive similar average annual rainfall, the differences in soil moisture are due to the ability of the soils to retain their complement of precipitation. Wetness variances are functions of slope, soil texture, porosity, and vegetational cover. For example, if all other factors are equal a site of modest slope retains its moisture more tenaciously than one of greater slope. Further, on a given gradient the lower portions of the slope stay wetter than points higher because of the downward seep. Porous soils (those of coarser granularity) are more easily drained than those of finer texture and, concomitantly, they are more easily leached of their minerals. Soils protected by a well-developed ground cover of vascular plants stay moister than like sites sparsely vegetated.

Among environmental forces, soil moisture, whatever its governing factors, has a particularly strong influence on the vegetation that might occupy a given site. So strong, in fact, is the selecting influence of moisture that, as a rule, different species of plants participate in the xerosere, the mesosere, and the hydrosere. Because the steepness of the slope is usually the critical factor in determining the soil moisture at a given Piedmont location, we can expect that most of the province's xeric sites will be found on hilltops and upper slopes, mesic conditions on the middle and lower slopes, and hydric conditions in the bottomlands.

To the foregoing we must add that occasionally in the Piedmont we find adjacent sites of very different soil moisture but only slight differences in elevation or slope. Fertile croplands can be found hard by forested wetlands which probably never have been farmed. The most frequent explanation is that certain rock formations weather and shrink beneath the soil, allowing the location to sink slightly, sometimes imperceptibly, below the level of neighboring land. Moisture retention in such depressions may be enhanced by the presence of imporous or impervious material beneath the surface. The vegetational import of these low places is to create islands of hydric conditions in an otherwise mesic countryside. Often the hydrophytic vegetation serves to identify such depressions more clearly than do the terrain features.

Historical Perspective

The eighteenth century saw the total occupation of the

Piedmont by European colonists. It was an era of forest clearing, and any virgin stand which survived that century was logged or farmed or both in the next. Prior to the present century, if a Piedmont farm became depleted, the depleters moved elsewhere. Farming methods were abusive to the soil and it was the expectation of a homesteader arriving on new land that he could eventually exhaust the soil and have to move westward. The idea of living with the consequences of one's agricultural practices was widely accepted only when there were no more western lands to be had, largely a phenomenon of the present century.

There were successive waves of occupants of the Piedmont. Many tracts, cut over, farmed out, and abandoned by the year 1800, would later be reclaimed by other immigrants who then contended with second-growth vegetation rather than with the climax forest. It is probably safe to suggest that many if not most Piedmont tracts went through several of such cycles before the twentieth century.

The exploitative ethic was not uniformly applied; conscientious husbandry of the soil was practiced by some from the time of the first Piedmont settlements. The existence today of fine old homes and barns dating from the nineteenth and even eighteenth centuries confirms that some who settled the province intended to stay and maintain the land for use by future generations. The northern Piedmont, particularly Pennsylvania, Maryland, and northern Virginia, is still graced with numerous working farmsteads of that nature—stone-sided barns embanked into a hillside, two-story brick and stone homes with a fireplace in every room. The architecture and well-kept fields bequeathed to the present by those farsighted freeholders have become the signature of the province.

The present century has been a time of extensive abandonment of the Piedmont's farmlands. The lure of new western lands has not been present, but economic and social forces have conspired to make agriculture in the Piedmont less profitable than in other parts of the nation. The flatter western farmlands, which are not naturally divided into small hilly fields, can better absorb the capital-intensive technologies of modern farming.

There has been for decades a continuing pressure to enlarge the unit size of farming operations to compensate for the rising costs of equipment and labor. Because the rolling, segmented terrain of the Piedmont does not lend itself to the efficient use of large machines, numerous families have liquidated their Piedmont assets in favor of flatter topography elsewhere. To continue in business, the surviving farms have been forced to increase productivity in some cases by more than 100 percent in the past three decades. This is particularly true of the Piedmont's beef, poultry, and dairy farms. The efficiencies have been achieved through the substitution of equipment, chemicals, and fossil fuels for horsepower and manpower. As a result there are fewer families now sustained directly by and on the Piedmont's farms than at any time since the eighteenth century. The number continues to decline.

For those who remain, the pressures are intense. Piedmont farmers face all the well-known cost–price squeezes which beset American agriculture, but they do not have open to them the remedy of enlarging their operations to achieve increased economies of scale. The proximity of large eastern cities has driven the price of farmland far beyond what can be amortized through working the soil, and because the land is now so valuable for nonfarming purposes, the ad valorem property taxes are higher. Few Piedmont farmers are not keenly—painfully—aware of these forces and of the options open to them.

To summarize the trend of farmland abandonment in the Piedmont, I make the following land use estimates, by state, based on plant succession patterns.

Table 3: Percent of Piedmont in Active Farmland

	1900	*1940*	*1960*	*Present*
New Jersey	70	60	55	40
Pennsylvania	80	70	67	65
Maryland	65	60	55	45
Virginia	70	50	40	35
North Carolina	75	50	40	30

Table 3: Percent of Piedmont in Active Farmland
[continued]

	1900	*1940*	*1960*	*Present*
South Carolina	75	50	35	20
Georgia	75	50	30	15
Alabama	75	40	20	10

The individual figures, particularly those more remote in time, are subject to the errors of estimation, but the trend they collectively depict is beyond contention.

Abandoned Farmland

Abandonment has become especially common in the southern Piedmont. From the Sumter National Forest in central South Carolina to the southwestern tip of the province, only about 15 percent of the land is under active cultivation in pasture or row crops. The northern half of the Piedmont, by contrast, sustains extensive agriculture, most notably in Pennsylvania. The causes of these trends we have partially examined; there are many other factors worthy of consideration by the interested observer. To further our understanding of the abandonment phenomenon, we will look briefly at the apparent fate of defunct farms in the northern and southern halves of the province, noting the contrast in the rates of abandonment in these respective sections during the twentieth century.

The northern Piedmont, subject to the pressures of the nearly continuous urban corridor along the fall line from New York City to Richmond, has seen an intense increase in suburban and exurban development. A wide range of commercial, industrial, and residential facilities have been built during the past few decades in the northern Piedmont countryside. So rapidly is abandoned farmland developed in some northern Piedmont communities that successional phases of plant growth have limited opportunities to appear. They most often occur at the edges of towns and cities, reflecting the blighting effect that urbanization typically has on adjacent agriculture where land is too highly taxed or valued to farm but is not yet ripe for development.

The Pennsylvania Piedmont, however, shows an encouraging near-absence of these circumferential zones of deserted land. The direct juxtaposition of cornfields and apartment buildings no doubt reflects some blend of enlightened land use taxation and a laudable tenacity on the part of the farmers.

The southern half of the province is characterized by wholesale abandonment across the countryside. The Virginia Piedmont between the Potomac and the James shows moderate recent abandonment, but between the James and the Roanoke the country is cloaked in nearly continuous successional woods, perhaps largely the result of failures of the 1930s and 1940s. The preeminence of successional woodlands slackens somewhat along the crescent of Piedmont towns anchored in the east by Raleigh, North Carolina, and in the southwest by Greenville, South Carolina. Proceeding southward from northern South Carolina, the province is effectively a single successional woodlot broken by occasional pastures and cropland. A significant fraction of the existing open space is in the early stages of succession, signifying recent abandonment. The land is sparsely occupied, and empty decaying homes and farmsteads are commonplace. Many large tracts are owned by wood products companies.

North and south, the overwhelming majority of the land in the Piedmont has at some time been cleared and farmed; the fraction probably exceeds nine-tenths of the province's total acreage. Exceptions are steep slopes, rocky outcrops, and permanently wet depressions and bottomlands. Even if a particular terrain seems too formidable for any modern farming application, it is usually wise to presume former cultivation because pasturage and orchards can be established on slopes up to 50 degrees.

Many abandoned fields well into forestation retain a washboard corrugation on the surface, the remains of old crop rows. Old terraces and drainage ditches are easily discernible regardless of present vegetation. Large piles of rocks were collected to rid fields of stones to facilitate plowing; these rock piles are visible in many tracts now wooded. Old fences of stone, wire, or rail in present woodlands probably enclosed or separated cleared fields at one time. Perhaps the most doleful reminders of previous farming are the erosion gullies which still gape in the Piedmont's woodlands. These washes are easily overlooked because the forest vegetation has the visual effect of unifying the topography and giving permanence to all terrain features; but they are nearly ubiquitous, particularly in the woodlands of the southern Piedmont. Such gullies do not, of course, occur naturally. They inscribe in the earth mute testament to the abusive farming practices of earlier times and to the human hardship which resulted.

Abandoned fields are overtaken by woody plants very soon after cultivation stops, and the aspect of open space is quickly lost. A field abandoned a century ago may have gone through several cycles or combinations of growth and cutting, reforestation, and timbering, with the result that the present stand may be uneven in age and may be composed of a mixture of forest plants from several stages in the sere. However, when we encounter an even-aged stand composed of what we understand to be plants associated with that seral stage, we are probably safe in concluding that we are seeing a textbook example of natural reforestation after abrupt abandonment. Attempts to decipher the recent history of a particular tract are not always successful, but some progress into the origins and interruptions of the sere usually reward the effort. The key lies in knowing the successional phases expected in a given part of the Piedmont.

Piedmont Succession Stages

In 1911, H. C. Cowles proposed the concept of plant succession and five years later F. E. Clements suggested that succession is the "mechanism of progress" in plant ecology. E. Lucy Braun, P. Dansereau, R. Dubenmire, and E. P. Odum expanded and refined Clementian successional concepts to include the following principles: (1) That succession is directional and predictable because (2) each species alters its environment to suppress itself and favor the requirements of its successor, and that (3) species and associations are attached to a place in the successional order. (4) That biomass (the aggregate of living tissue) and productivity increase as succession progresses. (5) That soil is developed as a concomitant of plant succession. Guided by these concepts, several researchers have, since the 1930s, devoted specific attention to the successional processes of the Piedmont. Notable among them are H. G. Wells, H. J. Oosting, and Catherine Keever, to all of whom we are indebted for their specific data on the composition of Piedmont seral phases.

The Clementian models are still embraced by many knowledgeable plant ecologists, but they have been questioned in the present decade by F. E. Egler, J. McCormick, W. H. Drury, and I.C.T. Nisbet, among others, particularly regarding the notion that each dominant assists its successor. Many successional ecologists question the predictability of a presumed sequence, and the traditional Clementian view of a world in which all sites, wet and dry, flat and steep, are headed toward a mesophytic stability is now thought by some to be naive.

Researchers who have focused on secondary succession in the Piedmont (particularly Wells, Oosting, and Keever) conclude that there is a rather rigidly predictable schedule whereby certain species emerge as dominants after a cultivated field is abandoned. Their conclusions, now at least a quarter-century old, are described in Table 4.

Table 4: Secondary Succession in the Piedmont

Year	*Dominant*
0	Crabgrass, *Digitaria sanguinalis*
1	Horseweed, *Erigeron canadensis*
2	*Aster* spp.
3	Broomsedge, *Andropogon* spp.

Table 4: Secondary Succession in the Piedmont [continued]

Year	*Dominant*
4	Broomsedge with emergent pine
5–50	Pine, *Pinus* spp.
50–90	Pioneer hardwoods
90–?	Climax hardwoods

My own field-observations show this sequence to be violated as often as followed, but I believe it is a good first approximation for mesic abandoned farmland in all but the most northerly parts of the Piedmont in Pennsylvania and New Jersey. We will use it as a basis and point of departure for our discussion of the herbaceous (early) mesosere.

Our task is to compare the ecological literature with the present appearance of the province and with the observable plant–animal associations in the existing seral stages and to draw our own descriptions of the life systems on the Piedmont's successional lands.

CHAPTER FIVE

The Mesosere, Habitats on Well-Drained Soils

The rolling terrain of the Piedmont places the great majority of the province's surface in the midrange of the drainage continuum. Only the tops of the taller hills and monadnocks and the badly eroded and otherwise barren places fall into the range we have defined as xeric. Hydric soils are principally limited to the alluvia of the bottomlands adjacent to flowing water. I estimate the distribution of Piedmont sites across the drainage scale as follows: xeric, 10 percent; mesic, 85 percent; hydric, 5 percent. Because the majority of Piedmont land is mesic and is cloaked in mesophytic successional vegetation, our first specific examination of a representative Piedmont sere will be that found on the well-drained soils and mild slopes typical of abandoned farmland.

The Herbaceous Phase

The last crop cultivated on well-drained soils (before abandonment) and the timing of the last cultivation affect the early secondary plant succession sequence, perhaps the entire sere. What was grown and for how many years and by what methods are all factors affecting the selection of plants which will appear. Another important selecting factor in determining the composition of the seral stages is the availability of seeds or the possible presence at abandonment of successional species in the form of viable root systems. The proximity of seed sources and the effectiveness of their transporting vectors are important. But the decisive factor in determining which plants will dominate or participate at subdominant rank in the seral stages is the natural history of each plant species, in particular the time of year when the seeds mature, when they can germinate, if they need a period of cold dormancy before germinating, and whether the after-germination life cycle makes the plant an annual, a winter annual, a biennial, or a perennial.

During the growing season, any row crop (corn, cotton, tobacco) must be cultivated repeatedly to prevent the incursion of unwanted plants. It is routine practice for a crop field to be plowed in late summer or autumn after the harvest and immediately replanted in a small-grain grass crop (wheat, barley, rye, oats) to prevent erosion during winter, meanwhile providing an additional forage crop which matures in spring. If the winter cover crop is not planted, that is, if the field is left fallow after the harvest, a weedy incursion will provide nature's own cover crop. Mats of ground-hugging herbaceous plants develop in the final weeks, even days, before frost—plants whose life cycle and tolerance for cold enable them to prosper in winter. Chickweed, *Stellaria media*, a seemingly frail herb, shows extensive, nearly solid mats of prostrate foliage which retain a soft texture and jade green hue through the harshest months. Henbit, *Lamium amplexicaule*, performs similarly but with less dense coverage and during warm periods in winter may send up floral stems to display the whorls of hooded pink and purple flowers that endear this delicate mint to anyone who takes the time to appreciate these exquisite diminutives. Cranesbill, *Geranium carolinianum*, shepherd's purse, *Capsella bursa-pastoris*, and winter cress (*Barbarea* spp.) are common in the first winter of abandonment. The turnip family (Brassicaceae) is especially well represented at this stage; many brassicaceous species are cold-adapted winter annuals that germinate in autumn and spend the winter as flattened rosettes of foliage, sending forth tall floral stems as soon as spring temperatures permit. Many are edible. This entire system of post cultivation ground cover is as readily observable in any Piedmont vegetable garden as in a neglected cornfield.

Crabgrass–Horseweed Dominance

Present among the low greenery of the first winter are

some of the plants which will dominate the vegetation during the following summer and others which may assume primacy in subsequent years. Crabgrass, *Digitaria sanguinalis*, may germinate after the last cultivation and develop in thick tufts which will soon send forth culms (flowering stalks) bearing the characteristic windmill-like arrangement of floral spikes. If the last cultivation is done in summer, crabgrass may dominate before the development of the decumbent winter herbs mentioned above. Horseweed, *Erigeron canadensis*, is a winter annual common in newly abandoned fields. The seeds of this plant mature by August first, arrive by wind and germinate without delay. The seedlings form low rosettes during autumn, and from this vantage they quickly erect head-high floral stalks to dominate the field in the first growing season after abandonment. The aspect of the field during this period is therefore one of an even-height stand of horseweed towering over matted clumps of crabgrass. The horseweed and the crabgrass are said to be dominants (in this case co-dominants) because they are the tallest plants and they occupy more of the field's area than do any other species.

These species dominate, but they are not the only plants present; they may not even be the most numerous. More than thirty additional species may be present by the first full summer of abandonment at a representative Piedmont site. Some of the taller and more obvious might include poke, *Phytolacca americana* (a frequent occupant of disturbed soils), Queen Anne's lace, *Daucus carota*, pineweed, *Hypericum gentianoides*, rabbit tobacco, *Gnaphalium obtusifolium*, and the berry briars, *Rubus* spp. Ragweed, *Ambrosia* spp., is probably present and may even dominate on poorer soils. The fescues, *Festuca* spp., are introduced weedy grasses which have come to share in the early mesosere. They may locally replace or mix with crabgrass in the first year of abandonment.

Most significantly, there are among the subdominants of the first summer plants which will later dominate. Several asteraceous species and broomsedge, *Andropogon* spp., all germinate during the spring following abandonment but wait until subsequent growing seasons before claiming the herbaceous canopy. Often the first woody plants (trees and shrubs) germinate at this stage also but do not ordinarily become visible, and certainly not dominant, until several years later. The more advanced thinking on woody incursion is that many if not all the plants which will later dominate were present at abandonment, a theory referred to as "initial floristic composition."

Eroded places in a field may have lost all topsoil and may therefore be dryer and poorer than surrounding areas. Erosion scars characteristically support vegetation much less demanding of moisture and nutrients than are plants found in ordinary mesic circumstances of first-year abandonment. Plants like *Plantago aristita* (a plantain with no common name), *Diodia teres* (no common name) and *Aristida dichotoma* (a three-awn grass) are vegetative indicators of previous erosion.

Aster Dominance

Botanists reserve the use of the term "aster" to members of the genus *Aster*, but in the looser sense applied here it encompasses the next taxonomic level, the family Asteraceae, which contains many genera, including the horseweed genus, *Erigeron*. Horseweed is usually the tallest and most conspicuous plant in the first year because it germinates as soon as the seeds mature in late summer and because it grows well in the low-light conditions of autumn and spring. By the first spring after abandonment it has the jump on all competitors. Its competitors are also in the family Asteraceae, but their seeds require a period of cold dormancy before germinating and they therefore do not sprout until spring, by which time the horseweed is already sending up columnar stems from the basal rosettes of winter. No plant species present in the first growing season has the genetic capacity to wrest dominance from the horseweed that year.

By the second summer, however, the perennial asters are in a position to make their bid for primacy. Having spent the previous growing season beneath the horseweed consolidating their purchase on life, expanding their root systems, and sending foliage to only two inches (5 centimeters) above ground, the asters begin a towering growth with spring's first warmth. Horseweed plants are still present, probably even in greater numbers than in the prior season, but they typically fail to achieve more than six inches (15 centimeters) of growth. The horseweed is its own nemesis, restricting its own second-year growth by crowding and by the chemicals introduced into the soil from its decaying roots. Shading by the competing asters is not held to be the cause of horseweed's second-year stunting, the species being shade tolerant. This moment in the sere provides evidence to confute the old contention that each species "paves the way" for its successor; the crowding and the chemical inhibitors work against the asters as well as the horseweed. The asters come to dominance in spite of, rather than with the help of, the horseweed.

The daisy fleabane, *Erigeron annuus*, and a daisylike composite (Compositae was an earlier name for the family Asteraceae, based on the structure of the inflorescence which is a composite of many small florets) with no generally accepted common name, *Aster pilosus*, are among the most common asters to dominate in the Piedmont's second-year fields. Ragweed, *Ambrosia artemisifolia*, another of the Asteraceae, though quite undaisylike, shares second-year dominance, showing its strongest growth where the soils were most abused during cultivation. All of these species can achieve chest-high growth by early summer; by then they have begun to flower, and their whites and purples become the descriptors of the landscape. A subdominant and very daisylike aster, *A. patens*, grows to perhaps three feet (1 meter), often less, and is frequently among the composites of second-year fields. In the New Jersey Piedmont, *Centaures vochinensis*, a waist-high composite with sky blue inflorescences, mixes with *A. pilosus*.

Again, the dominants are the tallest and most conspicuous plants, but they are not the only occupants of a field. Second-year subdominants might include Queen Anne's lace, *Daucus carota*, the cudweeds, *Gnaphalium* spp., (most often represented by rabbit tobacco, *G. obtusifolium*), the nightshades, *Solanum* spp., (such as horsenettle, *S. Carolinense*), and other Solanaceae, including jimson weed and ground-cherry, *Physalis angulata*, with its autumn seed cases resembling Japanese lanterns. The Solanaceae is a family of paradoxes. The fruits of many of its species are poisonous to man and livestock, yet the Irish potato, peppers, and tomatoes are solanaceous. Milkweed,

Asclepias spp., the thistles, *Carduus* spp., and pineweed, *Hypericum* spp., are typically prominent. Also at this stage there begin to appear some of the plants that make for rough going in the overgrown fields. Greenbriar, *Smilax* spp., multiflora rose, *Rosa multiflora*, and the berry briars, *Rubus* spp., are painfully thorny and can grow in thick tangles. Japanese honeysuckle, *Lonicera japonica*, forms low mats of tangled vines often covering the ground and climbing onto other vegetation. The seeds of cocklebur, *Xanthium strumarium*, beggar's ticks, *Bidens* spp., and the open-country species of beggar's lice, *Desmodium* spp., ripening in late summer, are adapted to clinging to the coats of animals to achieve distribution. They can encrust the clothing of a hiker. It has often occurred to me that acting as a seed vector and spending time at the ritual of deseeding my clothes are the dues exacted by the successional life scheme. "Stick tights" can test the sincerity of any naturalist.

As in first-year fields, the plantain *Plantago aristata*, the three-awn grass *Aristida dichotoma*, and *Diodia teres* indicate earlier leaching and erosion. Dominance by ragweed indicates less severe leaching. Of eroded sites in general it can be said that the sere proceeds more slowly than on better soils, and the initial succession plants are likely to be in evidence for several years after their good-soil counterparts have been replaced.

Also among the second-year subdominants are plants which will prevail or more fully develop in later years. Preparing to succeed the asters and lurking among them are the broomsedges, *Andropogon* spp., (principally *A. virginicus*), and the goldenrods, *Solidago* spp.

Broomsedge–Goldenrod Dominance

Broomsedge is a perennial grass (member of the family Poaceae), not a true sedge; sedges are in the family Cyperaceae. In all seasons the plant has a dry, shreddy, and straw-colored appearance which belies its vigor. The seeds mature in November and the wind tugs at the hairs, releasing and distributing them; the distance most seeds travel is not great, for they are relatively heavy in comparison with their supporting airfoil. Cold dormancy is required and the

7. Broomsedge, *Andropogon virginiana*

seeds do not germinate until the following spring. The seeds which arrive during the autumn following the last cultivation germinate in spring and the plants are visible among and beneath the horseweed during the first growing season after abandonment. That fall, when the horseweed stalks stand reedlike and leafless, the pioneer broomsedge arrivals release their seeds and establish the population which will dominate in the third year.

In the first year of a broomsedge plant's life it sends up only one or perhaps a few whispy stems. These stand throughout the growing season and winter. The following year the plant grows in an expanding clump to a diameter of six inches (15 centimeters) or greater, but remains about a yard (1 meter) tall. If enough seeds arrived at abandonment to produce an adequate parent crop by the first fall, then two years later, in the third growing season after abandonment, a heavy stand of broomsedge could result. The asters do not grow well in close proximity with broomsedge, probably because the robust *Andropogon* root systems usurp the moisture. The asters are intolerant of dryness, and the natural shortages of rain in the Piedmont in middle to late summer make the competition for moisture critical. Almost invariably the broomsedge prevails.

From northern Virginia northward, Indian grass, *Sorghastrum nutans*, also a tuft grass and resembling broomsedge but taller and less shreddy, may become dominant at about the third year. Both occur in older pastures and may dominate the aspect of even active pastures in late summer and fall if not mowed. When true abandonment occurs, the Indian grass and broomsedge—though the two do not usually share dominance—are in a position to assume immediate dominance. It is my observation that in newly abandoned pastures, as opposed to row-crop fields, broomsedge or Indian grass dominance occurs sooner and more completely and the number of subdominants is reduced. In the northern Piedmont particularly the present trend is for more pastures to be abandoned than plowed fields, so a large percentage of recently abandoned tracts show the straw brown color of *Sorghastrum* or *Andropogon* dominance.

The tuft grasses may come to occupy any percentage of a field between the onset of the third growing season and the closing of a woody canopy. This transition may span as much as fifteen years, though seven to ten years is more probable, during which, if the tuft grass density permits, there will be numerous other herbs on the land. Some of the asters of second-year dominance, notably *A. patens* and *A. pilosus*, are perennials and may retain enough vigor to produce significant flowering in the third year and for several seasons thereafter. Many of the important subdominants of the second year may also persist. They include rabbit tobacco, Queen Anne's lace, milkweed, the thistles, and the nightshades. Honeysuckle, greenbriar, multiflora rose, and the berry briars will probably advance prior to the woody takeover and may interlace to make many parts of the field impenetrable.

The most conspicuous newcomers of the period are the numerous and intermixed species of goldenrod, *Solidago* spp., which gild the mesic slopes of all sections of the

Piedmont in late summer and autumn (there are also *Solidago* species which prefer wetlands). Goldenrod, like the tuft grasses, may mix with its neighbors or may dominate a field completely or locally. It might be found mixed with the whites and purples of other asters, it may preempt the herbaceous canopy, or it may mix with the tuft grasses. Goldenrod is associated mostly with the middle and final years of the herbaceous phase of the mesosere, and in many fields it is quite prominent.

Animals of the Herbaceous Phase

Insects and Arachnids

In the mesosere's herbaceous phase the orthoptera are represented in strength, particularly soon after abandonment when crabgrass and the fescues are available. Grasshoppers (families Acrididae, the short-horned grasshoppers, and Tettigoniidae, the long-horned species) are prominent consumers of the grass foliage, and the crickets (family Gryllidae) are present to glean vegetable detritus nearer the ground. Numerous terrestrial beetles (order Coleoptera) inhabit the Piedmont's overgrown fields; prominent among beetle families of the habitat are the ground beetles (family Carabidae), of which most species, both as larvae and as adults, feed on other insects, principally moth larvae. Many carabids are nocturnal, as are the caterpillars they prey upon, such as the cutworms (some species in the family Noctuidae) and the cankerworms (a few species in the ubiquitous family Geometridae), which occupy the grassy tufts.

With the arrival of large numbers of asteraceous plants, particularly the daisylike types, *Aster* spp. and *Erigeron annuus*, and the goldenrods and thistles, there comes a rich variety of flying insects to collect their nectar and pollen and serve as pollinators. Numerous bees (order Hymenoptera) attend the aster inflorescences, the tiny halictids, the honeybees and the bumblebees being conspicuous. At least one, and probably several, species of wasps in the family Cynipidae lay their eggs in the stems of goldenrod and simultaneously inject a growth enzyme which causes a globular enlargement in the stem to protect and feed the larvae within. After defoliation these galls become conspicuous. Numerous species of hunting wasps search for spiders and insects on the asters to provision their nests. Among the coleoptera found on the aster inflorescences are soldier beetles (family Cantharidae), the net-winged beetles (family Lycidae), the shining flower beetles (family Phalarcridae), and the antlike flower beetles (family Anthicidae). Lightening bugs (family Lampyridae) rest during the day on the undersides of the umbels of Queen Anne's lace and probably feed there on wasted pollen and other detritus.

The capital insect pollinators in the aster fields are the lepidoptera. No list could exhaust the roster of moths, skippers, and butterflies that visit the Piedmont's asters and other flowering plants of abandoned fields. Some of the more important diurnal moths are the ermine moths (family Yponomeutidae) and several of the sphinx moths (family Sphingidae). Without question there are also numerous species of night-flying moths who would prefer to contend with the whippoorwills, bats, and screech owls than with the legions of bird, insects, and arachnid predators at large in the aster fields by day.

Among the butterflies, many members of the family Pieridae are important in the successional fields. Included are the whites, whose larvae feed on plants of the turnip family which are abundant immediately after abandonment (see page 89); the sulphurs, whose larvae eat herbaceous legumes such as the partridge pea, *Cassia fasciculata*, common in the habitat; and the orange-tips who feed on shepherd's purse as larvae. The milkweed butterflies (family Danaidae) include the monarch, *Danaus plexippus*, common in aster fields during the growing season and especially numerous in this habitat during autumn migration. Danaid larvae feed on milkweed, *Asclepias* spp. (see page 93), from which they take a taste disagreeable to birds, thereby gaining protection from avian predators as larvae and as adults. The adults feed on nectar but retain the offensive taste. Monarchs are among the few North American butterflies that migrate, and their southward movement through the Piedmont in September and October can be a spectacle. Numberless flocks sometimes fill the sky to the visual limit of 8X binoculars. The adults quit the continent in autumn, reproduce on their wintering grounds, and return with their offspring to the Piedmont in spring.

The brush-footed butterflies (family Nymphalidae) probably claim the largest number of species associated with the Piedmont's aster fields. This group includes the thistle butterflies (the red admiral and the painted lady, both of the genus *Vanessa* and fond of the flowers of thistles and other asters), the crescent spots, *Phycoides* spp., whose larvae feed on asters, the mourning cloak, *Nymphalis antiopa*, and a swarm of fritillaries, *Speyeria* spp. and *Boloria* spp., typically brownish to coppery with colorful spots. Included are two of the most common butterflies of the habitat, the red-spotted purple, *Basilarchia astyanax*, and the buckeye, *Junonia coenia*. Their larvae, feeding respectively on rosaceous leaves and plantains, are common caterpillars in the Piedmont's overgrown fields. Of particular interest is the viceroy, *Limenitis archippus*, smaller than the monarch but similar to it in shape and coloration, doubtlessly to trade upon the monarch's unpopularity with the predators.

The largest butterflies of the aster fields are the papilionids (family Papilionidae), the swallowtails. The tiger swallowtail, *Papilio glaucus*, is particularly fond of the floral heads of the larger asters, such as helianthus, and the thistles. The larvae feed on a variety of tree leaves. The larvae of the zebra, *Graphium marcellus*, and spicebush, *P. troilus*, swallowtails feed on the broadleaf shrubs of the mature deciduous bottomlands, pawpaw and spicebush, respectively, securing a faunal link between the two habitats. The larvae of the black swallowtail, *P. polyxenes asterius*, eat Queen Anne's lace in the successional fields, and the adults seek the inflorescences of the same habitat. In observing swallowtails we are cautioned that there is a dark phase of the tiger swallowtail which can be confused with the black and the spicebush.

The ambush bugs (family Phymatidae) and the assassin bugs (family Reduviidae), all "true bugs" (members of the order Hemiptera), abound as predators on the other insects in fields in the asteraceous phase. The ambush bugs are fitted to hunting from concealment, their yellow body colors blending with the inflorescences of this habitat, particularly goldenrod. I see nothing in the literature to suggest that the Phymatidae can change their colors, but I have seen on the white flowers of rabbit tobacco, *Ghaphalium obtusifolium*, whitish ambush bugs otherwise similar to the yellow individuals on the goldenrod.

The arachnids are represented in the overgrown fields by more species of orb-weavers (family Araneidae) than in pastures due to the availability of tall vegetation on which to string their vertical webs. Their prey is chiefly flying and leaping insects caught while moving laterally through the habitat. The argiope spiders, for example, subdue the largest grasshoppers with alacrity, but if an adult mantis runs afoul of the web, furious infighting ensues, with the spider usually cutting frantically at the strands to help free the mantis before it destroys the whole web. The sheet-web spiders (subfamily Linyphiinae) are also numerous in this habitat, their webs spread among the tufts, visible on dewy mornings like dropped handkerchiefs. The webs are horizontal, suggesting that the sheet-web spiders, sportsmanlike, take their prey on the rise. Wolf spiders (family Lycosidae) chase a variety of insects through the tall vegetation, though they usually remain near the ground. Resting in the tall grass late one summer twilight I watched a big lycosid scamper past through avenues in the crabgrass not evident to me and carrying a frantically flashing lightening bug. Lycosids, by the way, return an eyeshine exactly parallel to the source of light and are easily found if you hold the flashlight to your forehead and shine it into the grass. Standing in an abandoned field—or even your front yard—and musing thus with one's "headlight" will give an astonishing hint at the nocturnal life of that habitat.

Specializing in predation on the flying insects which visit the asteraceous floral heads are the green lynx spider, *Peucetia viridans*, a living form in translucent jade, and several species in the family Thomisidae, the crab spiders. I am convinced that at least one of the latter can change color to match their floral background, a "flower spider" I interpret to be in the genus *Misumena*. (Only a qualified arachno-taxonomist should attempt to place identities below the family level, because microscopic subtleties separate even the genera.) This thomisid I have seen ensconsed in the flowers of partridge pea, yellow-bodied at the precise shade of the petals and with the tips of its outstretched forelegs dipped in the same amber as the flower's sexual parts. The camouflage was so effective that from a foot away the spider could be seen and lost from view as the mind rearranged the scene to perceive only flower. The yellows paled when the spider was removed, and I have seen an apparently identical species matching shades of white and green in other backgrounds. The crab spiders wait on or below or inside the flowers, anchored with their diminutive two pairs of hind legs, and holding outstretched, crablike, their enlarged forward pairs to grab prey as large as swallowtails. The lynx spider makes a quick dash from within or beneath the inflorescence to fang its prey.

Amphibians and Reptiles

The upland chorus frog, *Pseudacris triseriata feriarum*, is seen leaping and climbing like a minute monkey in the weeds and grasses of the Piedmont from Virginia southward. The literature suggests that this species is restricted to moist areas, but my own experience includes too many sightings in mesic overgrown situations to ignore. It feeds on small insects.

The American and Fowler's toads, *Bufo americanus americanus* and *B. woodhousei fowleri*, occupy all parts of the province except for the extreme southeast. They range through nearly all habitats, wet and dry, open and wooded, showing a preference for mesic conditions except for breeding; the eggs and tadpoles require water. Food in herbaceous fields is the rich assortment of insects as well as snails and slugs. Young mice are recorded in the diet of both toads.

Caution is appropriate in handling toads, not because the urine causes warts as is commonly believed, but because secretions from their skin can be irritating or poisonous. Always wash your hands as soon as possible (in the next creek or puddle) after handling toads, and until you do, avoid touching mouth, eyes, or genitals.

There are many snakes to be found in a variety of living situations, including overgrown mesic fields, though there may be none peculiar to that habitat. Practically any snake not exclusively aquatic is a candidate for a sighting in an early successional field—in short, many snakes are where you find them. (See the animal tables in the appendix.)

The copperhead is the only venomous serpent likely to be encountered in the mesic Piedmont. The timber and canebrake rattlesnakes nominally include parts of the province in their ranges, but sightings are rare. Copperhead bites result in messy wounds, but fatalities are an uncommon result if the victim is taken immediately to a physician. The copperhead is said to be mild-mannered until aroused and certainly unlikely to attack without cause. I once stepped on a copperhead without incurring its displeasure, but I am frankly content not to enlarge my experience with the species. The seasoned field naturalist views any brownish snake from a respectful distance until identified, and handles no snake until familiar with the species and its disposition. Many of the nonpoisonous serpents can inflict painful bites or douse the handler with foul-smelling musk. Some thrash or fight violently when caught or cornered.

Beauty and fascination await the observer of snakes. The black racer moves with incredible speed when disturbed, disappearing as a whispered streak in the grass. The black rat snake can be of unpredictable disposition; I once picked up a six-footer (180 centimeters) sunning on a country road in North Carolina who was memorably docile, and another half that length who was unmanageable. The corn snake is predictably gentle and with its attractive patterns of red on orange, brown or gray is an appealing animal. The scarlet (*Cemophora coccinea*), docile but dazzling in its colors, mimics the lethal coral snake of the coastal plains of the deep South. Delicate and pencil thin, 2-foot (60 centimeters) long, the rough green snake (*Opheodrys aestivas*) allows itself to be lifted gently from a smilax vine and worn for a few minutes as a bracelet by the naturalist on his or her rounds. The hog-nosed snake, common in overgrown fields, puts on a show not to be missed. When approached, it hisses and spits like a cat, flattens and widens its head into the arrowhead shape of a viper's, and feints at striking. The display would be more intimidating if not so overdone that it arouses more curiosity than fear, in which case the hog-nose moves into step two of its act: convulsions, terminal gasps, and the belly-up stillness of obvious death. If the passerby rights the poor creature in an attempt to revive it, the hognose again rolls over on its back to emphasize that it is dead. After a cycle or two of what must be one of nature's most elaborate genetically-encoded melodramas (the killdeer's broken-wing display runs a weak second), the hognose becomes sullen and declines to perform.

Birds

The early phase of mesic succession harbors several bird species that nest almost exclusively in this habitat. Bold, noisy, and colorful, the yellowthroat, *Geothlypis trichas*, a warbler, nests on or near the ground in the weeds and briary tangles. During the breeding season the male sings a rollicking, explosively loud series suggestive of the Carolina wren. Wintering in the Piedmont's southern half, the yellowthroat's irritation note, likened to a single pluck on a loose banjo string, protests a human intrusion into its overgrown field. Its food in summer is insects, particularly caterpillars; winter fare shifts to small fruits such as honeysuckle berries. The continent's largest warbler, the yellow-breasted chat, *Icteria virens*, nests in this habitat, taking larger insects for itself and its young and favoring the later years of the herbaceous phase when woody incursion begins. The upward-percolating territorial notes of the prairie warbler, *Dendroica discolor*, rises from broomsedge and goldenrod. Nesting at knee level in the asters and berry briars, it feeds moth larvae and other soft-bodied insects to its young. The bobwhite, *Colinus virginianus*, and the meadowlark, *Sturnella magna*, are among the larger nesting birds in this habitat in the southern Piedmont. They are joined in Virginia and northward by the ring-necked pheasant, *Phasianus colchicus*, especially plentiful in the Pennsylvania Piedmont. The brown thrasher, *Toxostoma rufum*, and the mockingbird, *Mimus polyglottos*, use the overgrown mesic fields as nesting locations, particularly when there are woods and fencerows nearby.

Fringillids which nest in the weedy fields include the cardinal, *Cardinalis cardinalis*, the ground-nesting field sparrow, *Spizella pusilla*, and the chipping sparrow, *S. passerina*. They feed a variety of insects to their young and, except for the chipping sparrow, who retires to the southern Piedmont, remain in the same habitat in winter to feed on the seeds of grasses, asters and other weedy herbs.

In the Piedmont's overgrown fields, during the nesting season only, are two fringillids of very similar shape and color but different in size. The indigo bunting, *Passerina cyanea*, every feather quintessentially blue, and the blue grosbeak, *Guiraca caerulea*, twice as large and wearing a deeper, less iridescent blue broken by chocolate wing bars, occupy ecological niches supported, respectively, by small and large insects. The females of both species are drab brown. The grosbeak nests near the ground among the asters and briars; the bunting builds a cupped structure on a low twig in a pine or broadleaf sapling invading the field. Both also sometimes nest in a hedgerow bordering a weedy field. Rising to the highest perch in its territory, the bunting fills the summer days with a rich, elaborate song which the listener comes to associate with its overgrown habitat. The grosbeak's gentle slurred whistles and metallic *tinks* come subtly from the sumac and multiflora rose.

The swamp sparrow, *Melospiza georgiana*, is a bird of the wetlands in all seasons except during fall migration, when it feeds its way southward through the Piedmont's mesic abandoned fields taking seeds and insects. Arriving in autumn and remaining in the habitat through winter are the slate-colored junco,* *Junco hyemalis hyemalis*, the white-throated sparrow, *Zonotrichia albicolis*, the white-crowned sparrow, *Z. leucophrys*, and in far lesser numbers the vesper sparrow, *Pooecetes gramineus*. Their dietary staples are the seeds of grasses and asters. The rare Henslow's sparrow, *Ammospiza henslowii*, is periodically seen in the province's broomsedge fields on autumn migration, though it does not winter in the Piedmont. Autumn brings forth the song sparrow, *Melospiza melodia*, from its wetland and streamside nesting territories to feed in the mesic overgrown fields with the mixed flocks of juncos and white-throated and white-crowned sparrows. As far south as the James River, the tree sparrow, *Spizella arborea*, may join these wintering flocks.

*The author respectfully regrets his inability to employ the American Ornithologists' Union's newly sanctioned common name, "dark-eyed junco, slate-colored race."

The loggerhead shrike, *Lanius ludovicianus*, nests and winters in the overgrown fields and adjacent hedgerows, taking large insects when they are available and turning to small birds and rodents in winter. Weak of foot, and in that sense an unlikely raptor, the shrike is sometimes obliged to spike its prey on barbed wire or on a locust thorn while dismembering it, and the remains are occasionally seen still impaled. Shrikes are easily and commonly seen from the road perched on utility wires, fences, or bare twigs and scanning the overgrown fields they prefer as hunting habitat. The flight is a low swooping arc on wingbeats too rapid to count. White shows in flight at the base of the primary flight feathers in a pattern similar to that in the mockingbird's wings, the latter perhaps gaining an edge over its competitors for food through this mimicry because most small birds fear the shrike.

The American kestrel, *Falco sparverius*, a common sight on roadside wires or hovering on shallow wingbeats over a weedy field, is the continent's smallest falcon and one of the Piedmont's most numerous raptors. The prey is approximately that of the shrike, somewhat larger on the average and probably concentrating more on the rodents in all seasons. The kestrel winters throughout the Piedmont and nests regularly north of the Roanoke River. The male is our most brilliantly colored bird of prey, sporting a red back and tail, blue shoulders, and intricate markings on the head. The female has rusty plumage dappled in black.

Weaving among the fencerows, darting into the open to seize a small passerine, the sharp-shinned hawk, *Accipiter striatus*, smallest wood hawk in the province, has adapted well to man's opening of the Piedmont landscape. It especially favors newly abandoned fields where small birds abound and where visual barriers of vegetation afford the advantage of surprise. On short, rounded wings this bold and determined little raptor plunges into thickets to extract its prey. *A. striatus* has also adjusted to life in urban parks and residential areas where intermittent shade trees and shrubbery—and in winter, bird feeders—attract birdlife.

The marsh hawk, *Circus cyaneus*, winters throughout the Piedmont, quartering the lowlands and weedy fields in tireless powered flight. It sometimes rests on the ground, most often after a kill, but has been observed to remain

8. Sharp-shinned hawk, *Accipiter striatus*

in continuous flight for half a day. The marsh hawk does not use perches higher than a fence post. The female is brownish with dark striping which follows the contours under her wings; the male is a solid bluish gray with dark wingtips and a banded tail. Both sexes show a conspicuous white rump patch. The facial disc, an owl-like adaptation, aids in aurally tracking prey. No other bird of prey flies so low over the weedy fields for extended periods. The winter diet is mainly rodents.

The undisputed daytime ruler of the mesic open country in the Piedmont is the red-tailed hawk, *Buteo jamaicensis*, largest soaring hawk in eastern North America. It is present in all parts of the Piedmont in all seasons, reaching its greatest numbers in winter as the entire North American population shifts southward. The food in winter is almost exclusively rodents of the mesic fields, particularly the meadow vole and the hispid cotton rat. Summer fare includes reptiles and carrion, usually road kills, but rodents remain the staple.

The red-tailed hawk's hunting method is to scrutinize a grassy or overgrown area from a high perch such as a bare limb or the top of a utility pole and pounce on its prey after a direct glide. The bird is seen in large numbers along the Piedmont's major highways (especially in winter), where it hunts the grassy embankments from trees and utility poles. Training the eye to pick out the hulking form of this magnificent bird can make less tedious a long-distance drive on the interstate highways.

At sunset the red-tailed hawk yields the hunting of the Piedmont's open space to an even larger raptor, the great horned owl, *Bubo virginianus*. The great horned is a woodland owl by day, but it includes virtually all habitats in its nocturnal hunting lands and is the principal owl to take advantage of the wealth of rodents in the overgrown fields of the Piedmont. The prey, which can include nearly anything (up to the size of a fox), is mostly rodents in winter but in other seasons is more varied. To hunt the successional fields, the great horned owl watches and listens from a hunting perch at woods' edge or in a fencerow tree, launches and sails on muffled primaries to the kill. The same fields may be visited by the much smaller screech owl, *Otus asio*, in quest of mice or large insects, but without extreme care, the screech owl too becomes simply another morsel for the great horned, the largest owl south of the boreal forest.

Mammals

The rats, mice, voles, and shrews of mesic hayfields and pastures similarly populate abandoned fields in the herbaceous phase. The least and short-tailed shrews, *Cryptotis parva* and *Blarina brevicauda*, include the entire Piedmont in their ranges, and the southeastern shrew, *Sorex longirostris*, is present from central Virginia southward. None of these shrews is restricted to open country, but it is among their favored habitats. The shrews prey on a broad variety of invertebrates, such as the grassland orthoptera, snails, worms, and beetles, and they are known to include the young of certain vertebrates in their diet. They are active day and night and, though not often seen, are numerous and figure prominently in the flow of energy and nutrients. We are admonished, if the occasion arises to handle a short-tailed shrew, that the saliva is poisonous.

In the tall grass and weedy habitats there appears to be more than a casual relationship between the shrews and the rodents, whose runways the shrews utilize. Passage through the interlaced weeds and grass tufts would otherwise be difficult, and the shrews, being insectivores, lack the flat-bladed incisors needed for cutting vegetation. The young of rodents also constitute a portion of the shrews' diet.

As in Piedmont hayfields, the meadow vole, *Microtus pennsylvanius*, is prominent, holding perhaps a plurality of the rodents' numbers. The white-footed mouse, *Peromyscus leucopus*, and the deer mouse, *P. maniculatus*, are also numerous in overgrown fields, the habitat becoming more favorable for the white-footed mouse as grass height and weed incursion progresses. In the North Carolina Piedmont and southward, the hispid cotton rat, *Sigmodon hispidus*, prolific and voraceous, is also important in the grass-rich early years of abandonment. These rodents are significant first-consumers of abandoned-field vegetation, particularly the shoots and seeds, and grasses are among their favorite foods. Their habits include the construction of runways at ground level through the grass tufts, which serve as thoroughfares for other small mammals, reptiles, and for some ground-nesting birds. They may rear their young in aboveground nests or in burrows. Their principal predators are foxes, crows, hawks, owls, and snakes.

The eastern cottontail (rabbit) includes the entire Piedmont in its range and favors open, herbaceous habitats. It is often found in mesic abandoned fields, particularly during the growing season when young shoots of grass are available. In winter, when the diet shifts to twigs, buds, and bark, the cottontail is more often found in shrubby habitats of older seral advancement. The cottontail feeds mostly at night and avoids danger by hiding in brush during the day. Covered nests of woven grass (a rabbit weaving somehow stretches the imagination) containing four to seven young are concealed in tussocks and are visited by the mother to nurse, at intervals, until the third week. The young struggle to move through the thick grasses in their first week of independent foraging and, as an illustration of their vulnerability at this stage, are easily caught by a human on foot.

The woodchuck, a devotee of stable and well-managed grasslands such as pastures and hayfields, is not commonly found in sites of abandonment, regardless of the availability of grass. An unobstructed view of their surroundings and of approaching perils is evidently a requirement.

Present in the mesic abandoned fields are the mustelids which were described in the material on hayfields (pages 55–56), the long-tailed weasel, *Mustela frenata,* and the striped skunk, *Mephitis mephitis*. From Virginia southward they are joined by the spotted skunk, *Spilogale putoris*. Skunks occupy a variety of habitats and are not confined to herbaceous growths, but they find considerable reward in the rodent, reptile, amphibian, and insect populations in the overgrown fields. The weasel hunts rodents and other small mammals up to the size of rabbits; it occasionally takes birds. The mustelids are nocturnal and are active in all seasons.

Similarly, the gray and red foxes, *Urocyon cinereoargenteus* and *Vulpes fulva,* exploit the rodents of abandoned fields, and they are joined by the opossum, *Didelphis marsupialis,* in quest of the fruits of pokeberries, blackberries, and the fruits of the passion flower, *Passiflora incarnata,* also called maypops (although the fruit does not "pop" when stepped on until midsummer). The opossum is an omnivore whose appetite seems undiminished by the most unsavory food, regardless of the state of decay, but it is also a capable predator on birds, reptiles, amphibians, and small mammals. It must be emphasized that none of these predators limits its hunting to herbaceous mesic fields nor even to open country. It is the character of the province to be a mosaic of varied habitats, and those who occupy the Piedmont at a high trophic level are best suited who can exploit the numerous food sources the land offers.

Mesic Woody Succession

On the overwhelming majority of abandoned mesic Piedmont fields there are woody plants present by the time the asters and tuft grasses reach dominance, roughly by the third growing season after abandonment. Usually by the fifth year the woody seedlings are visible above the broomsedge, goldenrod, and briar tangles. The saplings may assume dominance (occupy more of the old field's area than do the herbaceous plants) by the seventh or eighth years and may close a canopy over the field by the twelfth to fifteenth year.

In the northern two-thirds of the province, from approximately the Savannah River northward, the woody seedlings may be deciduous or coniferous. South of the Savannah, initial woody plants are almost entirely coniferous, pines, *Pinus* spp., to be specific. North from the Savannah, deciduous saplings become increasingly important, as does a non-pine conifer, the eastern red cedar, *Juniperus virginiana*. As one moves northward in the Pennsylvania Piedmont from the vicinity of the 40th parallel of latitude, which bisects Philadelphia, coniferous participation becomes negligible in the early woody (or any other seral) stage. The same is true for all but a fringe of the southeastern New Jersey Piedmont. A variety of deciduous trees and shrubs are the characteristic woody pioneers in the northern Piedmont.

To facilitate our discussion of mesic woody succession, we will adopt two basic models representing this seral phase: a coniferous model applicable to the southern Piedmont and a deciduous model pertinent to the north. In the coniferous model we expect the initial dominant woody plants on an abandoned field to be conifers, either eastern red cedar or some species of pine. Similarly, deciduous saplings and shrubs are the initial woody plants to invade abandoned fields in the northern model. The northern extremes of the Piedmont are cloaked in deciduous trees at all woody stages of the mesosere, and are largely devoid of coniferous trees. The southernmost sections are occupied almost entirely by conifers as initial woody invaders. The ranges of these successional types are not equal in size, as shown in Figure 9. The coniferous zone occupies by far the larger portion of the province. As woody pioneers on abandoned land, the conifers effectively hold firm control over the Piedmont from central South Carolina southward. Coniferous saplings also dominate the majority of abandoned fields as far north as the Potomac River, locally sharing dominance with or deferring to deciduous pioneers. From the Potomac northward to the southern tier of counties in Pennsylvania, eastward across the northern tip of

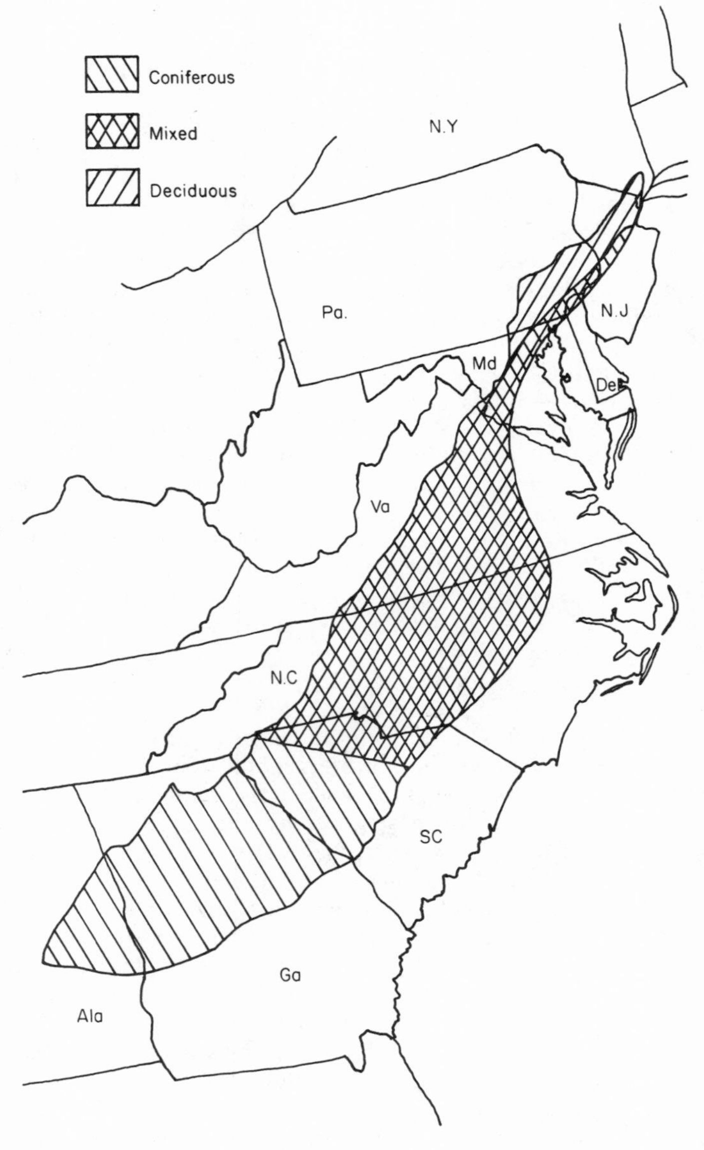

9. Deciduous/coniferous zones

Delaware and into the southern portion of the New Jersey Piedmont, the woody pioneers are about equally divided between conifers and deciduous saplings. There the pines and cedars end abruptly, and northward through the remainder of the province deciduous pioneers are the initial woody plants.

Southern Initial Woody Succession

Eastern Red Cedar, *Juniperus virginiana.* The eastern red cedar plays a significant role in early woody succession in the coniferous zone from southern New Jersey and Pennsylvania to northern South Carolina. Within this range and particularly from the James River northward, many of the abandoned fields are crowded with young cedar trees. The eastern red cedar may be the only woody plant visible, or it may share initial woody dominance with pine or with deciduous saplings. Cedar is particularly likely to dominate on old pastures because, unlike the pines, it can withstand cropping by mowing machines or by livestock. Pines that are bitten off or cut off at a few inches height typically die; young cedars seem to thrive under such croppings, often

10. Eastern red cedar, *Juniperus virginiana*

developing root stocks disproportionate to the foliage, which, when clipping is discontinued, nourish very rapid growth. Clipped cedar seedlings are commonly visible in active pastures.

Cedar saplings rise above the asters and tuft grasses often by the sixth or seventh year of abandonment in fields where crops were grown, sooner in abandoned pastures. Once established, the species grows, under favorable conditions, in a handsome, compact, conelike shape to a height of 50 feet (15 meters) and a dbh (diameter at breast height) of up to 24 inches (60 centimeters). However, in successional fields the cedars usually grow in close competition with one another or with competitive species, and their girth is commonly restricted to 6 inches (15 centimeters) or less. The foliage of seedlings is green to purple–brown and of bristly texture, contrasting with the rich green color and scaly texture of the foliage on older trees.

Pines, *Pinus* spp. There are three principal species of pine in the Piedmont. The Virginia pine, *Pinus virginiana*, is the most northerly, its range extending from the northern limits of the coniferous zone southward through most of the North Carolina Piedmont and southwestward onto the Dahlonega Plateau in Georgia. The Virginia pine is recognized by its short (1 to 1½ inches/25 to 40 millimeters) twisted needles, two to a follicle, its squat profile (as a young tree), and by the fact that it is slow to self-prune its lower branches, making a young or middle-aged stand of the species difficult to penetrate. The Virginia pine does not

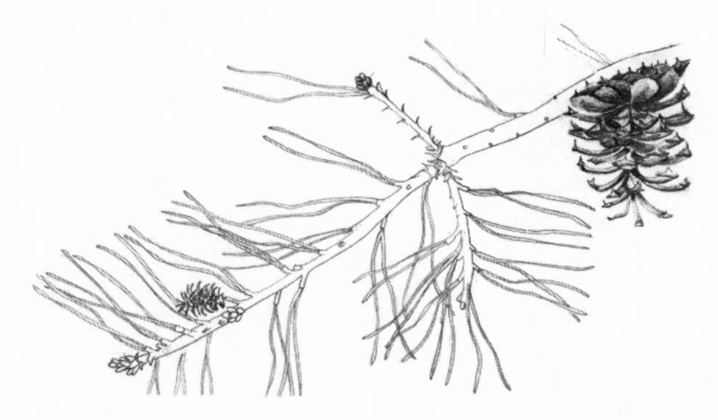

11. Virginia (scrub) pine, *Pinus virginiana*

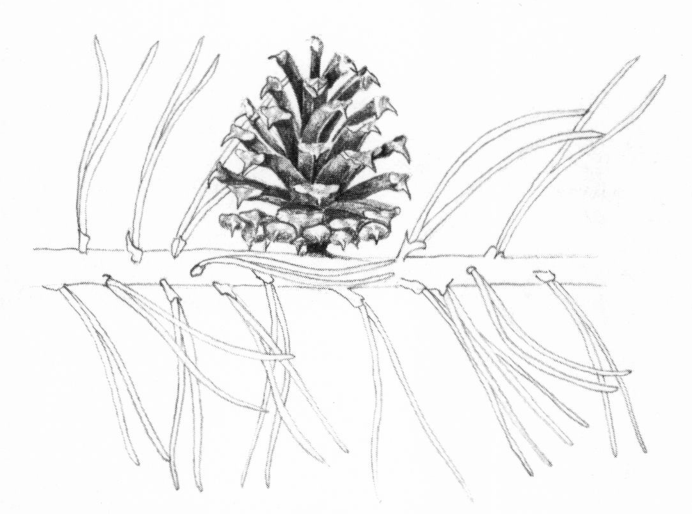

12. Shortleaf pine, *Pinus echinata*

achieve great height, 40 to 50 feet (12 to 15 meters) being average.

Shortleaf pine, *Pinus echinata*, has straight needles from 3 to 5 inches (75 to 125 millimeters) long in clusters usually of two each. Growing to perhaps 80 feet (25 meters), it is a valuable timber species as a mature tree. The range of the shortleaf pine includes the entire coniferous section of the Piedmont, but north of the James River it is substantially less common than the Virginia pine. The shortleaf is the most important successional pine in the Virginia Piedmont south of the James and in the western portion of the Carolinas Piedmont.

Loblolly pine, *Pinus taeda*, is the southernmost of the Piedmont's successional conifers. It is the dominant, and frequently the only, pine in the Alabama Piedmont and in the Georgia Piedmont outside the Dahlonega Plateau. Loblolly continues as the preeminent pine in the central and eastern parts of the South Carolina Piedmont and in the eastern third of the province in North Carolina. Along the Piedmont's outer margin, vigorous successional stands are seen as far north as Fredericksburg, Virginia; individuals and small clusters grow along the circumferential highway (Interstate 495) a few miles southwest of Washington, D.C. In the Virginia Piedmont, loblolly does not penetrate more

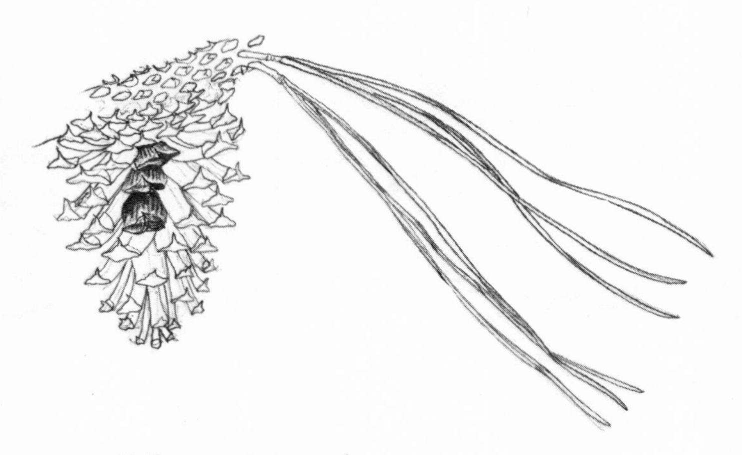

13. Loblolly pine, *Pinus taeda*

than a few miles westward into the province as a successional volunteer. However, because the loblolly is the most desirable pine for wood products, it is commonly planted in the Piedmont to the north and west of its natural range. The species is identified by the upward tilt of its twigs, the deeply furrowed bark on older trees, and the long (6 to 9 inches/15 to 23 centimeters) needles in clusters of three.

In addition to these three abundant, common, and wide-ranging successional pines, the pitch pine, *Pinus rigida*, is found as a mesic woody pioneer in the Piedmont of southern Pennsylvania, though even in that region the Virginia pine is the more common. White pines, *Pinus strobus*, are seen occasionally in the Maryland and southern Pennsylvania Piedmont but they characteristically appear later in the sere, after the climax hardwoods have become established. A disjunct stand of white pines exists on the south side of the Deep River in Lee County, North Carolina (see page 364). A final conifer, the Canadian hemlock, *Tsuga canadensis*, occurs sparsely and occasionally in the Piedmont of southern Pennsylvania. A few can be seen on the lower slopes of Big Round Top on the battleground at Gettysburg, and a luxuriant stand shades a north-facing slope to the south of Octoraro Creek in Chester County, Pennsylvania (see page 366).

14. Pitch pine, *Pinus rigida*

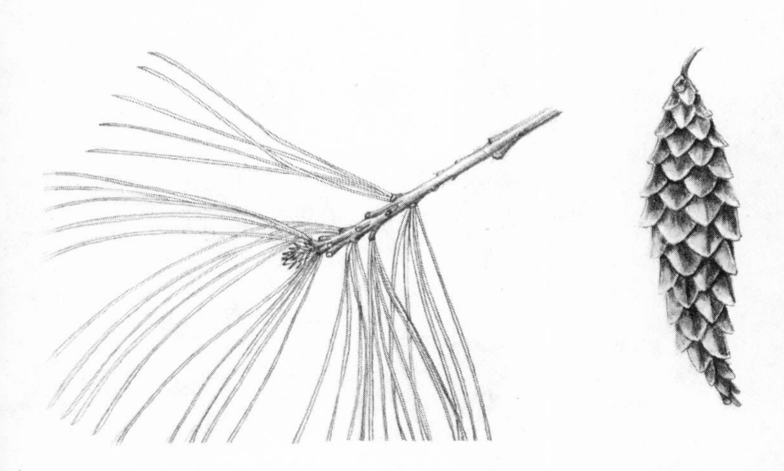

15. White pine, *Pinus strobus*

Northern Initial Woody Succession

Beginning at the Palisades, on the west bank of the Hudson River (the Piedmont's northern limit), and continuing southward into South Carolina is the region where deciduous trees and shrubs are among the first woody plants to invade abandoned land. Between the Hudson and the southernmost counties in the Piedmont of Pennsylvania and New Jersey, the woody pioneers are exclusively deciduous. South of the exclusive range and to the southern limit of the deciduous zone (see figure 9), the broadleaf trees and shrubs compete with pines and cedars to be the first woody dominants.

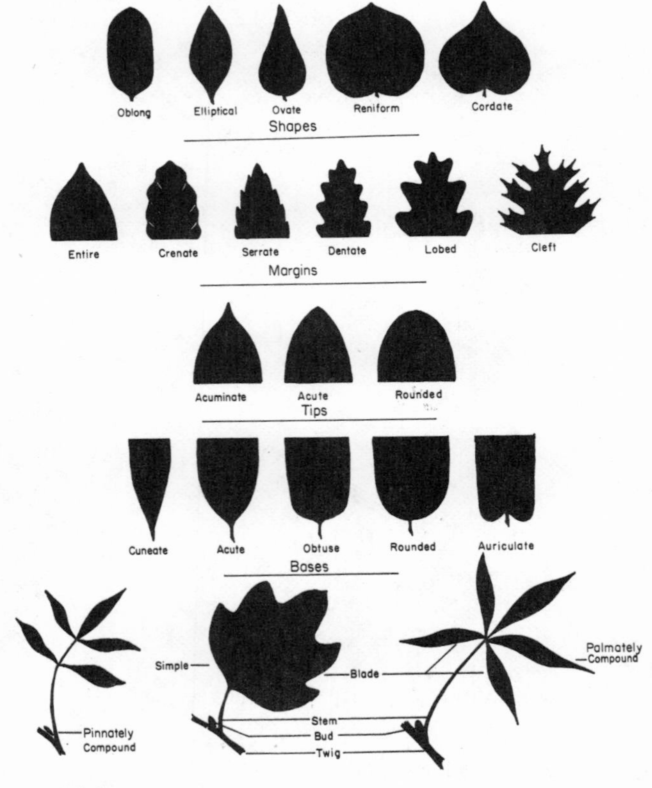

16. Leaf shapes

All combinations of broadleaf and coniferous saplings can be observed in this vast zone of overlap. Pure stands of every species of both groups exist, though more commonly the stands are mixed. Apparently every possible permutation by which the woody pioneers can be associated is now being explored by the forces that guide secondary succession. Enough patterns are detectable, however, to establish that the ranges and combinations in which the woody plants occur are not random. They are probably the result of local soil and hydrologic conditions, though not enough is known to sustain assertions regarding the preferences of each plant species for acidity, minerals, and moisture. We are mainly concerned here with the associations among the successional plants and between the plants and animals. The following are descriptions and ranges of the Piedmont's principal deciduous pioneers. The order of the descriptions is intended to suggest roughly their successional importance.

Red Maple, *Acer rubrum.* This is one of the most common trees in the Piedmont. It adapts to a variety of soil and moisture conditions and it is important at many locations

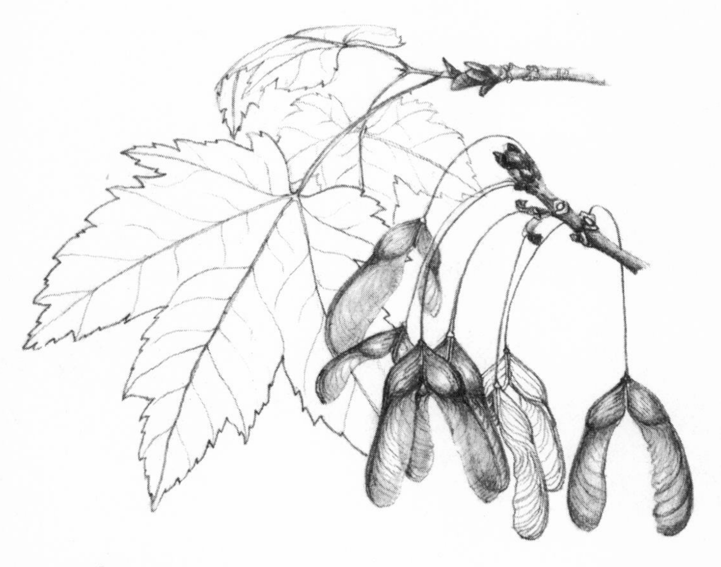

17. Red maple, *Acer rubrum*

throughout the sere. As a mesic woody pioneer, the red maple is most active in the Piedmont in the northern part of the province. Under mature oaks and poplars, the species commonly occupies a position in the subcanopy. In hydric soils it may dominate the climax canopy. The identifying characteristics of red maple are the three-lobed, coarsely serrate leaves, the crimson caste to the twigs lent by the flowers appearing in early spring before the leaves develop, and in winter the reddish twigs and buds, opposite in arrangement. The leaf stems and samaras (seed cases) arranged in opposing pairs and shaped like bees' wings are usually red, and in autumn the leaves of this maple are sometimes fiery red; indeed, the species is aptly named. The seed-bearing samaras when released spin rapidly, extending the duration of the trip to earth sufficiently for the wind to waft the little propellers some distance from the parent tree. The bark varies a great deal in color and texture and should not be relied on for identification.

American Elm, *Ulmus americana.* This species is particularly active as a mesic woody pioneer north of the Potomac, where it also thrives as a mature tree in fencerows and other relatively open environments. It does not compete well in a closed canopy except on damper soils, found

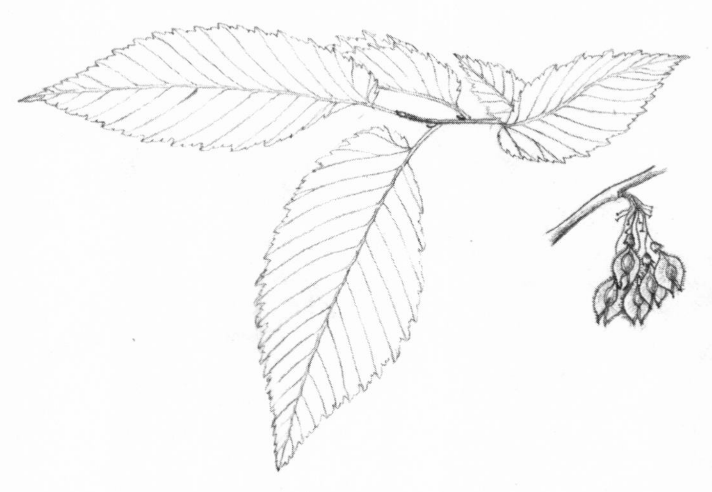

18. American elm, *Ulmus americana*

mostly south of the James River. Germinating in abandoned fields, the American elm often grows several main stems that develop into a cluster of equal-size trunks. With maturity, and particularly under competition, the number of trunks diminishes, usually to one or two. Most individuals grow vase-shaped in profile. The buds grow on alternate sides of the twigs. The leaves are 4 to 7 inches (10 to 18 centimeters) long, pointed, entire, doubly toothed, and asymmetrical at the base. The seeds form and are released in spring to be carried on saucer-shaped airfoils at the whim of the wind. Growing to 100 feet (30 meters) in height and up to 48 inches dbh (120 centimeters), this elm is favored as a shade tree, though many have been killed in recent years by Dutch elm disease, a fungus carried by a bark beetle.
Black Locust, *Robinia pseudoacacia.* If any shape could symbolize the abandoned fields and fencerows of the northern Piedmont it would be that of the black locust tree. It

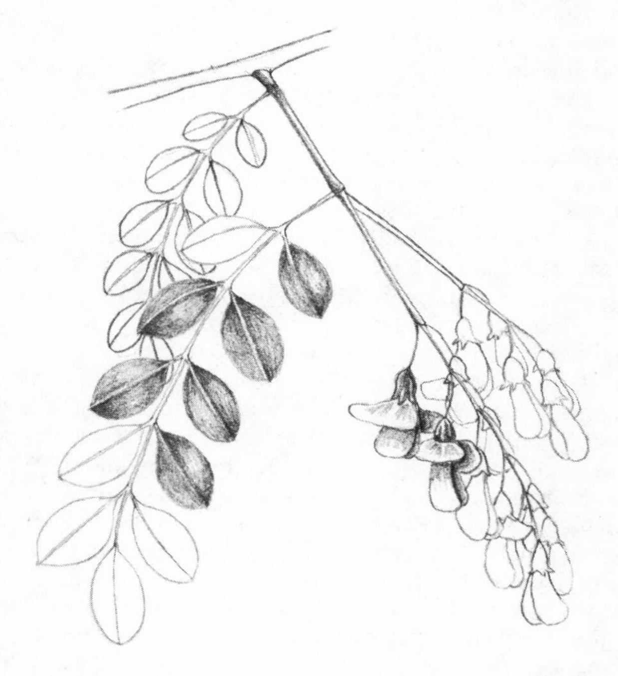

19. Black locust, *Robina pseudoacacia*

grows tall but keeps its limbs close, which creates the appearance of a cylinder of foliage. The twigs are spiny. The bark is deeply furrowed with intersecting trenches that make the ridges appear braided. The leaves are pinnately compound; the flowers, blooming when spring is well under way, are white and held in drooping clusters, the fruits encased in a beanlike pod 2 to 4 inches (50 to 100 millimeters) long. In autumn the pods split, revealing, but not ejecting, dense seeds that are anything but airworthy. They need the ripping, blustery winds that characteristically shake the Piedmont in late autumn to dislodge and distribute them. Birds and animals may also help in dispersing the seeds.

Tree of Heaven, *Ailanthus altissima.* Tree of heaven (also commonly called ailanthus) is an introduced species characterized by soft wood of no value to humans and by unbelievably rapid growth—10 feet per year is common. It is an urban invader frequently seen sprouting from sidewalks, fissures in pavement, gutters, rooftops, and a quietly implied threat that, if everyone were to leave the city for a summer, it would belong to *A. altissima* by fall. This species is equally aggressive on abandoned farmland, dominating the land in rapid spurts of growth, unquestionably retarding the native woody pioneers. Ailanthus spreads vegetatively (sprouting from the roots of older trees) as well as sexually. The seed-carrying mechanisms are samaroid, implying a wing, of sorts, for aerial transport. The leaves are alternate, pinnately compound, with twenty or more leaflets on each frondlike leaf. Tree of heaven resembles poison sumac (see poison ivy page 128).

The Ashes, *Fraxinus* spp. The ashes, with compound leaves and opposite arrangement of buds and twigs, are among the mesic woody pioneers most often seen rising above the goldenrod and broomsedge in the northern Piedmont. White ash, *F. americana*, and green ash, *F. pennsylvanica*, share the successional fields and at most seasons are devilishly hard to differentiate in the field. I have stopped trying. Literature suggests that the leaflets of white ash are larger. In spring, flowers of white ash appear before the leaves; those of green ash after the leaf buds have opened. The apex of young ash trees often show a cross-shaped twig configuration. The seeds are borne in single samaras which are longitudinally symmetrical (samaras of maple are doubled and, like a wing or propeller, have a thickened leading edge). Trees are straight and single-trunked, to 80 feet (25 meters) tall.

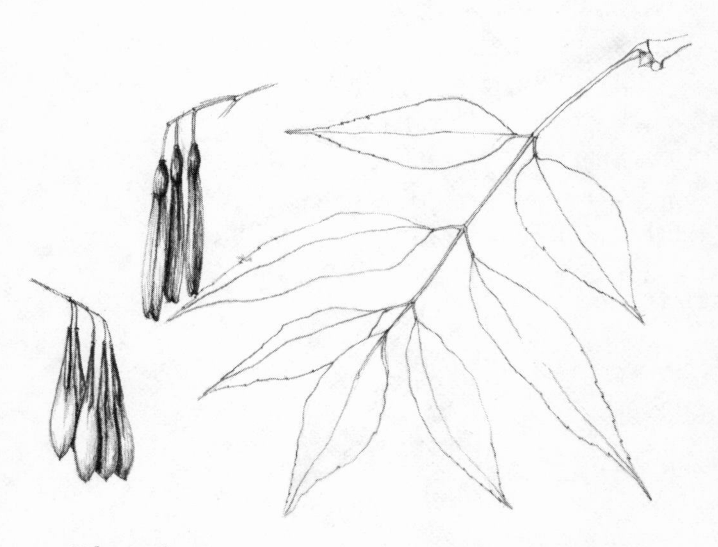

20. White ash, *Fraxinus americana* (top); Green ash, *Fraxinus pennsylvanica* (bottom)

The Cherries, *Prunus* spp. The cherries are important woody pioneers throughout the Piedmont. Black cherry, *P. serotina*, is native and ubiquitous in the province, although north of the Potomac sweet cherry, *P. avium* (a cultivated species now naturalized), may be locally more common. Black cherry can become a large tree—60 feet (20 meters) tall, 36 inches (90 centimeters) dbh. The leaves are 2 to 6 inches (5 to 15 centimeters) long, shiny on top and with brown tomentum (hair) along the midrib beneath. Flowers and fruit grow in grapelike clusters on racemes (elongated inflorescences with flowers, and later fruit, attached by stalks to the main stem). The drupes (fleshy, one-seeded fruit characteristic of the genus *Prunus*) are one-quarter inch (6 millimeters) in diameter, black and shiny. The bark on young trees of both species is ringed with ridges; on older individuals the bark becomes scaly, pronouncedly so on sweet cherry.

21. Black cherry, *Prunus serotina*

Northern Catalpa, *Catalpa speciosa.* Northern catalpa is important as a successional tree north of the James River, particularly in the Piedmont of Virginia. It is recognized by its heart-shaped leaves 10 to 12 inches (25 to 30 centimeters) long, shreddy bark, angular branches, and generally twisted shape. White flowers cluster at the twig tips in spring, developing into elongated (10 to 20 inches/25 to 50 centimeters) beanlike fruit pods in autumn. The seeds are enclosed in flattened, papery winglike structures that are exposed to the wind when the pods open. Although the seeds of the northern catalpa are encased in beanlike pods, the tree is in the family Bignoniaceae rather than the bean family (Fabaceae).

22. Flowering dogwood, *Cornus florida*

Flowering Dogwood, *Cornus florida.* Flowering dogwood is associated with the climax deciduous understory throughout the province but in many localities north of the James River the species is an important woody pioneer. Dogwood functions in this capacity very strongly in northern Virginia, where it commonly shares successional fields with cedar, often being the woody dominant for a decade or so after the close of the herbaceous phase. This small tree, rarely exceeding 30 feet (10 meters) in height and 8 inches (20 centimeters) in girth, is abundant in virtually all phases of the mesosere in the Piedmont. If it is not among the first woody plants, it is usually one of the first understory trees to appear. Heavy pioneer pine dominance, for example, often shows young dogwood in the understory as soon as the pine's lower branches are self-pruned. The "flowers" of dogwood are really compound inflorescences composed of compact clusters of small yellow florets surrounded by four large white bracts which give the appearance of being petals. Toward later summer the seeds develop within ellipsoid red berries which are important in the diet of many birds.

Tulip Tree, or **Yellow Poplar,** *Liriodendron tulipifera.* Tulip tree, or yellow poplar, is not a poplar (genus *Populus*) but is a member of the magnolia family. Its principal association is with moist soils, but its importance as a woody pioneer on mesic soils demands mention here. If the tulip tree becomes established as a woody pioneer it typically grows to great heights (more than 100 feet/30 meters is common) and continues as one of the dominant trees in the deciduous canopy as the site matures. In rich woods of some mesic postclimax sites whose soils retain moisture particularly well, the tulip may form part of the climax canopy. The trunk of the tree is, barring damage, arrow straight. Branches depart the trunk horizontally and arc vertically upwards to yield a pleasing, symmetrical shape. The leaves are large, to 10 inches (25 centimeters) across, and four-lobed. The cuplike flowers face vertically upward. There are six petals, greenish yellow distally and with an orange spot at the base. The fruiting structure is a conical arrangement of vertically compacted samaras which are released by the wind and by birds and squirrels in quest of the seeds.

23. Tulip tree, *Liriodendron tulipifera*

24. Sweet gum, *Liquidambar styraciflua*

Sweet Gum, *Liquidambar styraciflua.* Sweet gum is important as a woody pioneer throughout the deciduous zone and, like the tulip tree, commonly remains on the site as a canopy tree as the tract matures. In the coniferous zone and on sites in the zone of overlap where conifers are the woody pioneers, sweet gum is among the first deciduous trees to invade and wrest control from the pines later in the sere. The leaves are star-shaped, 5 to 7 inches (13 to 18 centimeters) wide with pointed lobes. Twigs often have corky ridges along the sides similar to those on the winged elm. The seeds are borne in prickly, burrlike spheres 1 inch (25 millimeters) in diameter which open and spill the seeds into the wind. Birds take many of the seeds before they are released.

Persimmon, *Diospyros virginiana.* The black, checkered bark appearing on mature trees in adjacent cubes is a reliable identifier. The alternate, glabrous leaves are simple, entire, elliptical, 3 to 6 inches (8 to 16 centimeters) long and half as wide. The fruit is a sweet, pulpy, spherical, berry, pumpkin-colored and 1 inch (2½ centimeters) in diameter. The seeds—tan, elliptical and flattened—are commonly seen in the droppings of foxes, raccoons, opossums and other mammals, confirming that these fruits are important wildlife food. They are also edible by humans when very ripe; eaten before ready, they pucker the mouth unpleasantly. Persimmon is thinly scattered across mesic sites in the early woody phase, as would be expected from the relatively small number and distributional methods of the seeds. They do not dominate in any habitat but are present in virtually all forest phases in all drainages.

25. Persimmon, *Diospyros virginiana*

Winged Elm, *Ulmas alata.* Winged elm is the southernmost deciduous woody pioneer of importance, being most significant in the Carolinas, where it may locally dominate or may mix with pines or cedars. The winged twigs are more delicate than those of sweet gum. The leaves are similar in shape to those of the American elm but are much smaller, usually about 1 inch (25 millimeters) in length. The winged elm blooms in early spring and samara-borne seeds are released soon after.

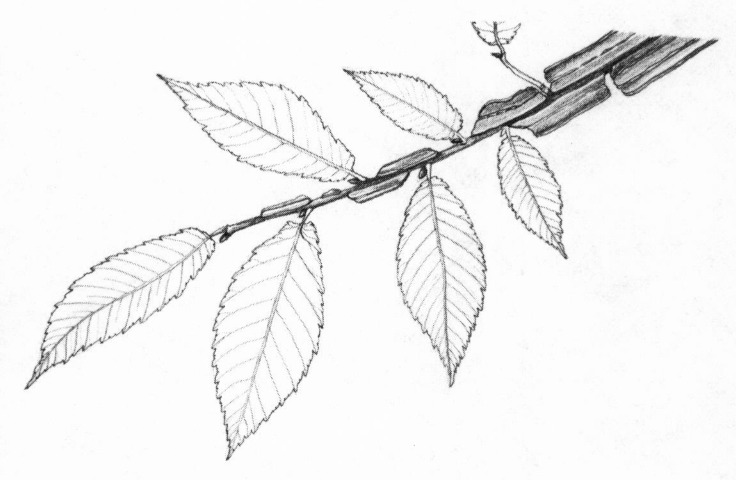

26. Winged elm, *Ulmus alata*

Shrubs and Vines

Sumac, *Rhus* spp. Sumac is a weedy shrub occurring throughout the province. Staghorn sumac, *R. typhina*, occurs principally north of the James; shining and poison sumac, *R. Copallina* and *R. vernix*, can be found throughout the mesic Piedmont. All occur as woody pioneers on abandoned fields, roadsides, and waste places. The fruits of staghorn, hairy red drupes 1/10 inch (3 millimeters) in diameter, are packed in a pointed cluster at the plant's apex; the leaves are pinnately compound with hairy rachises (stems) from which a milky fluid seeps when crushed. Shining sumac (also called winged sumac) is similar but is distinguished by winged stems. Sumac seeds are distributed by birds, but once established on open ground a plant may reproduce vegetatively by sprouting from its root system, with the effect that dense compact islands of sumac occur at intervals on abandoned fields.

Black Haw, *Viburnum prunifolium.* Black haw has, as the name implies, leaves similar to cherry but opposite in arrangement; it is chiefly associated with the hardwood understory, but in the northern Piedmont is among the woody pioneers on mesic fields.

27. Trumpet creeper, *Campsis radicans*

Trumpet Creeper, *Campsis radicans.* Trumpet creeper is rigid enough to stand waist high but often climbs on other vegetation as a vine, its many branches curving toward maximum sunlight. The flowers have stunning orange corollas 4 inches (10 centimeters) long, from which protrude the yellow sexual parts. Leaves are pinnately compound with seven to fifteen leaflets 1 to 2 inches (25 to 50 millimeters) long. Like its close relative, the catalpa tree, trumpet creeper produces winged seeds in a long, beanlike pod. The plant contributes to the weedy appearance of the late herbaceous phase of abandoned fields and is found in fencerows, clearings, and along roadsides. It is entirely associated with open spaces and does not survive under a closed canopy of trees.

Poison Ivy, *Rhus radicans.* Poison ivy, with its characteristic three shiny leaflets, appears at many sites very early in the sere as an upright woody spike to 36 inches (1 meter) tall. Later in the sere it climbs on trees, typically capturing its ration of the sunlight in the crowns of the dominant hardwoods. Older plants cling to trees with an abundance of short tendrils that give the vine the appearance of a fuzzy

28. Poison ivy, *Rhus radicans*

snake climbing the tree. The seeds, contained in small drupes, are important in the diet of many birds. The foliage and twigs contain a virulent poison to which probably no one is totally immune; I thought I was until at age 36 and wearing only hiking shorts I shinnied up a tree festooned with the glabrous foliage. The toxin is not released unless the plant is broken (or burned) so there is little danger of infection from merely brushing against the leaves. It is well to be able to identify the plant at a glance and avoid it. Repeated latherings with soap and water soon after contact can reduce the danger of allergic reaction.

Without exception, the initial woody plants so far discussed have in common the property of being distributed by wind or by birds. The seeds are specifically adapted by the nature of their encasement to be carried either on the wind or in the digestive tracts of birds. Accordingly, they are small and numerous. Those which are encased in a drupe or other fruity form tend to pass through the birds undamaged or, better, scarified to facilitate germination. The major point to observe is that squirrels—which are exclusively arboreal and therefore not active on abandoned fields—are not employed in distributing the first wave of trees across the landscape. Mammals that are partially arboreal (the gray fox, raccoon, and opossum come to mind) may assist in distributing some of the drupe-borne seeds of the initial woody plants discussed and may be instrumental in the dispersal of several which are common but which occur too sparsely to be considered of primary importance as woody incursives. Among the latter are persimmon, crab apple, and sassafras. The droppings of the three omnivores mentioned are rich in fruit-borne seeds during late summer and autumn. Moreover, I have noticed that foxes and raccoons use their feces as territorial communications, often placing them at conspicuous, well-traveled locations such as paths, old roadways, barren spots—in general, places devoid of tree cover and suitable to the needs of woody pioneers.

Initial Woody Succession Habitats by State

Having introduced the principal initial woody plants of the Piedmont mesosere, we may now describe the general appearance of the landscape at this phase and examine some of the relationships which comprise the life system. Generally, the mix of woody plants moving onto abandoned land is more complex in the northern Piedmont than in the south to the extent that in the coniferous zone there is a tendency for cedar or for one species of pine to dominate. The deciduous newcomers of the northern Piedmont tend to be more diverse.

NEW JERSEY

The New Jersey Piedmont as far south as Morristown, Plainfield, and Perth Amboy came under the ice of the Wisconsin Glacier, which receded from the area about 10 thousand years ago. The glaciation created new highlands in the forms of eskers and moraines and new lowlands where the mountains of moving ice scooped the soil down to bedrock. In some places they melted in repose to form glacial lakes which sediments and vegetation have subsequently filled (see page 308). Beyond creating new lowlands and highlands, the ice sheet brought in soils and minerals from other landscapes, some from as far north as Canada, locally altering the mineral mix and acidity of the then-present soil. Precisely how these alterations are expressed in the modern vegetation is uncertain; no obvious demarcations in the floral composition divide the New Jersey landscape at the limits of glaciation. We are most confident in observing the effects of glaciation on plant life in terms of high ground and low, with the expected results in soil moisture. The vegetation on New Jersey's glaciated mesic slopes appears similar to that on comparable slopes not touched by the ice.

On abandoned fields in the Piedmont of northern New Jersey, the first woody plants to rise above the asters, ragweed, and multiflora rose are likely to be black locust, catalpa, red maple, ash, or ailanthus. A given field might be invaded by all and either become zoned into groves of individual species or covered by a random integration of the woody pioneers. White birch, *Betula populifolia*, is locally dominant on a few mesic sites in the glaciated zone. Climbing buckwheat, *Polygonatum cuspidatum*, which grows in dense shrubby masses, and swamp dogwood, *Cornus ammomum*, both of which characteristically favor the lowlands, are active in the mesic abandoned fields of northern New Jersey. In this region ailanthus and black locust are the most numerous woody pioneers.

Pines, which are relatively rare and not important as first woody plants in the northern New Jersey Piedmont, become significant in the Piedmont of central New Jersey south of the Great Swamp. They become visible as one travels southward on Interstate 287 past the Watchung Mountain between Pluckemin and Exit 13. Virginia pine is the most evident conifer; it is joined locally by Austrian pine (*P. nigra*), a naturalized species. Virginia creeper, *Parthenocissus quinquefolia*, can be seen climbing in the pines as they mature; in autumn the five-lobed leaves of this liana present a brilliant red contrast with the pine's green needles. Red maple is also important in the central New Jersey Piedmont as a woody pioneer, along with ash and black cherry. It is typical of young woodlands here to be composed of the pine–maple or ash–maple mixtures encountered southward through Virginia and, in places, the Carolinas.

Cedar is scattered throughout the New Jersey Piedmont but the species are significant in early woody succession only in the southern and southeastern sections. A site offering a representative view of the mesosere in southern New Jersey is the Hutcheson Memorial Forest, one mile east of Millstone on the south side of Route 514. This tract, in addition to being cloaked in original growth forest (which we will discuss in conjunction with the climax mesosere), includes 71 acres (28.7 hectares) of fields in which segments apparently have been "abandoned" at intervals for study purposes. A large field, lying between the forest and the road, contains juxtaposed sections whose age since the last mowing I have estimated and whose plant populations my notes record as follows:

2–4 years–Daisy fleabane, *Erigeron annuus*, goldenrod, the dazzling purple aster, *Centaures vochinensis*, thistles, *Carduus* spp., multiflora rose, milkweed, *Asclepias viridiflora*

5–10 years–Same asters, cedars to 10 feet (3 meters) tall

10–15 years–Same asters, cedars to 15 feet (5 meters) bearing heavy fruit, American elm, box elder, *Acer negundo* (actually a maple), flowering dogwood, islands of staghorn sumac, sweet cherry (*Prunus avium*)

With the stepped sequence of early succession adjacent to the mature forest, the Hutcheson tract no doubt offers a true picture of both ends of the mesosere in the Piedmont of southern New Jersey. The tract is owned by Rutgers University, from whom permission to visit should be obtained. Hutcheson Memorial Forest is now a National Natural Landmark.

Twelve miles (20 kilometers) to the south, Princeton University maintains a tract equally instructive in comprehending the New Jersey Piedmont's natural vegetative pattern. Herrontown Woods, on the northern outskirts of the attractive town of Princeton, is bordered on the west by Mount Lucas Road and Poor Farm Road and on the north by the Herrontown Road. The main entrance is on the east side of the 400-acre (160-hectare) tract on Snowden Lane. On the left side of the access road at the entrance is a field which appears to have been abandoned in about 1970. Its rank growth of broomsedge and purple top, *Tridens flavus*(a grass whose sparsely branched floral heads lend a purple cast to many overgrown fields throughout the Piedmont), is punctuated by shrubs of multiflora rose and by saplings of cedar and crab apple laced with Virginia creeper.

The area on the right of the access road was abandoned perhaps five years earlier. By late 1977 the cedars had reached 15 feet (5 meters) and were interspersed with flowering dogwood, swamp dogwood, black cherry, red maple, and black haw. Pin oak, *Quercus palustris,* seedlings are present and there are islands of staghorn sumac. In the few remaining open spots, goldenrod nod in the autumn gusts. Virginia creeper grows with the trees that offer sunny exposures. Herrontown Woods contains excellent acreage representing the transition between the initial woody and mature phases.

There are numerous other examples of early woody succession on private lands in the vicinity, but it is fortunate that these two windows on the process are open to the naturalist. For a further description of both sites see Part Four, Special Places.

PENNSYLVANIA

The preeminence of deciduous woody pioneers continues through the northern half of the Pennsylvania Piedmont, the likely woody pioneers being American elm, black locust, ash, ailanthus, silver maple, and red maple. As a pioneer, the American elm shows a mushroomlike profile composed of several trunks.

Within 20 miles (30 kilometers) of the Maryland border, pines and cedars appear on the abandoned fields. Virginia pine is the most numerous but there are many groves of pitch pine as well. Cedars are scattered throughout the Pennsylvania Piedmont, but occur in significant numbers chiefly toward the Maryland border. White pines are planted in some abundance, but they appear to occur naturally only in well-developed hardwood stands rather than as pioneers.

Pin oak occurs thinly but consistently as a woody pioneer on the old fields of the Pennsylvania Piedmont. Sweet cherry, *Prunis avium*, also occurs thinly but in numbers great enough to be of significance to the life system. Silver maple, mostly associated with the lowlands in the Piedmont north of the James River, occurs persistently on mesic sites in the Piedmont of southern Pennsylvania. Its density in mesic conditions is much lower than in the river bottoms, where it may dominate in shimmering groves.

The majority of the northern Pennsylvania Piedmont is cloaked in deciduous early succession on its mesic sites. Conifers are few. Into this deciduous dominance, scattered conifers and coniferous groves are introduced toward the Maryland border.

MARYLAND

The mesic abandoned fields of Maryland's Piedmont display the same general aspect as those of southern Pennsylvania. Most sites are dominated by ash, American elm, and black locust, but this deciduous majority thins toward the Potomac. Conversely, the importance of pine and cedar increases as the observer moves southward across Maryland. Pitch pine becomes sporadic to nonexistent except in the western extremes.

In profile, a representative mesic field in the Maryland portion of the province is likely to be dotted with cedar but dominated by the hardwood saplings typical of the northern Piedmont. Virginia pine and eastern red cedar are the important conifers, and where they occur they are likely to dominate in solid groves.

VIRGINIA

Between the Potomac and the James there occurs a transition between the deciduous primacy of the northern Piedmont and the primacy of conifers to the south. The groves of Virginia pine become larger and less widely spaced. Shortleaf pine, *Pinus echinata*, appears as a coniferous pioneer. Toward the fall line these pines are joined by the loblolly pine, *P. taeda*, south of Fredericksburg. On the whole, the northern half of the Piedmont in Virginia is divided equally between initial woody dominance by pine, cedar, and deciduous saplings. Many tracts are mixed; many are solid stands, particularly where pines are concerned.

While the conifers come into prominence in this region, deciduous pioneers are still important. In the vicinity of Manassas and Warrenton, flowering dogwood spreads vigorously over abandoned mesic slopes. On many fields in the vicinity of Bull Run Mountain it shares early woody dominance with cedar, dappling the landscape in autumn with contrasting reds and greens. Sassafras, *Sassafras albidum*, persimmon, *Diospyros virginiana*, redbud, *Cercis canadensis*, tulip tree, *Liriodendron tulipifera*, and ash, *Fraxinus* spp., are also important between the Potomac and the James. Catalpa is locally vigorous, as is black locust.

South of the James River there begins a large region which approximates the character of the great pineries of the southern Piedmont, but which is too rich in deciduous pioneers to be identical. Virginia and shortleaf pine share dominance, but the groves are usually enriched with deciduous coevals including red maple, tulip tree, sassafras, and sweet gum, *Liquidamber stiraciflua*, an important woody pioneer of the central Piedmont (sweet gum is also an important *deciduous* pioneer, arriving soon after and displacing the pines throughout the province south of the James). Cedars are considerably less common between the James and the Roanoke than are pines. Loblolly comes into prominence in this region, particularly toward the fall line, extending its influence inland toward the Piedmont's centerline near the Virginia–North Carolina border. Black locust and catalpa persist as woody pioneers south of the James only along the Piedmont's western margin.

NORTH CAROLINA

The initial woody phase of the mesosere in North Carolina approximates that of Virginia south of the James River. Loblolly pine continues its march westward across the Piedmont, comfortably dominating early woody succession in the Piedmont's eastern third in this state. Shortleaf and Virginia pine are the principal woody pioneers in the central and western segments, respectively. The majority of abandoned fields in all but the eastern and western extremes are usurped by a mixture of pines, usually shortleaf and loblolly. Cedar is present and occasionally dominates, but it is most frequently of secondary importance. In the Piedmont of North Carolina, as in the more northerly states, cedar is apt to gain dominance on abandoned pastures.

Coniferous initial dominance gains momentum in North

Carolina; many of the deciduous pioneers characteristic of the northern Piedmont are absent altogether. Gone is the black locust. The American elm ceases to function as a mesic pioneer. Ash and ailanthus are infrequent. Dogwood, redbud, and sassafras are common, but less so than in Virginia.

Although the trend of initial woody succession in the Piedmont is for the prominence of deciduous pioneers to be reduced with lower latitude, several important deciduous pioneers assert themselves in North Carolina. Sweet gum, of nominal influence north of the James River, is probably the most important deciduous pioneer in North Carolina. It is also among the first to replace the pines on sites of initial dominance by conifers. Winged elm becomes abruptly important on abandoned fields from about the Roanoke River southward and exerts local dominance throughout the Piedmont of North Carolina and northern South Carolina. Black cherry also rises to new importance as a woody pioneer. Honey locust, *Gleditsia triacanthos*, with its long-thorned trunks, replaces black locust.

SOUTH CAROLINA

In South Carolina a major shift occurs in the pattern of initial woody succession, the last significant variation one finds in traveling southward through the Piedmont. North of the northeastern unit of the Sumter National Forest (there are three noncontiguous tracts bearing this name, all in the South Carolina Piedmont), we find a considerable portion of the newly abandoned farmland, of which there is no dirth, occupied by winged elm. This region is the southernmost outpost of deciduous woody pioneers. Here the winged elm and, on some sites, sweet gum and black cherry (occasionally sassafras) claim the open fields alone or in company with conifers. Black cherry dominates in patches on some fields. Red maple is prominent as a woody pioneer. Ash is in evidence, cross-shaped apices rising above the broomsedge and goldenrod. Trumpet creeper covers the ground and some of the woody ascendants in some parts of the old fields. This is not to say that pines are of secondary importance in the Piedmont of South Carolina. Except for an area in the vicinity of York and Rock Hill where winged elm and cedar predominate, the initial woody mesosere in northern South Carolina is principally in the control of shortleaf and loblolly pine. Loblolly is prevalent in the eastern Piedmont here; shortleaf to the west. The two species share dominance in the central zone.

Southward and westward of central South Carolina and continuing through Georgia to the southwestern tip of the province just east of Montgomery, Alabama, there exists an immense region where pine dominance of the early woody mesosere is nearly absolute.

GEORGIA

There are few abandoned fields in Georgia which do not fall under the control of pine. Loblolly is dominant over the majority of the Georgia Piedmont, mixing with shortleaf to the east and south of the Dahlonega Plateau. Deciduous pioneers are uncommon. The same deciduous trees which act as pioneers within the southern part of what we have called the deciduous zone of the Piedmont—black cherry, red maple, winged elm, and sassafrass—typically become established under the pines in the Piedmont of Georgia, later to replace them.

The overwhelming majority of the Georgia Piedmont is now devoted to growing pine timber and pulpwood. Many tracts which were abandoned earlier in this century have already grown one or more crops of pine. Sometimes the pines are replanted; sometimes the landowners hope for natural seeding after a cutting. On such sites, the presence of young deciduous trees becomes more apparent than that of pines because the deciduous trees are often spared or, if they are cut, send up sprouts from the stump. Thus, many apparently young stands of mixed pine and hardwood or of hardwood predominance give the false impression of having been populated initially by deciduous pioneers.

On the Dahlonega Plateau, which occupies roughly the northwestern quarter of the Georgia Piedmont, the average altitude is approximately 1600 feet (500 meters). Prominences on the plateau probably exceed 2000 feet (600 meters). Since 1000 feet (300 meters) of altitude is said to be the climatic equivalent of a change of 600 miles (1000 kilometers) in latitude, we expect and find that the vegetation on the Dahlonega Plateau, 800 to 1000 feet (250 to 300 meters) above the rest of the Georgia Piedmont, shows a more northerly aspect. Virginia pine is the dominant woody pioneer. Scattered groves of deciduous pioneers dominate locally, mostly sweet gum, dogwood, and mimosa. Here black locust makes it southernmost appearance in the Piedmont.

ALABAMA

The Piedmont in Alabama experiences initial woody dominance by pines similar to that in Georgia and the southern half of South Carolina. Along the southern margin, in the vicinity of Auburn, cedars are more in evidence than in any locality south of northern South Carolina. Willow oak, *Quercus phellos*, not infrequently participates in the woody takeover, though I am not aware of any tract on which it asserts dominance. The loblolly pine controls the initial woody mesosere in Alabama's Piedmont with scant help from cedar and willow oak.

Animals of the Mesic Woody Succession

Insects and Arachnids

Invasion of the landscape by woody plants results in the reduction of herbs and grasses and in the elimination of many plants with large showy flowers adapted to pollination by insects. Many, though not all, of the trees, shrubs, and vines which replace the herbs are wind pollinated; that is, pollen is produced in great quantities and is transported by the wind rather than by animal vectors. Thus, there is a reduction in the number of adult insects such as bees, moths, skippers, butterflies, and certain beetles that are associated with plants bearing conspicuous flowers. Similarly, as the trees take over and shade out the grasses, the grasshoppers, crickets, and ground-dwelling beetles of the herbaceous phase are reduced.

We might venture the generality that, so far as the lepidoptera are concerned, a vegetational shift from herbs to woody plants changes the land from a foraging ground for the adults to a nursery for the larvae. Table A-1 (in the appendix), Selected Lepidoptera of the Piedmont, shows food preferences of some of the common and conspicuous lepidoptera, but by no means for all; the number of species is vast. A majority favors the leaves of woody plants, many of which are among the pioneers we have been discussing. Black cherry, for example, is a food plant for the caterpillars of the tiger swallowtail, the red-spotted purple, the wild cherry sphinx, the twin-spotted sphinx, the blind-eyed sphinx, the cecropia moth, the promethea moth, the eastern tent caterpillar, and the wild cherry and cherry scollop shell moths. At any time in the growing season, the bedraggled cherry foliage reflects these appetites. Elm is eaten by the question mark butterfly and by the four-horned sphinx, elm leaf caterpillar, and polyphemus moths. Sassafras is attacked by the black and spicebush swallowtails and by the promethea and imperial moths. The latter is also fond of the leaves of sweet gum as are the larvae of the luna and royal walnut moths. The great ash sphinx feeds on ash leaves. The silver-spotted skipper eats locust; the cecropia and polyphemus eat maple. The alien ailanthus apparently brought with it from Asia a direct heterotroph, the Cynthia moth. When mimosa arrived, it found a waiting forager, the (then unnamed) mimosa webworm that had previously limited its enthusiasms to honey locust. A species within a group known as the hand-maid moths, the larvae recognizable by their unique behavior of raising both ends of their bodies when disturbed, eats sumac. It is called the sumac caterpillar.

The lepidoptera seem to be less concerned with the foliage of the Piedmont's conifers than with the broadleaf trees, but interest is not entirely lacking. The larvae of the olive hairstreak butterfly as well as those of the bagworm moths eat eastern red cedar foliage. Caterpillars of the banded-elfin butterfly are fond of pine in the Piedmont, notably pitch and Virginia pine, as are the bagworms.

At probably no other stage in the sere are the caterpillars, the larvae of lepidoptera, so numerous or at least so observable as when woody plants first dominate. The number of species inhabiting the Piedmont has not been reckoned, but more than 10 thousand have been identified north of Mexico. Not all of the larvae have been matched to the adults, so one should not be discouraged by difficulties of identification. It must also be remembered that many caterpillars undergo drastic changes in form and coloration as they grow. The caterpillar of the spicebush swallowtail, for example, begins life resembling a bird dropping and develops ultimately a fierce, dragonlike visage.

A plenitude of caterpillars heralds an army of predators, many from the order Hemiptera, the true bugs. Even more important in the control of the lepidoptera are the parasitic hymenoptera, particularly the many wasps of the family Braconidae. Adult females oviposit in or on the caterpillars' bodies, the eggs hatch, and the wasp larvae devour the host internally. The importance of braconids as caterpillar killers can hardly be overstated. While the yellow-billed cuckoo idles in the catalpa branches taking a few "catawba worms" (said to make excellent fishing bait) in late summer, the ground below is covered with the dead and near dead, each black and yellow caterpillar festooned with the white cocoons of braconid larvae. The larvae of many flies, many from the family Tachinidae, also parasitize caterpillars. Robberflies (family Asilidae) prey upon the adult moths, skippers, and butterflies. Alexander Klots, whose works are a major contribution to the popular understanding of the lepidoptera, summarizes the struggle in *Butterflies of the World*. "Only the enormous reproductive powers of the moths, butterflies, and skippers enable them to survive the attacks of so many hostile organisms." Predation and parasitism on the lepidoptera conveys captured solar energy and nutrients from the plants to the higher life forms.

Before leaving the insects we must observe that, although the insect-pollinated herbs have been displaced by the trees and shrubs, many of the woody pioneers are insect pollinated. Conspicuous flowers hum with insect activity on catalpa, dogwood, cherry, black locust, and honey locust. These arboreal flowers sustain communities of insect pollinators and their associated predators and parasites, much as do the flowers of the forbs (non-grass herbs).

Spiders who can engineer their webs to span the distance between branches or between trees are favored in the wooded environment. The white micrathena, *Micrathena mitrata* (0.2 inches/5 millimeters), and the spined micrathena, *M. gracillis*, man their webs in the young trees in place of the arrowhead micrathena, *M. sagittata*, who sifted the wind for flying insects when the land was in asters and thistles. All, of course, are harmless to humans, but a micrathena web across the face is uncomfortable, particularly with the frightened occupant seeking refuge inside one's collar. Gone is the argiope of the open spaces, replaced by *Araneus* spp., who wait in rolled-up leaves for a tug at their webs.

Ticks, *Dermacentor variabilis*, are common and numerous in virtually all Piedmont habitats. The folk name "wood ticks" suggests greater vigilance while walking in the woods, but in my own ramblings I seem to pick up more in the grasslands and weedy places. Ticks evoke in humans responses ranging from revulsion to terror, probably a combined result of the appearance of these hard-shelled arachnids, their association with disease, and their blood-sucking feeding habits. Ticks are monumentally undesirable but they are not legitimate cause for panic—it takes several minutes for them to find a suitable cranny or surface on the mammalian body and to slowly insert their sucking mouth parts. The best defense by far is short pants and hairy legs. A tick or insect can travel unfelt up the inside of clothing or on bare skin, but levering onto a hair, it immediately betrays its presence. Relief is but a finger-flick away.

The attack of the chigger, *Trombicula* spp., is more insidious. These minute, red mites, usually too small to be seen, burrow into the skin and cause severe irritation. Symptomatic relief becomes the victim's first concern. The tiny arachnids are encountered in a variety of habitats in the Piedmont south of Virginia, particularly woodlands. The only effective defense is a thorough and vigorous scrubbing with soap and water immediately after returning from a walk during the summer. Chiggers can be washed off before they become entrenched, a process which apparently requires several hours.

Reptiles and Amphibians

The advent of forestation signals the arrival of numerous reptiles whose preferences are wholly or partially arboreal. The lizards and snakes who hunt and bask in trees and other exposed places above ground level are much more easily noticed than those who slither in the deep grass or under rock and soil. Therefore, fields newly overgrown with trees and shrubs are considered more "snaky" than herbaceous tracts. What has happened is that some of the reptiles, following the trophic flow vertically with the rise of tall vegetation, have begun to exploit the height dimension of the habitat and have become more visible. The northern fence swift, a lizard, prowls the bark and branches of young pine and deciduous trees, and remains as the canopy closes. In the southern Piedmont the green anole, the "chameleon" sold in pet stores, basks and hunts insects in the shrubs and vines in the early woody mesosere. The five-lined skink loses ground as the habitat becomes more woody, but will reappear later among the fallen logs and stumps if the first growth of trees is timbered.

The eastern garter snake, the rough green snake, the eastern coachwhip, and the northern black racer are the predominant snakes that arrive with the trees. South of the Roanoke River they are joined by the scarlet king snake, a harmless harlequin who mimics the deadly coral snake. (The latter is a denizen of the Coastal Plain and is not reported in the Piedmont.) These serpents are all at least partially arboreal and are reliably found in young forests, feeding on amphibians, birds, eggs, other reptiles, and rodents. Reptiles of the Piedmont (see the appendix), may assist in identifying snakes and lizards of the young woodlands; however, we must recognize that snakes, as a group, are notably unfaithful to any given habitat. Herpetologists are fond of saying, "Snakes are where you find them."

Birds

The early waves of spring migrants which overrun the Piedmont include the yellowbilled cuckoo, *Coccyzus americanus*, who will feast on the outbreak of tent caterpillars. The blue jays turn their attention to these and other hairy caterpillars of the developing woodlands. The wood warblers (family Dendroidae) attack the loopers, larvae of the geometrid moths, and other hairless, soft-bodied caterpillars in the young trees at greater heights. The king bird hunts flying insects in the very earliest phase of woody dominance, yielding to the crested flycatcher and the eastern phoebe after the canopy grows closed. By day, the barn swallow laces into the treetops after flying insects and the chimney swift flies at still greater height. Both fly tirelessly and feed while on the wing. At night the whippoorwill hunts a variety of wooded and open habitats in the same manner as the swifts and swallows.

Nesting in the young trees we find cardinals, rufous-sided towhees, catbirds, brown thrashers, loggerhead shrikes, white-eyed vireos, prairie warblers, yellowthroats, yellow-breasted chats, orchard orioles, indigo buntings, and blue grosbeaks. After the trees have reached 30 feet (9 meters) or more, nesting species include the bluejay, the common crow, the mourning dove, and the pine warbler.

Once a stand of trees has developed, a number of definite relationships becomes observable between the trees and the birds. The pine warbler is not commonly found far from its namesake in any season, though it sometimes feeds on the ground in open places adjacent to pine woodlots. The brown-headed nuthatch limits its breeding and wintering ranges in the Piedmont to pine woods, arriving as the trees close into canopy and remaining until the hardwoods replace the pines. The preference of the red-breasted nuthatch for pine stands, while wintering in the Piedmont, is only slightly less exclusive than that of the brown-headed nuthatch. The brown creeper, *Cerphia familiaris*, often joins the nuthatches, golden-crowned kinglet, *Regulus satrapa*, and the pine warbler in loose companies foraging in the winter pines.

When the sweet gums and tulip trees reach sexual maturity, an event which occurs long before full height is attained, they attract numerous fringillids as winter foragers. Hanging on the sweet gum balls and prying out the seeds, we commonly see purple finches, house finches, and goldfinches, along with black-capped and Carolina chickadees, and pine siskins. Evening grosbeaks and purple finches extract the seeds from the artichoke-like structures on tulip trees, a food source for which they must compete with the eastern gray squirrel.

The sharp-shinned hawk and the loggerhead shrike are the principal diurnal birds of prey of the developing mesic woodlands. The red-tailed hawk, essentially a hunter of the open lands, often feeds eastern gray squirrels to its young, so we must presume that the Piedmont's largest *Buteo* visits the woods. At night the screech owl and the great-horned owl take small mammals and large insects from the young woodlands. The secretive barn owl, rarely seen or heard but known to be present in significant numbers, probably hunts this habitat. Crows and jays also take some small rodents and birds. Overhead, the black and turkey vultures constantly scan, and in the case of the latter, sniff for carrion.

Mammals

At this stage we observe a transition between occupancy by open-country mammals and those of the forest. The meadow vole of the grasslands yields to the pine vole, the latter gaining prominence as the woody canopy closes, actually preferring deciduous woodlots to conifers. The semiarboreal golden mouse, *Peromyscus nutalli*, appears with the adolescent ashes, maples, and tulip trees. The presence of many caterpillars eating at the deciduous leaves implies heavy nighttime visitation by moths, accounting in part for the great variety of bats (see Mammals of the Piedmont in the appendix) which patrol above and within the young woodlands.

There are numerous mammalian species who prefer access to a variety of adjacent habitats, including young woodlands. Both of the resident foxes, the white-tailed deer, the cottontail, the opossum, and several of the mice and shrews inhabit a variety of open, weedy, and wooded places, seemingly to widen their feeding opportunities. The theme of generalism, rather than specialization, is clearly favored among the mammals that occupy the mesic Piedmont.

The major zoologic shift to be marked within the young woodlands is the appearance of the strictly arboreal mammals. The flying squirrel, *Glaucomys volans*, is likely to arrive first and, due to its shyness and nocturnal habits, to be noticed last. Unlike the fox squirrel, *Sciurius niger*, and the eastern gray squirrel, *S. carolinensis*, the flying squirrel resides comfortably among the conifers and may therefore occupy a tract a seral step ahead of the larger squirrels if the initial woody invasion is by pines. The flying squirrel dens and rests by day in the excavations of medium-size woodpeckers such as the hairy, the red-headed, and the red-bellied. The excavations of the downy woodpecker may also be adequate. The flying squirrel does not propel itself through the air—bats are the only mammals capable of powered flight—but leaps from the trees and glides with grace and agility, usually ending a glide with an upward swoop to alight on a tree trunk. The fox squirrel and the much more common gray squirrel typically show no interest in a woodlot until there are sexually mature deciduous trees present to provide flowers, fruit, nuts, and fungi. Thus, a site pioneered by, say, the ashes, maples, and tulips will host the diurnal squirrels sooner than a neighboring pine-cloaked site abandoned in the same year.

The Growing Forest

There are few cases, if any, in which the first species of trees and shrubs to appear on the land remain indefinitely and ultimately mature to dominate the climax vegetation. The typical Piedmont mesosere features an initial woody crop, as the previous section describes, which yields dominance in a few years to another set of vegetation. These successors may, with modifications, comprise the climax stand or they may be replaced by another plant community or perhaps a third before the vegetative mix stabilizes and a mature climax forest comes into being.

In this section we will regard all seral stages between the initial woody plants and the climax stand as the "growing forest." After posting the generalities of the northern and southern Piedmont vegetative models as they appear in the growing forest, we will, as in earlier sections, identify the principal trees, shrubs, vines, and herbs. We will then examine the important plant–animal and animal–animal relationships which complete the trophic flow.

The Coniferous Growing Forest

The pines and cedars that comprise the dominant initial woody vegetation in the coniferous zone usually reach their maximum height within thirty to forty years after germination. As they grow, they compete vertically for the sunlight. They maintain verdant crowns of foliage which capture the majority of sunlight and cast shade on the lower branches. As the light reaching the lower branches is reduced, they die and, with a little help from the fungi, insects, and woodpeckers, decompose. To appreciate the phenomenon of self-pruning and the resulting opening of the growing forest's understory, we must bear in mind that the only place on a woody plant where vertical or new growth occurs is at the tips of the twigs. Trunks and branches thicken laterally, but a tree grows vertically only by adding new wood at the top. A given branch, therefore, remains at its original level; it does not move upward as the tree grows. On many species, coniferous and deciduous, the unmarked surface of the trunk at maturity obscures the former existence of branches at the lower elevations, perhaps giving the impression that the branches moved upward and a span of clear trunk lengthened beneath them.

Thus, as the conifers mature, the conical shape of their adolescence gradually, after decades of close competition, becomes that of a long-stemmed mushroom. A single pine growing in the open would be functionally absurd with eighty feet of clear trunk capped by ten feet of foliage, and open-grown trees do, in fact, assume a lower, fuller profile. But in the growing forest, no other shape is successful.

Another characteristic of the coniferous growing forest is the high and steady mortality of the pines. They die suddenly and totally as a result of numerous diseases which tend to attack the situationally least-favored individuals or groups of trees. The effect over the first two or three decades is to gradually thin the growing pines and cedars and to give an increasingly open aspect to the understory.

Into the void left by the pines' fallen lower branches grows a generation of deciduous trees that will eventually replace the pines. A logical question arises: why don't pine seedlings enjoy the same robust germination under the maturing and fertile pines as the deciduous successors? The answer is that the pines change the conditions of moisture and acidity in the soil, and they create shade—conditions their own offspring cannot tolerate. Young pines cannot grow in dense shade, and they are best adapted to dry conditions. The growing pines increase the amount of moisture in the soil by creating a layer of decomposing litter and by reducing the evaporating effects of wind and sun.

The First Wave of Deciduous Trees and Shrubs, Coniferous Zone

The traditional view concerning the introduction of the first deciduous trees is that the seeds arrive and germinate after the pines have become established. The more recent theory of initial floristic composition contends that the seeds of everything which will grow on a site were present at abandonment. Either way, the deciduous transgressives (young trees presently at a lower level than they will eventually achieve) germinate and grow into the open branchless space beneath the coniferous crowns. This invasion by deciduous transgressives occurs more quickly under shortleaf and loblolly pines than under Virginia pine because the latter is more reluctant to shed its lower branches. The following are the principal broadleaf transgressives.

Sweet Gum, *Liquidambar styraciflua.* Throughout that portion of the Piedmont where conifers are the woody pioneers (that is, throughout the coniferous zone), sweet gum is one of the first deciduous trees to appear under the coniferous canopy. It is more abundant in the southern Piedmont than in the north, but it plays an important successional role as far north as the southern part of New Jersey. Sweet gum is recognized by its star-shaped leaves, spherical (1 inch/25 millimeters in diameter) spiny seed capsules, and corky, winglike projections on the bark of twigs. The seed balls should not be confused with the similarly sized and shaped, but smooth, structures on sycamore.

Red Maple, *Acer rubrum.* Red maple may act as a true transgressive or it may remain in the subcanopy indefinitely. The relative height it achieves in comparison with its competitors appears to vary directly with the moistness of the soil. Red maple is present in important numbers in most growing forests of the Piedmont and is particularly active as a mesic transgressive south of the Potomac. There are many mesic sites, however, on which the red maple never penetrates the canopy but remains in the substory. The leaves of red maple are three-lobed and have red stems. Bark is extremely variable in color and texture and is not reliable as an identifying character.

Tulip Tree or **Yellow Poplar,** *Liriodendron tulipifera.* This is a tree of many roles in Piedmont secondary succession. Where it occurs with or under young pines it is always a transgressive. It is usually the first to reach the canopy and to begin eliminating the pines. The chances are good that the tulip tree will remain on the site and ultimately participate in the climax canopy, especially in the more moist locations. Its growth is rapid, equaling or exceeding that of the pines. The leaves are four-lobed. The fruiting structure, an upright cone of tightly packed samaras, persists into winter and provides a conspicuous identifier. The trunk is straight and symetrically branched.

Trees and Shrubs of the Understory

Sourwood, *Oxydendrum arboreum.* Sourwood characteristically grows in the understory beneath pines and other dominants of the growing forest. It does not attain canopy stature. It may remain as a subdominant as the site matures. The tree flowers conspicuously in summer, the floral

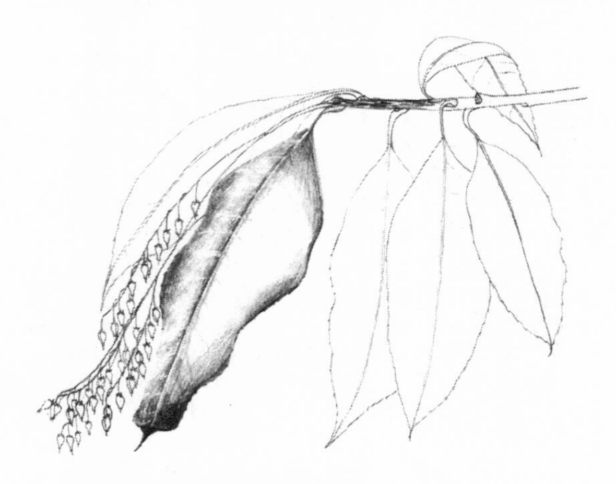

29. Sourwood, *Oxydendrum arboreum*

sprays, and later the fruit, arranged like fingers on a skeletal hand. Branching is irregular, twisted and angular. Diameters greater than 8 inches (20 centimeters) are uncommon in the Piedmont. The leaves are alternate, simple, narrowly elliptical, 5 to 7 inches (13 to 18 centimeters) long.

Flowering Dogwood, *Cornus florida.* The observer of succession in the Piedmont finds that dogwood may make its appearance at any stage of the mesosere. In the coniferous zone it normally appears under growing pines. It remains a member of the subcanopy, its crown typically nested under red maple or sourwood. The leaves are oval and pointed. The small yellow flowers are clustered within white showy bracts. Shiny red berries are grouped at the tips of the twigs in autumn.

Black Cherry, *Prunus serotina.* If black cherry germinates with the pines, it is likely to be overgrown by them and to remain on the site as a subcanopy tree. It commonly germinates under the pines after the understory is open, in which case it will probably remain a stunted member of the shrub layer or lowest tree stratum. The flowers and quarter-inch (6 millimeters) black cherries grow on racemes of 4 to 6 inches (10 to 15 centimeters). The leaves are elliptical and pointed, with tufts of brown tomentum (fuzz) beneath the midrib.

30. Redbud, *Cercis canadensis*

Redbud, *Cercis canadensis.* Redbud, or Judas tree is a legume (member of the family Fabaceae) found growing under pines and, later, under the initial deciduous and climax deciduous canopies throughout the mesic Piedmont. It does not reach the canopy, being restricted to a height of 30 feet (10 meters) or less. Leaves are heart-shaped. Pink flowers and beanlike fruit pods grow along the twigs, branches, and trunk.

Winged Elm, *Ulmus alata.* Winged elm is frequently coeval with the pines south of the James River. It is sometimes overtopped by the pines or it may share the canopy with them. Whether the species has the genetic capacity to germinate under the pines then grow to penetrate the canopy is questionable. Leaves are 1 inch (25 millimeters) long, elliptical and serrate. The slender stems (one-quarter the thickness of sweet-gum stems) have corky projections.

Black Haw, *Viburnum prunifolium.* As the scientific name suggests, the leaves of this shrub resemble those of cherry. They are arranged in pairs on opposite sides of the twigs. Black haw sometimes germinates during the herbaceous phase shortly after abandonment and remains in the shrub layer of the growing forest. In most cases it is excluded from the mature forest. Frequently, it makes its initial appearance under the coniferous canopy after the understory opens. Black haw is endemic to the Piedmont, being infrequent in the Coastal Plain and rare in the Blue Ridge Mountains.

Downy Arrowwood, *Viburnum rafinesquianum.* As with all viburnums, the leaves are opposite. They are downy above and beneath, oval to elliptic in shape, and serrate. The plant grows to head height in clusters of a few stems, the lower sections of which commonly include an arrow-straight segment of perhaps 3 feet (90 centimeters). This shrub occurs almost exclusively in woodlands. It is often the first to arrive under the developing canopy, and it may remain to become part of the climax forest.

Squaw Huckleberry, *Vaccinium stamineum. Vaccinium* is the genus of the blueberries, of which several species are common in the mesic growing forests of the Piedmont. Perhaps because it is usually the first to arrive under the maturing pines, squaw huckleberry may be the most common. It may grow to 15 feet (5 meters), but waist height is the more frequent stature. The elliptic leaves are at alternate positions on the stems and are of varying sizes. Flowering specimens (May and June) show clusters of white, five-lobed, bell-shaped corollas from which the yellow stamens protrude. Berries are tart to bitter, but are eaten enthusiastically by birds.

Silverberry, *Elaeagnus umbellata.* In the pine stands of the Piedmont south of the James, this escaped shrub occurs locally in profusion. The leaf is elliptic, green above, silvery below, leathery, 2 to 3 inches (5 to 8 centimeters) long and of wavy margins. Long, unbranched twigs characterize the shape. The flowers and berries are clustered on short stems along the twigs.

Deciduous Holly, *Ilex decidua.* This congener of the evergreen holly responds to the changing seasons with normal deciduous behavior. One or more trunks may develop from a single root system. Height is generally limited to 15 feet (5 meters). The leaves are elliptic to elliptic–spatulate, 2 to 3 inches (5 to 8 centimeters) long, ½ inch (1 centimeter) wide, the margins serrate. The white flowers and lustrous orange-red berries occur along the stems and at leaf nodes. Plants are frequent but widely spaced on most sites; the species is common from the Potomac southward, usually making its first appearance under the growing pines. Deciduous holly is in many ways similar to black haw, *Viburnum prunifolium*, with which it shares the shrub layer of the growing forest, but is distinguished by its alternate leaves; the viburnums all have opposite leaves.

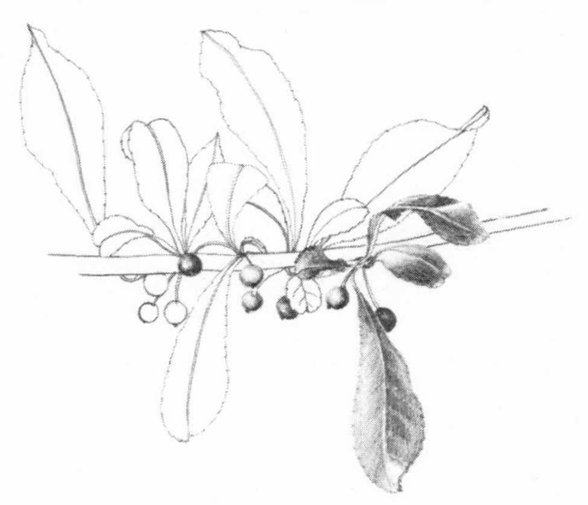

31. Deciduous holly, *Ilex decidua*

Shadbush, *Amelanchier arborea.* North of the James, shadbush is a common shrub in the growing forest. The white flowers are conspicuous in the bare woods in early spring on lax drooping racemes. The purplish plumlike fruit is edible. The leaves are elliptic with cordate bases, serrate, 1 inch (25 millimeters) wide, 2 inches (5 centimeters) long. The plants commonly grow to head height or higher. Multiple stems are uncommon.

American Holly, *Ilex opaca.* In the growing forests throughout the Piedmont, American holly frequently occurs as a very small tree of shrub height. Occasionally, attrition in the canopy permits the tree enough sunlight to achieve its handsome genetic potential, a densely foliated slender cone of 40 feet (12 meters) or more. The leaves are elliptic, 2 to 4 inches (5 to 10 centimeters) long, 1 to 2 inches (2 to 5 centimeters) wide, leathery, glossy green above and armed with spines at the margin. The leaves of all hollies are alternate. In this species, the leaves are retained throughout the winter to be shed the following spring after new leaves have developed; a given leaf lives approximately one year—not indefinitely. Male and female flowers occur on separate trees. Only female trees bear fruit, glossy red berries on short stems along the twigs, which persist well into winter and yield a significant food source for many birds. Humans use this holly for decking halls.

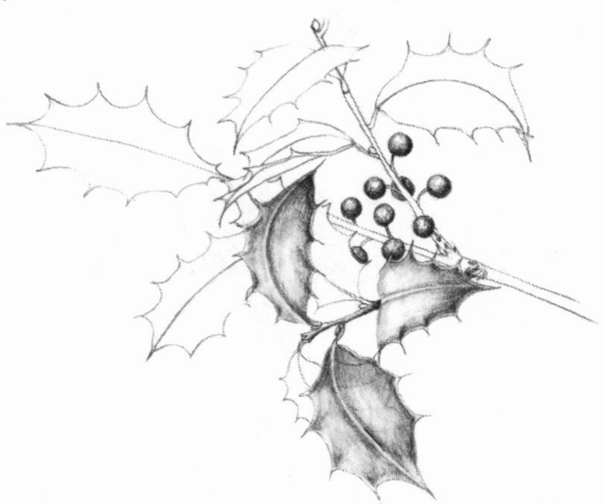

32. American holly, *Ilex opaca*

The Second Wave of Deciduous Trees and Shrubs, Coniferous Zone

As the sweet gums, red maples, and tulip trees begin to share the canopy with the maturing, and doomed, pines, a second group of transgressives begins to move into the open understory. At most mesic locations in the coniferous zone (from southern Pennsylvania and New Jersey southward), this wave of young hardwoods will eventually dominate the mature canopy.

Willow Oak, *Quercus phellos.* Willow oak may be found anywhere in the Piedmont, but it is most prevalent south of the Roanoke River. In its primary range (the heart of the coniferous zone), it commonly is the first of the oaks to arrive, often appearing with, or within a decade of, the sweet gum. It is occasionally a woody pioneer, arriving with the pines. On mesic sites the willow oak does not, as a rule, remain as a member of the climax canopy, although it may do so in the extreme southern Piedmont. It does, however, participate in the climax communities of more moist sites from Virginia southward. It is a paradox that a species which favors moist soils should appear on mesic sites early in the woody sere, when the soil is relatively dry, and should later disappear as the site matures and becomes more moist. The leaves of this oak are willowlike, linear, ½

33. Willow oak, *Quercus phellos*

inch (1 centimeter) wide and 2 to 4 inches (5 to 10 centimeters) long, entire, dark green above, pale below. The acorns are very small (¼ to ½ inch/6 to 12 millimeters in length and diameter), permitting them to be eaten and distributed by birds. This may account for the species' appearance very early in the sere at some sites. The bark is dark and shallowly furrowed.

Spanish or **Southern Red Oak,** *Quercus falcata.* Throughout its range (in the Piedmont, northward to the Potomac), Spanish oak remains as a participant in the climax canopy wherever it occurs. The first appearance is usually under the maturing pines at about the time the sweet gums and tulip trees begin to penetrate the canopy. The species bears a dimorphic leaf which changes as the tree ages. Young trees are foliated with wedge-shaped leaves ending in a three-forked tip; leaves on older individuals end in a single, narrow, pointed lobe and have similar lobes lateral to the midrib spaced with deep sinuses. The acorn is compressed, ½ inch (1 centimeter) long.

34. Spanish, or southern red oak, *Quercus falcata*

Black Oak, *Quercus velutina.* Black oak seedlings appear as the pines mature and the tulip and sweet gum assault the canopy. As a sapling, its leaves may be oversized, perhaps a foot (30 centimeters) in length, while retaining their characteristic shape of an oval outline cut by two or three deep sinuses on each side. The five to seven lobes each have sev-

35. Black oak, *Quercus velutina*

eral points. Acorns are ¾ inch (2 centimeters) long, slightly elongated. Bark is black, ridged, the furrows widely spaced. On upper portions of individuals, the bark may be shiny. Black oak is the dominant oak in the climax canopy at maturity on many mesic slopes as far north as the Delaware. It grows to be a very large tree; 3 feet (90 centimeters) dbh is not uncommon on mature sites.

36. White oak, *Quercus alba*

White Oak, *Quercus alba.* White oak is another of the climax hardwoods that makes its debut as the first wave of deciduous trees reaches the pine canopy. It grows to be a very large tree and, along with the black oak, dominates a majority of the mature mesic sites in the coniferous zone—perhaps in the entire Piedmont, for white oak is common throughout the province. It is not principally a southern tree, as are the Spanish and willow oaks. The acorn is nearly 1 inch (25 millimeters) long, oblong. The leaves are 5 to 9 inches (13 to 23 centimeters) long, 3 to 4 inches (8 to 10 centimeters) wide with rounded lobes and with sinuses of varying number and depth.

Post Oak, *Quercus stellata.* Post oak prefers soils toward the drier end of the mesic scale, and throughout the mesic Piedmont may be seen slightly uphill from its close relative, the white oak. Saplings appear as the pines mature. The leaves are leathery with rounded lobes. The central lobe is enlarged to lend a cruciform appearance to the leaf. Acorns are ¾ to 1 inch (20 to 25 millimeters) long, oval.

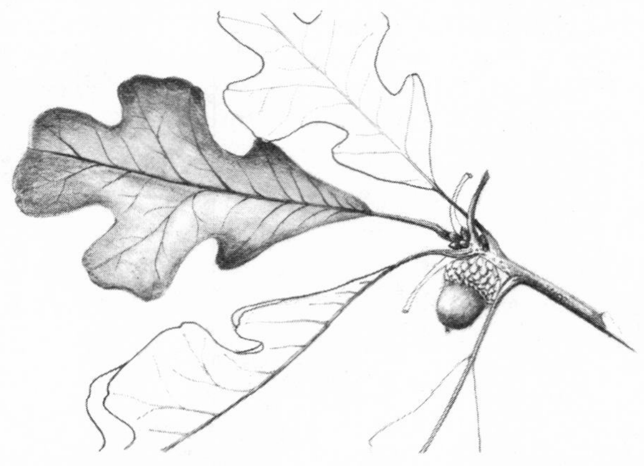

37. Post oak, *Quericus stellata*

The Deciduous Growing Forest

We have observed that in the Pennsylvania Piedmont north of the latitude of Philadelphia and in the northern two-thirds of the New Jersey Piedmont, the initial woody plants of the mesosere are principally deciduous. To the south lies a region where deciduous early succession alternates on neighboring sites with conifers—or mixes with them on the same site. This zone where deciduous and coniferous initial woody succession overlap extends southward through northern South Carolina. In the material which follows, we will examine the succession process in the northern reaches of the Piedmont, where the initial woody plants are almost wholly deciduous, and in areas of the south where a similar pattern appears.

The ashes, elms, locusts, maples, black walnut, and ailanthus (tree of heaven), which invade the abandoned land in the northern Piedmont, grow almost as rapidly as their coniferous southern counterparts. By the end of the first decade after abandonment, the deciduous pioneers usually have formed a closed canopy. Later, as these trees approach their maximum height (this may be achieved well in advance of the midpoint of their lives), the same process of self-pruning which opens the coniferous understory causes the lower branches of the deciduous pioneers to be shed. Into the understory beneath the first wave of trees in the deciduous zone grows a successor group. These transgressives, with a few later additions, will ultimately dominate the climax canopy.

Thus the northern growing forest differs essentially from the southern by bypassing the coniferous stage, though we would be lured into oversimplification in viewing this as the only difference. The first wave of trees holds dominance longer in the deciduous case than in the coniferous perhaps by a factor of two. The composition of the climax canopy is somewhat different, as is the understory. In addition, the postclimax vegetation, should this phase be achieved, can be different in the northern and southern zones. Withall, the similarities uniting the two types of growing forest are more remarkable than the differences, once we view the pines as an omissible step.

Page references in the following list refer to more detailed descriptions given elsewhere.

Red Oak, *Quercus rubra*. This species is similar in form and in seral function to the black oak, with which it hybridizes where the ranges overlap, principally north of the Potomac. Neil Jorgensen, in *A Sierra Club Naturalist's Guide to Southern New England*, makes the following comment on

38. Red oak, *Quercus rubra*

the taxonomic confusion resulting from hybridization among oaks:

> [The red oak] is easily recognized by its tall stature, its leaf shape, which is intermediate between the stout leaves of black oak and the lacy leaves of the scarlet oak, and its bark, which is gray with wide reddish furrows. The red, scarlet, and black oaks hybridize readily; the ensuing progeny may hybridize again. As a result, there are many southern New England oaks showing characteristics intermediate between one species and another, thus making identification uncertain.

Hybridization is no less widespread among members of the red oak group (those species having leaves with pointed lobes) in the Piedmont. The red and the black are fundamentally similar and, as Jorgensen's passage implies, intermediate gradations exist. Most red oak leaves have nine lobes; those of black oak, seven. Trees of the northern Piedmont the majority of whose leaves are nine-lobed are at least partly red oak. The species makes its initial appearance under the ashes and locusts as the understory opens. By one hundred years after abandonment, *Q. rubra* is probing the canopy.

Black Oak, *Quercus velutina.* See page 154. On lands pioneered by deciduous trees, the seral timing and function of black oak are similar to those on coniferous lands—the species appears as the initial deciduous understory opens. It grows to be a large tree and a major participant in the growing and mature forests.

White Oak, *Quercus alba.* See page 155. White oak is likely to be one of the first oaks to join the growing forest in the deciduous zone. This sequence contrasts with that of the coniferous zone, where the black and willow oaks are likely to precede the white. Still the white oak arrives at about the same number of years after abandonment in both zones, being preceded in the south by other oaks and in the north by ash, locust, and their allies. The range includes the entire Piedmont.

Basswood, *Tilia americana.* Basswood, or linden, transgressives occupy a significant position in the growing forests of the northern Piedmont. The individuals are generally too scattered, however, to dominate at maturity. The leaves are heart-shaped and finely toothed. The flowers and fruit are born on modified leaves of spatulate shape, from which the midrib detaches midway to become the main pedicel. On mesic sites, basswood is more often seen under the red and white oaks than in the canopy.

Scarlet Oak, *Quercus coccinea.* The scarlet oak arrives on mesic sites in the northern Piedmont as the deciduous pioneers reach maturity. It is most common north of the James but can be found as far south as the Dahlonega

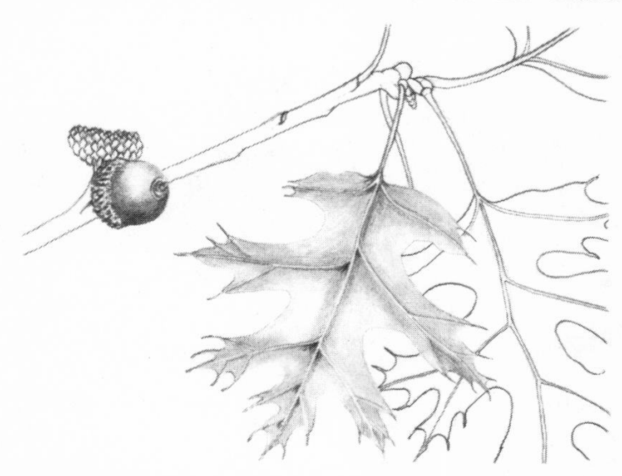

39. Scarlet oak, *Quercus coccinea*

Plateau. It does not characteristically reach the great height of its seral coevals, the red oak and the white, but is restricted to 75 feet (23 meters) or less even under optimum conditions. It rarely ventures off the mesic slopes. The leaves have the same number of lobes as the red oak (usually nine), but the sinuses cut almost to the midrib and are wider than those on red oak leaves. In autumn they turn scarlet (those of red and black oak become brown). The acorn is a good identifier: the tip is nipple-like and is surrounded by concentric circles.

Cherry Birch, *Betula lenta.* Principally a tree of the Appalachians and regions north of our concern, the cherry birch extends southward into the Piedmont, encompassing the province in Pennsylvania and New Jersey. In this region (the purely deciduous zone of the Piedmont) cherry birch is active as a participant in the canopy of the growing forest. The leaves are similar to those of American elm, being ovate, 3 inches (8 centimeters) long and finely serrate. The bark is shreddy on the twigs, but on the main trunk it bears the lateral striations characteristic of cherry. The twigs have the taste of licorice.

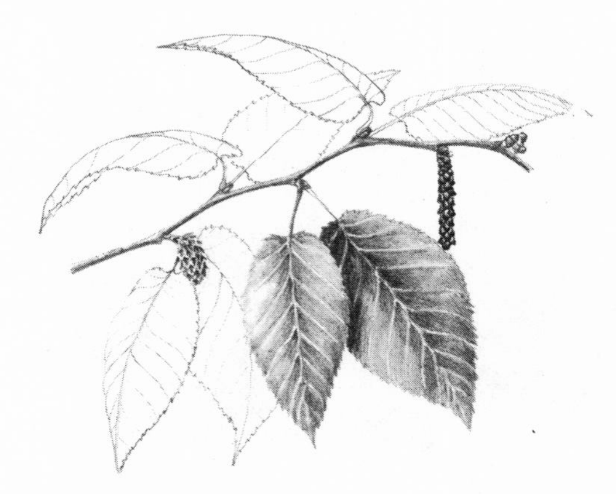

40. Cherry birch, *Betula lenta*

Black Gum or **Black Tupelo,** *Nyssa sylvatica.* Black gum is significant to dominant in the understory of many growing forest stands in the Piedmont north of the James. It may arrive soon after the canopy is established or it may wait until much later in the sere. If grown in the open or in a hedgerow, it may achieve 2 feet (60 centimeters) dbh, but as a member of the deciduous understory, to which it is characteristically confined, the girth is usually less than one foot. On most sites where it occurs in the deciduous zone, black tupelo remains to participate in the understory of the mature forest. The leaves grow in fanned clusters at the twig tips; they are entire and elliptic to obovate in shape. Fruit is an ellipsoid blue drupe ½ inch (13 millimeters) long and of great importance as a food source for birds.

41. Black gum, or black tupelo, *Nyssa sylvatica*

Flowering Dogwood, *Cornus florida.* See page 123. In the deciduous growing forest, flowering dogwood is content to restrict its height and accept a subcanopy role. It may first appear with the woody pioneers and be outgrown by them or it may arrive somewhat later. In either case it is likely to remain on the site indefinitely as a member of the climax understory.

Redbud, *Cercis canadensis.* See page 149. All that is said above regarding dogwood applies to redbud except that on many sites redbud becomes less frequent as the forest matures. Redbud is commonly among the woody pioneers, and it is a prominent tree of the understory of the northern growing forests.

Sassafras, *Sassafras albidum.* Sassafras is a common tree of the understory of the growing forests throughout the

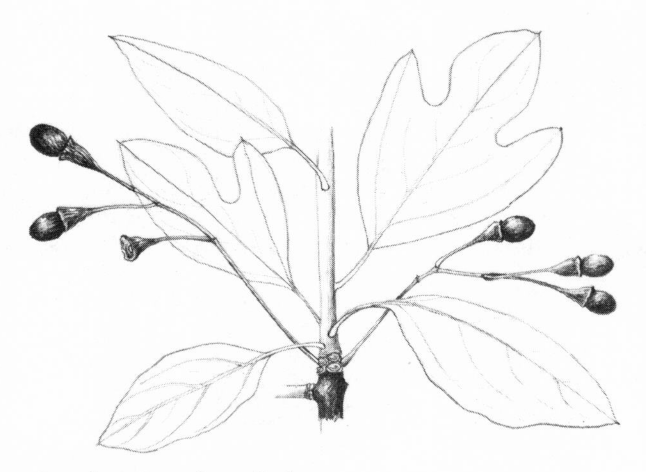

42. Sassafras, *Sassafras albidum*

Piedmont, though less so in the northern extremes than in the south. It arrives beneath the deciduous pioneers of the northern Piedmont as they reach maturity; it thus occurs later in the northern mesosere than in the southern, where it sometimes functions as a woody pioneer, germinating with or shortly after the pines.

Witch hazel, *Hamamelis virginiana.* Like spicebush, witch hazel is associated with the bottomlands in the southern Piedmont, but is common on the mesic slopes in the northern part of the province. A close relative of the sweet gum, witch hazel is a small tree with leaves which are alternate, broadly elliptic, asymmetrical at the base, and with large rounded teeth. The flowers, with threadlike yellow petals, may appear any time during autumn, winter, or early spring; blooming usually precedes the opening of leaf buds. Witch hazel becomes established only after the deciduous forest is well into its developmental sequence, when the ashes and walnuts are 18 inches (45 centimeters) or more in diameter, and the black oaks have penetrated the canopy.

Spicebush, *Lindera benzoin.* Spicebush is a shrub of head-height. The leaves are thin, entire, obovate and 2 to 4 inches (5 to 10 centimeters) long. The twigs are thin and spicy tasting, in winter studded with sessile spherical buds which open in early spring to produce quarter-inch (6 mil-

43. Spicebrush, *Lindera benzoin*

limeters) yellow flowers. The plant arrives beneath the deciduous pioneers of the northern Piedmont's mesic slopes after the canopy is established and the understory opens. From the Potomac northward, the presence of spicebush under the older ashes and locusts is strongly probable. It is active in the shrub layer of certain sites in the southern Piedmont as well, but these tend to be hydric rather than mesic; in light of its hydric preferences in the south, it seems remarkable that spicebush's participation in the northern mesosere is essentially limited to the midrange of woody succession. As the mesic sites mature, the presence of spicebush diminishes in spite of the enhanced capacity of the soils to retain moisture.

Black haw, *Viburnum prunifolium.* See page 150. This shrub is endemic to the Piedmont; it grows scantily in the Blue Ridge and in the Coastal Plain. Black haw commonly appears late in the herbaceous phase of the mesosere and remains through the phase we call the growing forest. The cherry-like leaves are opposite in arrangement, but the plant is otherwise easily confused with young black cherries.

Downy Arrowwood, *Viburnum rafinesquianum.* With its downy, 1½-inch (4 centimeters), coarsely serrate leaves (opposite, as in all viburnums), downy arrowwood appears under the developing deciduous pioneers of the northern mesic Piedmont at a seral moment comparable to that of its appearance in the coniferous Zone—as the understory opens. It is likely to remain on the site and participate in the shrub layer of the climax forest.

Viburnum dentatum (no common name). This species is similar to downy arrowwood, but as the epithet suggests, the leaves are dentate; that is, the notches at the margins are substantially larger than on the arrowwood, as are the leaves themselves. In addition, they lack the "downy" feel of *V. rafinesquianum's* leaves. The flowers and fruits are arranged in a circular, flattish inflorescence called a cyme. *V. dentatum* grows in the open hydric soils of the northern Piedmont and on the mesic slopes after the initial canopy of cherry, locust, or ash is established. On northern mesic sites it tends to arrive and disappear before downy arrowwood; however, it may remain and grow under the oaks as the site matures.

44. *Viburnum dentatum*

Shadbush, *Amelanchier arborea.* The arrival of shadbush coincides with that of the oaks in the deciduous zone.

Mesic Growing Forest Habitats by State

The initial woody phase of the mesosere is more complex and of greater duration in the deciduous zone than in the coniferous. The floristic complexities are approximately equalized, however, by the time the conifers of the south yield to their deciduous successors. The diversity of both models increases as the understory opens and the members of the communities sort themselves into the several subdominant levels—the subcanopy, the shrub layer, the herb layer, and the dwellers at shoelace height and below. As the growing forest progresses, the occupants of this seral phase are augmented by the transgressives that will eventually dominate.

NEW JERSEY

The northern two-thirds of the New Jersey Piedmont is dominated by the deciduous pattern of succession. Here ash, red maple, catalpa, black locust, and ailanthus rise to close a canopy by two decades after abandonment; beneath this the understory opens two or three decades later. These pioneers, particularly the ash, maple, and locust, are long-lived in comparison with the more southerly conifers, and they may be expected to control the canopy for a century or more. During their tenure, it is easy to observe beneath them the transgressive red and white oaks reaching for the canopy and the black gum and cherry birch, which probably will remain in the understory. Tulip trees may be present and, with silver maple, may be locally common in the early mesic canopy. Dogwood and black cherry are common in the understory; both may have arrived as part of the initial woody complement, later to be overgrown. Spicebush is expected as the prevalent shrub. Black haw and downy arrowwood are generously distributed in the shrub layer. Jack-in-the-pulpit, true and false Solomon's seal, wind flower, May apples, and grape and Christmas ferns, in their seasons, populate the herb layer.

The conifers are not wholly excluded as woody pioneers even from the most northerly reaches of the Piedmont, but they figure significantly only south of the Raritan River. Islands of pine dominance can be seen along Interstate 287 as far north as New Vernon, but generally pines and cedars are uncommon north of the vicinity of New Brunswick. But there, in the neighborhood of the Hutcheson Memorial Forest (see page 384), cedars and American elms share initial woody dominance. At the Herrontown Woods (see page 384) near Princeton, the decadent cedars can be seen overgrown by sweet gum (one of this species' more northerly outposts in the Piedmont), tulip trees, and red maple. These, in turn, are being pressed by vigorous ranks of red oaks, white oaks, and pignut and shagbark hickories (*Carya glabra* and *C. ovata*, respectively). The time span which separates the waning of the woody pioneers and the arrival of the climax dominants is too great for the two ends of the woody sere to coexist on many sites. In the Herrontown Woods, however, a few skeletal cedars stand beneath red oaks of 18-inch (45 centimeters) diameter and shagbark hickories one foot (30 centimeters) dbh. Old sweet gums and tulip trees more than two feet (60 centimeters) dbh are the largest trees present.

PENNSYLVANIA

On July 3, 1863, the third and decisive day of the battle of Gettysburg, an incident occurred which draws little attention from military historians, but which might figure prominently in helping us understand the life forces of the Piedmont. Longstreet's massive assault on the Union center had just failed. Jubilant couriers raced to inform the Union flanks, "We turned the charge. Nine acres of prisoners!" To most Federal officers, this was good news beyond belief. But to Union Cavalry Commander Kirkpatrick, the tidings represented fading opportunity for glory. Presuming that the Confederate right, which he faced, was demoralized and vulnerable, he ordered Brigadier General Elon J. Farnsworth of the First Vermont Cavalry to lead a charge.

Farnsworth rode out to reconnoiter. He watched the Georgians and Texans, firing disciplined volleys from behind stone walls, decimate several infantry regiments sent to test their strength. The terrain was boulder-strewn. The path of the charge would be uphill toward the stone walls. Farnsworth concluded the venture would be ineffective and suicidal, and so advised his commander.

The latter responded by questioning Farnsworth's courage. The young brigadier, breveted for gallantry only days earlier, stiffened, fought the impulse to issue a challenge, saluted and rode, as he fully expected, to his death.

Farnsworth fell in an open field, the exact spot recorded in a photograph taken shortly after the battle and, at present, by a marble monument. Today the site is a splendid example of a northern Piedmont growing forest. It has grown undisturbed since the day of the battle. Few other sites offer an opportunity for dating this precisely an uninterrupted successional growth.

Surrounding the Farnsworth marker are the decadent black locusts and white ashes which, as an interim topographic map shows, had begun to claim the field by 1894. The ashes are two feet (60 centimeters) in diameter and still growing vigorously. The locusts average about 15 inches (40 centimeters) and many are dead or dying. Occasional black walnut trees of two feet (60 centimeters) dbh are present; they were among the woody pioneers. Most significantly, there are many young red and black oaks in the tract; some have already claimed a place in the canopy. Together with the ashes and a few tulip trees, the oaks are destined to hold the canopy exclusively for a time, later to share it with the scattered white pines and Canadian hemlocks whose transgressive saplings have already appeared in the understory. The subdominants include redbud, dogwood, and American elm. Witch hazel and a robust growth of spicebush occupy the shrub layer. Black haw, a relic from the period of early woody dominance, is still present. Herbs include jack-in-the-pulpit, loose-flowered phacelia, false Solomon's seal, windflower, mayapple, and Christmas fern.

To the north of Gettysburg the successional pattern generally fits our northern, or deciduous, concept, and examples of coniferous succession are sparce. French Creek State Park south of Reading and Pottstown offers many examples of the transition between the deciduous pioneers and their successors. But in the vicinity of Gettysburg and southward there is a considerable amount of coniferous pioneering. Cedar is common, as is Virginia pine. Pitch pine, *Pinus rigida*, is less common. The same civil war photograph of the Farnsworth site shows Round Top in the background, intermittently cloaked in pine and cedar. It is a maturing forest today, as we shall explore in the next section.

Where coniferous pioneers occupy the Piedmont of Pennsylvania, they yield in forty to sixty years to red maple, pin oak, *Quercus priniodes*, tulip tree, and ash. American elm, a prominent deciduous pioneer in the area, does not appear to germinate under the pines and cedars but does succeed the conifers on many sites where they germinated simultaneously. Thus, in Pennsylvania one finds groves of decadent pines and cedars yielding to red maple (occasionally silver maple), pin oak, and American elm. On advanced sites of this category, the saplings of black and red oak probe the understory, foretelling a future of oak–hickory–white pine dominance similar to what the site would ultimately experience had the woody pioneers been deciduous.

MARYLAND

The growing forests of Maryland exhibit about equally the working of the coniferous and deciduous models. Many sites blend the two. Where the pioneers had been pines or cedars, the deciduous successors are commonly red maple, ash, and pin oak. Black oak follows in a few decades. In the case of deciduous pioneers, the black oak and, to a lesser extent, red oak, follow the locusts, ashes, ailanthus, and elms by about half a century. Dogwood and redbud seem likely to invade the understory sooner as one proceeds southward from the northern extremes of the Piedmont; indeed, they are more likely to be among the woody pioneers. Spicebush loosens its absolute grip on the shrub layer and in Maryland begins to share that level with more viburnum (generally, downy arrowwood and black haw), and shadbush. Common herbs include true and false Solomon's seal, jack-in-the-Pulpit, mayapple, wild yam, Ladies fern, Christmas fern, lily-leaved orchid, and several violets, *Viola* spp.

VIRGINIA

Proceeding southward from the Potomac, the greater part of the successional lands are pioneered by conifers. Virginia pine is most prominent in the northern counties, gradually diminishing toward the Roanoke River where there is approximate parity between Virginia, shortleaf, and loblolly. Under these pine canopies develop the red maple, tulip, ash, and sweet gum successors who typically reach the canopy by the fifth decade. It seems evident that the climax hardwoods appear sooner in the latitude of Virginia than, say, one hundred miles to the north in Pennsylvania. Virginia creeper, poison ivy and grape vines may appear in the canopy of Virginia's mesic growing forests. The understory beneath the decadent pines and ascendant maple and tulip trees may contain not only the red and black oaks, but the hickories, principally the mockernut, *Carya tomentosa*, and shagbark, *Carya ovata*, as well. It might also contain the mountain holly, *Ilex ambigua*, dogwood, redbud, and sassafras.

Deciduous-pioneered stands in Virginia are typically dominated by black locust, tulip tree, red maple, ash, or some combination of these. Black walnut is scattered on many sites. After growing for several decades, the understory opens in these stands and soon the oaks make their appearance. Dogwood, redbud, and sourwood typically dominate the understory above a shrub layer rich in *Viburnum* and *Vaccinium* spp. It is interesting that in the growing forests of the middle Piedmont some of the woody pioneers frequently are overtaken by taller deciduous trees and continue to exist as understory trees. Species exhibiting this behavior include cedar for a time and, on a more permanent basis, red maple, dogwood, and redbud.

Bull Run Mountain is a site which must be mentioned as a splendid example of the Virginia Piedmont's mesic growing forests and of other forest habitats as well. Just north of Thoroughfare Gap, a historic passage through which migrants and armies have marched, the ridge encompasses a tract owned by the Natural Heritage Foundation called the Bull Run Mountain Natural Area. A canopy of mature black locust and occasional black cherry loom over red oak and tulip tree transgressives with an understory of redbud and red mulberry. Spicebush dominates the shrub layer. Here a notable display of ferns including Christmas, royal cinnamon, and grape (*Botrychium* spp.) shares the herb layer with a seasonally shifting variety of forbs, such as *Hieracium* spp., and other composites.

The region where one is likely to encounter deciduous woody pioneers in dominance lies principally to the north of the James River. Hence, nearly all of North Carolina is initially controlled by conifers. The stands are more likely to be pine than cedar, and very frequently they are interspersed with ash, winged elm, and tulip tree coevals. By their thirtieth year, the pines have thinned and pruned themselves sufficiently to accommodate a vigorous understory of sweet gum, tulip, and red maple. These deciduous transgressives spend the next three to five decades invading the canopy and slowly wresting control from the pines, which eventually all succumb to competitive pressures and disease. Flowering dogwood and scattered red mulberry dominate the understory beneath the pines and deciduous ascendants; later, as the canopy becomes more predominantly deciduous, redbud and sourwood arrive in the understory. The secondary layer under the tulips, ashes, and maples will likely contain some winged elm and sassafras, residuals from the initial woody invasion that was overgrown by the pine canopy. Nor will all individuals of red maple ascend to the canopy; some, perhaps most, will remain in the understory during all woody phases of the mesosere.

The inevitable death of the pines in this area is not popularly understood. There are numerous diseases which overtake the pines as their deciduous successors weaken them by usurping the sunlight and moisture, and since they do not replace themselves their ultimate disappearance is a natural feature of the succession process. There is much concern about destructive outbreaks of the Southern pine beetle, whose infestations can kill large stands of pine timber and valued shade trees. Although the pine beetle gives the appearance of being a rising threat due to the recent vast increase in pine acreage in the southern Piedmont, it should comfort some landowners to bear in mind that the beetle is a native North American evolutionary partner of the deciduous forest, hastening the removal of the pines. If the beetle doesn't take a given pine, some other pathogen will, probably before its fiftieth year. In the context of the region's natural successional pattern this is of little consequence because the forest is naturally diverse and replacements are immediately available. However, when a monoculture of planted pines is attacked, the results can be economically disastrous. More enlightened forestry practice attempts to achieve some species diversity either by planting deciduous trees with the pines or by letting natural secondary succession provide them. In general, the problem of the Southern pine beetle is a natural consequence of massive, sudden reforestation. No known measures are apt to be broadly effective against it so long as a disproportionate part of the southern Piedmont is devoted to growing pines.

Beneath the growing pines of the mesic Piedmont of North Carolina a rich shrub layer develops, many members of which will remain as the sere progresses. Black haw, *Viburnum prunifolium*, is likely to appear first and to be followed shortly by downy arrowwood, *V. rafinesquianum*. The naturalized shrub silverberry, *Elaeagnus umbellatus*, which escaped from cultivation in the nineteenth century, is well established in the shrub layer of the growing forest. The deciduous holly, *Ilex decidua*, technically a tree because it has only one stem for each root system, typically does not rise above the shrub layer. Its obovate leaves are easily confused with those of black haw except that they are alternate in arrangement. The evergreen American holly, *Ilex opaca*, also a small tree, spends a decade or more in the shrub layer before rising to the low understory. Common under maturing pines are several members of the blueberry clan. They are mostly shrubs of the lowest stature, waist-level and below, including in descending order: squaw huckleberry, black highbush blueberry, *Viburnum atrococcum*, and lowbush blueberry, *V. tenellum*. Sparkleberry, *Vaccinium arboreum*, a shrub commonly attaining 10 feet (3 meters) in height, is associated with xeric woodlands and with mesic sites during the pine phase, as is the much shorter (8 to 24 inches/20 to 60 centimeters) *V. vacillans* which has no common name although it is quite common under growing pines in the Carolinas. In the herbaceous layer, the Christmas fern, *Polystichum acrostichoides*, grape ferns (*Botrychium* spp.) and the ebony spleenwort, *Asplenium platyneuron*, are common, the latter seeking rocky crevices. Running cedar, *Lycopodium clavatum*, a club moss and one of the most primitive plants in the Piedmont, is a rhyzomatous evergreen commonly outlining the texture of the old crop rows beneath growing pines. Flowering plants include pipsissewa, *Chimaphila maculata*, rattlesnake orchid, *Goodyeara pubescens*, and sourgrass, *Oxalis* spp.

By the time the sweet gums and tulip trees become well established in the canopy, the black oak and willow oak saplings are moving through the shrub layer. Spanish oak, *Q. falcata*, and white oak generally arrive a few years later. The leaves of the sapling Spanish oak are shaped quite differently from those of the mature tree (see page 154), and at this phase in the growing forest it is a generality that the leaves of the seedling oaks tend to be enlarged.

SOUTH CAROLINA

The growing forest of South Carolina is similar to that of North Carolina where the pines were the woody pioneers, but in the north-central portion of South Carolina's Piedmont, in the vicinity of Rock Hill, there is a region where winged elm is exceptionally active as a woody pioneer (see page 150). In this circumstance the winged elms, often mixed with cedar, black cherry, honey locust, and sweet gum close a canopy within a decade. The winged elms, like the red maple in other growing forests, may be successful in the subsequent vertical competition with the sweet gums and tulip trees or be relegated to the understory where it may survive for some decades, but without replacing itself. Within a quarter-century or so the sweet gums and tulip trees succeed the winged elm, black cherry, honey locust, and red cedar pioneers just as they do the pines in other sites in South Carolina's mesic Piedmont. Meanwhile, the growing forest displays a mixture of the deciduous pioneers and their successors, or, as on the majority of all mesic sites south of the James River, features a canopy of loblolly and shortleaf pines over a distinct layer of deciduous transgressives, notably sweet gum and tulip tree. Black cherry, honey locust, and winged elm are generally present only where they had been among the woody pioneers.

After either deciduous or coniferous pioneering, and as the deciduous successors overtake the canopy, a shrub layer develops which usually includes black haw, downy arrowwood, squaw huckleberry, strawberry bush, *Euonymus americanus*, and highbush and lowbush blueberry. Deciduous and American holly may be present. The herb layer is likely to include several of the beggar's ticks, *Desmodium* spp., numerous composites (family Compositae), pipsissewa, rattlesnake orchid, and Christmas and grape ferns.

GEORGIA

Outside the Dahlonega Plateau, all of the Georgia Piedmont experiences pioneering by loblolly and shortleaf pine, with loblolly predominating. Even on the plateau the initial woody plants are likely to be pines, probably Virginia pine, although stands of ash and black locust are common. Both successional types result in the closing of a pine canopy and subsequent elevation of the canopy to 80 feet (25 meters) or more as the understory opens beneath and accommodates the deciduous successors. The understory opens more slowly beneath the Virginia pines than under the loblollies, ubiquitous outside the plateau.

Sassafras plays a major role as a transgressive beneath the pines in Georgia's better-drained mesic sites. A stunted, gnarled occupant of the understory in more northerly parts of the Piedmont, sassafras makes straightway for the canopy at many Georgia locations, overtaking the pines in a quarter-century or less. Red maple, tulip tree, and sweet gum quickly follow the sassafras or, in its absence, comprise the initial deciduous assault on the pine canopy. Where winged elm or black cherry are present beneath the pines, they are probably overgrown coevals of the conifers rather than transgressives. Dogwood may appear in the understory at about the time the canopy is being overtaken by the first deciduous transgressives or at any subsequent time.

Oaks follow the sweet gum and sassafras by about two decades in Georgia, as in most locations south of the James. Spanish oak is first on many sites; black oak on the rest. White and post oaks may soon follow, the quercine saplings moving into the understory at about the time the deciduous takeover of the canopy is complete. Redbud, white mulberry, dogwood, and sourwood commonly move with the oaks to their place in the understory.

The shrub layer develops perhaps more slowly on Georgia's mesic Piedmont than to the north. Strawberry bush, lowbush blueberry and downy arrowwood may arrive with the oak saplings. As in most parts of the southern Piedmont at this seral stage, the herb layer is not well developed. Beggar's ticks, pipsissewa, rattlesnake orchid, and Christmas fern are often present, as may be any of several composites which prefer the growing forest habitat.

ALABAMA

Being at approximately equal latitude and of comparable climate and phytosociological history with the Georgia Piedmont (outside the Dahlonega Plateau), Alabama's mesic Piedmont differs little from that of Georgia. The difference is perhaps least during the growing-forest seral stage.

Throughout the mesic Piedmont of Alabama, loblolly pine is the predominant woody incursive. Cedar is locally active, mostly along the province's southern margin in Alabama. As in most parts of the Piedmont south of the James, the pines form the initial woody canopy of Alabama, dominate for three to five decades, then yield to deciduous transgressives, mostly sweet gum, tulip tree and sassafras. Winged elm and honey locust, where present beneath the pines, have probably been overgrown by the conifers.

Spanish oak and black oak invade the subcanopy beneath the decadent pines and the victorious sweet gum and tulip trees. White oak and post oak (on drier mesic sites) will follow. There is extensive hybridization among the oaks.

The subcanopy of Alabama's mesic growing forests commonly contains white mulberry, dogwood, sourwood, and redbud in addition to the transgressive oaks and the overgrown winged elms and honey locusts. French mulberry, *Callicarpa americana*, appears occasionally in the shrub layer with the squaw huckleberry and lowbush blueberry.

Animals of the Growing Forest

Insects

The black cherry, black locust, honey locust, catalpa, and dogwood have flowers designed for insect pollination.

Many species of flies, bees and ants, and moths and butterflies visit these arboreal flowers just as they, or their close relatives, attend the herbaceous flowers earlier in the sere.

In the discussion of insects and arachnids of the early woody succession (see pages 138–41), are detailed the principal lepidopteran larvae and other insects associated with the trees of the early woody mesosere. As these trees mature, the same caterpillars continue to eat the foliage, the most conspicuous of these relationships being that in which the eastern tent caterpillars, *Malacosma americanum*, ravish the black cherry, often taking foliage and flowers, thereby preventing the development of fruit. This devastation takes on alarming dimensions and would without doubt arouse popular concern were the black cherry of any perceived economic consequence to man. The origins of the damage coincide with those of the depredations of the southern pine beetle—wholesale recent abandonment and reforestation of the land resulting in unnaturally high populations of these woody pioneers. There are probably more black cherries in the Piedmont now than at any other time since the retreat of the ice.

When the oaks arrive beneath the tulip trees and sweet gums of the southern Piedmont or the locusts and ashes of the north, they bring their specializing caterpillars. Eminent among the eaters of oak leaves are the caterpillars of the oakworm moths, *Anisota* spp., the variable oakleaf caterpillar, *Heterocampa manteo*, and the rough prominent, *Nadata gibbosa*. The buckmoth, *Hemileuca maia*, eats sweet gum and oak.

The woody lianas (vines) grow with the ascending canopy. Virginia creeper and poison ivy rise with their structural hosts, the woody pioneers, and remain as a permanent part of the canopy, often being joined by wild grapevines, *Vitus* spp., in the crowns as the site matures. The lianas, like the oaks, attract their own caterpillars, mostly the larvae of sphinx moths of the genus *Pholus*.

A subfamily of the wasps, the Cynipinae, is responsible for many of the galls seen on the leaves and twigs of oak. The signatures of this group include the oak apple gall, egg-size, brown, and papery, and the wooly red gall, both caused by growth hormones injected into the leaf or stem during ovipositing. The larvae eat the abnormal plant tissue growing inside the gall and are protected by it. They pupate inside the gall and emerge as adults. Other wasps are gall-makers on other trees, shrubs, and herbs of the growing forest and other habitats.

There is an immense number of wasp species in North America, and the list of those occupying the Piedmont's growing forests is by no means small. Some live as larvae in the leaf litter, some in decaying wood, some in living plant tissue. Their specialties also include parasitizing arachnids, insects, and vertebrates and feeding on dung or fungi. The biotic importance of wasps is known to be critical in the forest community. They are particularly vital as controllers of the lepidoptera.

The developing forest hosts many beetles whose adults and larvae feed on decaying wood. The horned passalus beetle, *Popilius disjunctus*, 1½ inches (4 centimeters) long, black, shiny, and with large cutting mandibles, lives in rotting logs. The family Cerambycidae, the long-horned beetles, chew audibly on decaying wood—their munching is one of the subtle sounds to touch the naturalist sitting quietly in the woods. One cerambycid, the locust borer, *Megacyllene robiniae*, is destructive to living locust trees. The adults are gatherers of pollen common on goldenrod in autumn.

The flies and gnats (order Diptera) have many representatives who have adapted to life in the growing forests. The larvae of numerous families live in the leaf litter eating decomposing plant material or fungi or parasitizing other insects; the xylophagids, xylomyids and stratiomyids are notable in this regard. The larvae of small-headed flies (family Acroceridae) parasitize spiders. Some species from the families Trixoscelididae and Lauxaniidae spend their larval stages in the nests of birds. A dipteran family of major importance in the growing forest is the Agromyzidae, the leaf-miner flies, whose tiny larvae eat pathways through the interiors of the leaves of many deciduous plants, sometimes destructively. The gall gnats (family Cecidomyiidae), like many wasps, inject growth hormones during ovipositing. Any gall which is not perforated by an exit hole is likely to contain its maker, and the larvae are easily identified as to order by the characteristics of the head. The larvae of dipterans are wormlike with no well-developed head; those of hymenopterans have discernable heads. The larvae of the dipteran suborder Nematocera, which embraces the mosquitoes, midges, and craneflies (which look like enlarged mosquitos but do not bite) live in water, wherever it accumulates in the growing forest, or in moist soils.

Arachnids

Any wooded habitat may harbor chiggers (see page 141), but experience teaches that the red mites attack most enthusiastically from the growing forest, particularly where pines are present. Ticks (see page 141) show a clear preference for the grassy and weedy accommodations of early succession, but they are also ubiquitous in the growing forest. In fact, they are ubiquitous in the Piedmont during the frost-free months. The white and spined micrathena spiders (see page 140) are common at this seral stage, their webs often containing prey items wrapped and stored in a vertical line as if to resemble the droppings of birds. Numerous species of jumping spiders (family Salticidae) are present, pouncing on small insects on the forest floor and at all layers above. They are prominent in the diets of several passerine birds of the growing forest in all seasons, including winter, when they may be pried from their hibernacula in the bark crevices. In the leaf litter, wolf spiders (family Lycosidae) dash after beetles and other insects.

Araneus is a large and cosmopolitan genus of woodland spiders, well represented in the Piedmont's growing forests. *Araneus diadematus* is prominent, spanning a yard or more with orb-shaped webs and waiting patiently in a rolled leaf near one of the web's anchor points, usually an upper corner. Seen with the light behind her, a mature female (the one with the big abdomen) glows with brilliant reds on her bulbous body and bands of yellow on her legs; she is probably the most colorful spider in her habitat. Occasionally in Araneus' webs in the adolescent woods we meet *Argyrodes*, a tiny silver-bodied scavenger spider who feasts on the remains of Araneus' captives.

Reptiles and Amphibians

The growing forest, with its cool moist soils and abundance of decomposing wood and associated fungi, is a favored habitat of the eastern box turtle, *Terrapene caroline carolina.* The five-lined skink, *Eumeas fasciatus,* is fond of terrestrial haunts in the open spaces and, after a tract becomes thoroughly forested, will be found in the sunny ecotones at the woods' edge. The similar broad-headed skink, *E. laticeps*, however, is strongly arboreal and is common in the upper story of the mesic woodlots. The much smaller ground skink, *Lygosoma laterale*, is common on the forest floor. The three-inch (8 centimeters) body and tail, golden brown coloration, and dark dorsolateral stripe of the ground skink fit well in the leaf litter where the diminutive reptile eats insects, detritus, and fungi.

Snakes of the habitat include the northern brown snake, the red-bellied snake, the ringneck snake, the worm snake, the northern black racer, the eastern coachwhip, the black and yellow rat snakes, the eastern and scarlet kingsnakes, the scarlet snake, the copperhead, and the timber rattler. Only the last two are harmful to man and both behave in such a way as to mitigate the likelihood of danger. The copperhead is notably forgiving of inadvertent molestation, and the rattlesnake avoids man when possible. Neither attacks gratuitously.

The life cycles of most amphibians include an aquatic larval stage and, commonly, a water-oriented adult life. The well-drained forests of the Piedmont are therefore not rich in amphibian life, although the woodlands may be cleft by streams and seepages where amphibians abound. In thinking of the forest, "tree" frogs come to mind, but that group shows a clear preference for alluvial and swamp forests, so they will be discussed with those habitats.

Salamanders of the genus *Plethodon* are an exception. Their eggs are laid in moss and moist earth and the larvae do not require water. They are commonly known as the woodland salamanders. The red-backed salamander, *Plethodon cinereus cinereus,* occupies the northern Piedmont to the Roanoke River, hunting at night in the leaf litter for insects that would seem to be unmanageable by a fragile, three-inch amphibian. The food includes biting ants, hard-shelled beetles, and hemipterans which spray foul odors in their attackers faces. A subspecies called the Georgia red-backed salamander, *P. cinereus polycentratus*, claims an isolated range on the Dahlonega Plateau. In the more moist, north-facing slopes in the growing forest habitat, the slimy salamander, *P. glutinosus glutinosus*, can be found. The slimy salamander is large (to 7 inches/18 centimeters), and its body is a lustrous ebony, sometimes salted with white flecks. Secretions from the skin are harmless but difficult to remove. The spotted salamander, *Ambystoma maculatum*, also black-bodied and large (to 8 inches/20 centimeters) but colored with yellow or orange spots, burrows in rotting logs and hides under debris throughout the Piedmont.

The eastern narrow-mouthed toad, *Gastrophryne carolinensis*, prefers similar quarters. It is most frequently found burrowing in rotting wood, in leaf litter, in sawdust piles, or under stones in pursuit of ants, termites, and beetles. The American and Fowler's toads, old friends from habitats earlier in the sere, are active in the growing forest, indulging their tastes for flying insects. The eastern spadefoot toad is distinguished from the true toads by its vertically elliptical pupil (the pupils are horizontal in the American and Fowler's toads) and by an elongated, sickle-shaped black spade (used in burrowing) on each hind foot. The spadefoot, *Scaphiopus holbrooki*, principally an amphibian of the forest, makes its appearance at this seral stage.

The leopard frog, *Rana pipiens*, and the pickerel frog, *R. palustris*, represent the true frogs in the mesic woods, with occasional help from the wood frog, *R. sylvatica*. All are aquatic as tadpoles and they retain an aquatic orientation in adulthood, though many individuals wander far from water and explore a variety of mesic life styles. The spots on the pickerel frog are rectangular; those on the leopard frog are rounded. The wood frog sports a dark mask suggestive of the loggerhead shrike's. Glands on the skin of the pickerel frog secrete fluids which snakes find distasteful. Other predators, including humans, take gustatory delight in the Ranadae, the family which produces the frog legs *au table*.

Birds

Deadwood becomes plentiful as the forest grows and the woody pioneers die. Dead pines are soon riddled with the chiselings of downy, hairy, and red-bellied woodpeckers. The common flicker is present as the first deciduous trees mature. South of the Roanoke River, the brown-headed nuthatch is abundant in the pine woodlots, nesting and roosting in the abandoned cavities of the downy woodpecker, or excavating for himself.

In winter the red-breasted nuthatch is common in the middle-aged woods throughout the Piedmont. It is joined by feeding cliques of golden-crowned kinglets, brown creepers, and yellow-rumped warblers. Pine warblers are abundant throughout the Piedmont's pinelands in summer, retreating south of the Roanoke River in winter. The yellow-bellied sapsucker taps the cambium of many species of deciduous trees including tulip, red maple, oak, pecan, walnut, and fruit trees. Other wintering migrants include evening grosbeak, purple finch, house finch, and pine siskin, all fond of the seeds of the sweet gum and tulip tree. Tree and fox sparrows feed at forest edges. Where the growing forest borders on pastures and brushy fields, wintering flocks of juncos, white-throated sparrows, white-crowned sparrows, and song, field, and chipping sparrows fly to it for refuge when flushed.

Many species use the growing forest for nesting but do much of their feeding elsewhere. Examples include the blue jay, common crow, cardinal, rufous-sided towhee, catbird, robin, bluebird (a cavity nester), and white-eyed vireo; their nests are often near the woods' edge. Nesting and feeding mainly in the forest, though not exclusively the mesic growing forest, we encounter the black-capped chickadee, Carolina chickadee, tufted titmous, great-crested flycatcher (a cavity-nesting bird), eastern wood peewee, yellow-billed cuckoo, summer tanager, and scarlet tanager (Roanoke River northward). This behavioral grouping also includes the wild turkey, now becoming common where there are extensive deciduous woodlands, but still reclusive and not often seen.

The wood warblers, family Dendroidae, are strongly associated with the forest. Although the mature woodlands are by far richer in resident and migrant warblers (see page 213), than any other seral stage, the growing deciduous woods host several species. The Kentucky warbler commonly nests on the ground amid the maples, sweet gums and tulip trees. The hooded warbler does so frequently and the redstart appears to prefer the growing forest to other mesic habitats. Significant numbers of nearly all warblers of the eastern United States pass through the habitat in the spring and fall. The chestnut-sided, blackpoll, and black-throated blue warblers are among those more commonly seen. The naturalist who takes an interest in the migrations can not but see an important association between the northward movement of warblers in spring and the hatching of the caterpillars that eat the leaves of trees, on which they feed voraciously. The warbler–caterpillar–growing forest relationship is short but intense. It spans approximately the vernal aspect of the year (see page 239).

45. Cooper's hawk, *Accipiter cooperii*

As the forest grows, it becomes a hunting habitat more suitable for the Cooper's hawk than for the sharp-shinned, a bird of the hedgerows and sprouting woodlots. Both nest in the Piedmont's growing forests, showing a preference for pine. The sharp-shinned hawk is common, the Cooper's much less so, but as a nesting raptor in the Piedmont, perhaps not so abysmally uncommon as some authorities suggest. I usually enjoy several sightings in a summer.

Throughout the Piedmont, the red-tailed hawk nests in pines, often old pines in woodlots all but overtaken by deciduous trees, constructing a platform of twigs two feet across, at fifty feet or higher. The habitat's most common nesting raptor, the broad-winged hawk, *Buteo platypterus*, is strictly a woodland bird. It selects accommodations similar to the red-tailed's. Hawk nests tend to be reused and to be traded in successive years between species. They are also used occasionally by the great horned owl, chief of the habitat's nocturnal raptors. Perhaps to avoid becoming a meal for the horned owl (see page 106), the barred owl selects deeply sheltered nesting sites such as hollow trees, a commodity not common in the growing forest because pines rarely become hollow. Both these larger owls and the diminutive screech owl include the mesic growing woods in their hunting territories. The black vulture, with its relatively short wings, is better adapted to penetrating the woods to attend a carcass than is the turkey vulture.

Mammals

The opossum is present in virtually all Piedmont habitats and is certainly abundant in the growing forests. It is mainly nocturnal and in a night's prowl will encompass a variety of adjacent habitats. Following its nose, the province's most notable scavenger and omnivore takes fruit, carrion, insects and rodents. One can create good opportunities for observing opossums, as well as skunks, raccoons and gray foxes, by placing table scraps on a platform in a tree. The platform discourages neighborhood dogs, and, unfortunately, red foxes. An interesting point about opossums is that they have expanded their range to more than double their pre-Columbian size, generally toward the north and east. A close look will show that some individuals lose portions of their ears and tail to frostbite in colder climes; the phenomenon is well noted in the Piedmont at least as far south as North Carolina.

The short-tailed shrew hunts the woodlands of the entire

Piedmont, dashing furiously through the leaf litter and the tunnels and burrows of small rodents. The long-tailed shrew is found in the mesic growing forests and other wooded habitats in the province in Pennsylvania, New Jersey, and Maryland. South of the Potomac, the southeastern shrew is found in woodlots. All prefer moist conditions where snails, earthworms, and insects are plentiful.

At night, the air is cleft by a copious variety of bats darting among the treetops of the growing forests, though probably none could be said to be limited to this habitat. Their food is flying insects, principally moths, soft-bodied beetles such as lightening bugs, and dipterans. Bats are notably difficult to identify, even when captured. In-flight identifications are attempted only by experts. The following list is probably not comprehensive: little brown myotis, keen myotis, small-footed myotis (Potomac northward), silver-haired bat, hoary bat, and evening bat.

Mustellids of the habitat include the least weasel (preying on small birds, rodents, and mammals) and the omnivorous striped skunk. The spotted skunk is present south of the Roanoke. The raccoon and red and gray foxes hunt the mesic growing forest, in conjunction with adjacent habitats of varied moistness and vegetation.

By far the most important mammal in the scheme of forest regeneration is the eastern gray squirrel, *Sciurus carolinensis*. This arboreal rodent's nests, two-foot globes of twigs and leaves, appear in the growing forest as the deciduous trees overtake the canopy. The nests are uncommon in pines, as are sightings of the squirrel itself. The seed structures of ash, maple, and tulip trees are among the foods which attract the squirrel to the growing forest. From nearby stands of oak and hickory, the gray squirrel brings nuts and acorns which it buries for storage. The rodent possesses an amazing ability to recover the cached nuts, and the forest floor in winter is pocked with small excavations which are economical and direct, not random trenches made in an uncertain search. Still, many remain buried, planted at optimum depth for germination. The seeds of the oaks and hickories appear to be transported exclusively by squirrels (they are too large to be carried by the wind), and it is safe to assert that the final phase of plant succession in the mesic Piedmont is initiated by the industry of squirrels.

The eastern fox squirrel may participate, but its numbers are so reduced that the contribution may be slight. The southern flying squirrel is plentiful at this seral stage throughout the Piedmont, but the food is mainly smaller seeds, and such nuts as are stored are cached in nests and in the crotches of trees rather than in the ground.

The white-footed mouse and, south of the James River, the semi-arboreal golden mouse forage in the growing forest for seeds, fruits and tender shoots. The pine vole is at home in both the deciduous and coniferous seral phases. The eastern cottontail favors the twigs and buds of the shrubs and transgressives of this habitat, particularly in winter. The whitetail deer, steadily increasing in many parts of the Piedmont, takes the same fare from higher levels on the plants. The growing forest, thicker than the woods at maturity, may be the whitetail's favorite haunt.

The Mature Forest

As the length of time after logging or abandonment increases, so does the likelihood that the progress of the sere will be disturbed. Although flooding is not a problem on mesic sites, fires are. Disease constitutes a successional disturbance only when an alien organism is involved, for the native microbes, fungi, and insects (such as the southern pine beetle), serve to expedite the progress of the sere. Man, of course, is the principal disturber. From the growing woodlots and from the mature forests man harvests firewood and pine and hardwood timber and, in the process, retards and confuses the successional process. There is no reason he shouldn't—the practice of devoting a portion of one's land, usually the less tillable acreage, to the satisfaction of one's needs for wood is economically sensible, and it is consistent with good stewardship of the soil. To the naturalist, this simply means that relatively few of the Piedmont's mature tracts have developed to their present condition without the distortions and scars of intermittent logging.

One recognizable result of cutting is the presence in the mature forest of plants associated with earlier stages in the sere. Pines standing amid oaks and hickories are a common example. Another is the presence in the canopy of trees normally confined to the subcanopy; where the oaks have been cut, the red maples and black gums may gain and retain dominant status. Sawed-off stumps, mangled survivors, and gaps in the canopy point to recent logging. Sometimes one finds large trees which bear the kinks and deformities of cuttings half a century gone.

With all its probable irregularities, its departures from the norms of undisturbed growth, how do we recognize the mature forest? In the mesic Piedmont, a forest is mature when its canopy becomes dominated by oaks, although considerable seral advancement still lies ahead. The mesic oaks are the black, white, red, willow, scarlet, and Spanish. Where tall, straight-trunked individuals of these oaks hold a majority of the canopy, we can be confident that the forest grew as a unit from a common start. This contrasts with woodlots, which are characterized by widely spaced oaks with low, spreading branches and by the presence of obviously younger trees between the oaks. These spreading oaks achieved their initial growth in the open, perhaps as shade trees in a pasture, and their shape is quite distinct from that of forest-grown trees that have had all their lives to compete vertically for sunlight. A forest interspersed with spreading oaks is hardly more mature than it would be in the absence of the former shade trees, for the soils lack the depth and moistness of maturity and the subdominant plants characteristic of the advanced forest are not yet present.

Successional progress does not cease when the oaks assume dominance and the forest is said to be "mature." The advent of oak dominance marks the beginning of another prolonged era of development. Under the aegis of the oaks, the canopy, and later the subdominant strata, will be transformed—perhaps over centuries—from the passage vegetation of the growing forest to the permanent populations of the climax condition. Some might argue that the changing never ceases, that the soils continue to deepen and the flora continues to shift, though at an ever-slowing rate. It seems practical, however, to view forest soils as seeking an equilibrium, as they mature, between organic accumulation and loss to erosion and leeching. Similarly, the plantlife moves toward a stability that will be altered only over geologic time.

There are very few sites at their climax condition in the Piedmont, and of those that are, perhaps the fewest are mesic. Mesic lands are the most valuable from nearly every commercial aspect, and are the least likely to be abandoned. They make the best croplands, the best pastures and, when abandoned, the best commercial forests. These forests are rarely left undisturbed. The volunteer, work-free crop of growing forest trees provides a lucrative return on no greater investment than the taxes on the land. The ancient trees of mature forests are not so free of internal rot and blemishes as are the prime populations of the growing woods; therefore, almost invariably, the woods are cut before they mature.

Do we pursue an illusion in discussing the climax forest as a reality in the Piedmont? I think not. There are a few sites presently in the climax condition (see Part IV, Special Places) and there are many more which will approach that condition in the coming generations, particularly if human concerns for the natural environment continue to mature. The significant point is that numerous sites have recently crossed the threshhold into maturity. Our challenge is to identify the vegetation characteristic of the era of maturity as the forest approaches the condition of climax.

Maturity in the Canopy

A site enters maturity when the oaks take control of the canopy. They are then the most numerous tree at that stratum and are about a foot (30 centimeters) in diameter at breast height. They are sexually productive; acorns pepper the leaf litter each autumn. The oaks are somehow more vigorous than the sweet gums against which they press aggressively. Some tulip trees retain their positions, individuals competing successfully with the oaks by retaining a few feet advantage in height. (Altitude is not a guarantee of survival—spires in the forest crown invite the attentions of wind and lightening.) Some tulips persists well into the forest's maturity and even into the climax. They are commonly the largest trees in a mature tract; tulips of 4 feet (1 meter) dbh amid oaks half that girth are common at this stage in the sere.

Hickories follow the oaks. They may follow by only a few decades or by a century or more, but I have yet to see a mesic site on which hickories preceded the oaks into the canopy. The mockernut hickory, *Carya tomentosa*, occupies the entire mesic Piedmont and is the province's most common hickory. The pignut hickory, *C. glabra*, thrives in the drier mesic soils, and on such sites may precede the mockernut. The bitternut and shagbark hickories, *C. cordiformis* and *C. ovata*, prefer more moist sites. A half-century or so after germinating, the hickories gain the canopy and are sexually mature; hickory nuts are added to the fallen mast at the end of the growing season. A mesic forest dominated by oaks of 24 inches (60 centimeters) dbh and sexually mature hickories perhaps a foot (30 centimeters) dbh is well into maturity. Such a growth requires a minimum of 150 years, more on many sites.

Throughout the Piedmont the American beech, *Fagus grandifolia*, is associated with the mature mesosere. Beech may appear with or before the hickories; it prefers the damp end of the mesic regime. On drier sites it occupies the lower slopes, gradually ascending as the decades pass. Beech is commercially useful but in restricted quantities, so it is not so ravenously lumbered as are the oaks and pines. Although *Fagus* species sometimes dominate lower slopes from which the oaks have been cut, their more typical behavior is to occupy a secondary position with the red maples beneath oaks and hickories. Occasionally north-facing slopes will pass out of the period of oak dominance, perhaps with the woodcutter's help, and later stabilize under the combined dominance of maple and beech. Such maple–beech associations are common in the north of the Piedmont. Their frequency decreases with latitude and with distance from the mountains. Maple–beech dominance in the Piedmont is sometimes viewed as a postclimax condition, signifying that the site has become more mesic (more moist) than it was at simple maturity. This level of moistness becomes possible as soils deepen.

Basswood, *Tilia americana*, is another latecomer in the sere. It arrives on mesic soils in the Piedmont about when the hickories do, but it is thinly distributed. Its locations are scattered; they tend toward the northern and western parts of the province. Like beech and maple, basswood usually fails to reach the canopy.

The Subcanopy

The secondary layer undergoes fewer changes in reaching maturity than does the dominant level. Mature subcanopies commonly contain red mulberry, red maple, sourwood, sassafras, and flowering dogwood. Witch hazel and American hornbeam, *Carpinus virginianus*, are two small trees that show clear preferences for hydric conditions but which commonly venture onto mesic sites at maturity. The American chestnut, *Castenea dentata*, extinct as a mature tree, valiantly rises to the subcanopy on a few mature sites in the Piedmont, inevitably to be struck down by the blight. Perhaps the most definitive indicator of maturity in the subcanopy, at least north of the Roanoke River, is the black gum, or black tupelo, *Nyssa sylvatica*. The black gum dominates the secondary level on some sites almost to the exclusion of all else save the canopy transgressives. It is revealing to see black gum and mockernut hickory side by side at 12 inches (30 centimeters) dbh under oaks, the hickory having secured its station in the canopy while the black gum reaches only to the substory. The respective assignments of stature seem clearly to reside in the genes of each plant.

The shrub layer is marked at maturity by the presence of the mapleleaf viburnum, *V. acerifolium*. Rarely does this shrub become established other than under the oaks; it almost always signals forest maturity. Also present in the shrub layer is the tallest *Vaccinium*, sparkleberry, *V. arboreum*, an occupant of the upper, drier slopes. Sweet shrub, *Calycanthus floridus*, occasionally lends its heady scent to the lower slopes. Shadbush, *Amelanchier arborea*, spicebush, *Lindera benzoin*, downy arrowwood, *Viburnum rafinesquianum*, and several *Vaccinium* species of lower stature than sparkleberry also characterize the mature shrub layer, *V. tenellum*, and deerberry, *V. stamineum*, in particular.

There are, of course, regional peculiarities in the mature forest. Owing to the northern Piedmont's proximity to regions which at maturity are cloaked in conifers, Canadian hemlock, *Tsuga canadensis*, and white pine, *Pinus strobus*, are common participants in the mature forests north of the Potomac. These conifers grow sparsely but consistently beneath old oaks and hickories and, after many decades, take their place in the canopy. Sites of such maturity are uncommon, however, and where the white pines and hemlocks are found they are typically of transgressive stature. In the southern Piedmont, the mature oaks and hickories are joined on some sites by cucumber tree, *Magnolia acuminata*, a deciduous magnolia of the shrub layer and subcanopy. Except for the regional presence of these trees, mature forests throughout the Piedmont are amazingly consistent in their composition. The very great majority of the mature mesic acreage is cloaked in some combination of oak, hickory, and tulip tree, though maple and beech reach for the canopy on certain very mature north-facing slopes.

In spring, several herbs are common on the mature mesic forest floor, notably jack-in-the-pulpit, *Arisaema triphyllum* and mayapple, *Podophyllum peltatum*. Small seeps and bogs in mesic tracts often show the green tongues of skunk cabbage, *Symplocarpus foetidus*. The lily-leaved orchid, *Liparis lilifolia*, and the Catesby's trillium, *Trillium catesbaei*, grow on selected slopes. Later in the growing season the cranefly orchid, *Tipularia discolor*, sends forth a spike of flowers but keeps its leaf below ground until autumn. The single leaf then appears at shoelace height, dark green above, purple below. Partridge berry, *Mitchella repens*, grows locally in prostrate mats. The elegant whorled leaves of Indian cucumber root, *Medeola virginiana*, catch the eye in late spring. Common mesic ferns include the Christmas and lady ferns, *Athyrium aspenioides*. From the seeps frequent in mature and otherwise mesic forests grow waist-high fronds of royal fern, *Osmunda regalis*, cinnamon fern, *O. cinnamomea*, and the smaller maidenhair fern, *Adiantum pedatum*.

Associated with the mesic forests at maturity, and to some extent with the bottomland hardwoods (see Chapter Seven) as well, is a group of herbs called the spring ephemeral wildflowers. These forbs, representing several botanical families, have adapted very specifically to life beneath the mature deciduous giants. They send forth their leaves and flowers after winter has done its worst but before the leaves in the canopy emerge to block the sunlight. In a brief span, hardly more than a fortnight for some individuals, the spring ephemerals manufacture the entire year's food supply, store it in an underground bulb or tuber, bloom, and set seed. By the time the forest floor is shaded by canopy foliage, the aboveground portions of the ephemerals wither and decompose. No vestige of the plants is then visible until the prevernal aspect of the following year. We overlook this striking event in the biological calendar unless we visit the mature woodlots earlier than we would ordinarily expect herbs to appear. The ephemerals are extraordinary herbs in their tolerance of cold; they do not hesitate to push through a late snow to bloom. Trout lily, *Erythronium americanum*, and toothwort, *Dentaria* spp., are the most common ephemerals. They appear in profusion on selected lower and middle slopes, usually those facing north. Spring beauty, *Claytonia* spp., prefers moist conditions but often climbs the slopes. On older, richer sites grow Dutchman's breeches, *Dicentra cucullaria*, and occasionally golden corydalis, *Corydalis micrantha.* To the same degree as the oaks, hickories, and mapleleaf viburnum, the spring ephemeral wildflowers signify seral maturity.

Blooming simultaneously with the ephemerals, but differing functionally from them by retaining their foliage after the canopy is foliated, is another group of prevernal forbs. Windflower, *Thalictrum thalictroides*, bloodroot, *Sanguinaria canadensis*, and hepatica, *Hepatica americana*, continue photosynthesis in the dim light that filters through the mature woods' multiple layers of foliage.

The herbs of the mature forest include many species with close functional ties to the dominant plants, some even more direct than the chronological kinship linking the ephemerals with the oaks. The false foxglove, *Aureolaria pedicularia*, is parasitic on the roots of black oaks. It is profusely branched and waist high, flowering in midsummer with yellow corollas up to 2 inches (5 centimeters) in length. Beechdrops, *Epifagus virginiana*, gains its entire food supply by parasitizing the roots of beech. The plant therefore lacks chlorophyll and even during the growing season appears dead and twiggy, thrusting its brown to purplish racemes as much as a foot (30 centimeters) above the leaf litter. Indian pipes and pinesap, *Monotropa uniflora* and *M. hypopithys*, also lack chlorophyll because they gain their food from decomposing material in the leaf litter. Such plants are called saprophytes. Indian pipes send up clusters of white, translucent stems to a few inches height, each capped with a single nodding flower. Pinesap has a single, 4 inch (10 centimeter) stem which bears numerous flowers. The name probably originates from the plant's color, a rich translucent amber with reddish accents.

Trees of the Mature Forest

Page references refer to the more detailed descriptions of these trees presented in earlier sections.

White Oak, *Quercus alba.* See page 155. White oak is the mainstay of the mature canopy throughout the Piedmont. It is not usually the first oak to germinate but it is the species that on the majority of sites eventually holds a plurality in the canopy. White oak is the common denominator uniting the mesic canopies at maturity at all latitudes in the Piedmont, from the white oak–red oak–scarlet oak alliances of New Jersey to the white oak–willow oak–Spanish oak associations of Alabama. Look for the ashy-white bark, which becomes shaggier as the eye follows the trunk upward.

Red Oak, *Quercus rubra.* See page 157. This is the northerly, pointed-lobed oak much given to hybridizing with the other oaks in its taxonomic group (i.e., those with pointed lobes). From the Potomac northward it is the most common partner of the white oaks. It is uncommon south of the James.

Black Oak, *Quercus velutina.* See page 154. Black oak is to the central and southern Piedmonts what red oak is to the north—the principal partner of the white oak in the domination of the mature forest. It is the most common pointed-leaf oak between the Potomac and the Savannah, south of which it yields gradually to the Spanish oak. Still, black oak is a significant presence in the mature forests of all sections of the Piedmont. The leaf is broader than that of the red oak, and it has fewer (usually seven) lobes.

Spanish Oak, *Quercus falcata.* See page 154. The influence of this oak is strong as far north as the Potomac; however, it is mainly a southern species. South of the Savannah River it is the main associate of the white oak in mature mesic canopies. The lobes are long, narrow, and pointed. The terminal lobe is asymmetrical.

Post Oak, *Quercus stellata.* See page 156. In the mature mesic forests of the Piedmont, post oak is the second most common round-lobed oak after the white oak. It is most numerous toward the southern end of the province, but it is a numerically significant canopy species throughout. In Georgia and Alabama its populations on some tracts may exceed those of white oak. The cross-shaped leaf is this tree's main identifier.

Scarlet Oak, *Quercus coccinea.* See page 159. The scarlet oak is an important canopy species of the northern Piedmont, though it is generally not so large a tree nor so numerous as the red oak. The leaf is similar to that of the red oak, but the sinuses are cut more deeply, almost to the midrib. The coloration in autumn accounts for the common name.

Beech, *Fagus grandifolia.* The American beech generally does not germinate until the oaks are well established. It grows thinly on drier mesic soils, more densely toward the lower slopes. Beech may be either a tree of the canopy or of the subcanopy. Its height is characteristically less than that

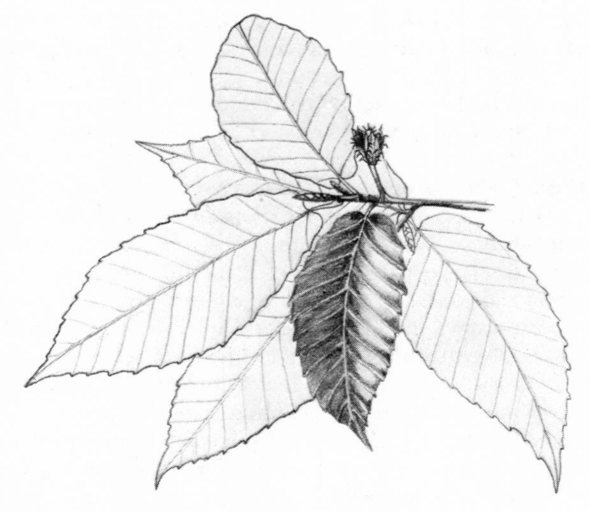

46. American beech, *Fagus grandifolia*

of the oaks and hickories, but it is a long-lived tree and may accede to the canopy if a neighboring oak falls. Self-pruning is very slow; the lower branches tend to remain vital and, in winter, to retain their dead leaves—legend says the Indians and early European settlers favored the easily climbable beech as a means of escaping irate bears.

The buds are sharply conical to 1 inch (25 millimeters) long; the leaves are alternate, 2 to 6 inches (5 to 15 centimeters) long, elliptic, dentate. The seeds are three-sided, ½ inch (12 centimeters) long and compacted in a marble-size capsule armed with spined bracts. The bark is light gray and bloched with darker areas, and when wounded it forms dark scars that persist for decades. This quality invites amorous, vandalous, and territorial inscriptions that speak of long-past human visits to the old woodlots.

Chestnut, *Castanea dentata.* This tree once occupied the entire Piedmont as well as the rest of the Appalachian highlands, holding the key position in the mature deciduous canopies. In other words, the oak–hickory forests were formerly chestnut–oak–hickory. For its wood and for its fruit, a protein-rich nut of hearty flavor, the American chestnut was once valued more highly than any other North American tree. Early in the present century, the species was extirpated by a blight caused by an introduced fungus, *Endothia parasitica.* At present, saplings rise occasionally to the understory before being killed by this persistent disease. The lanceolate leaves 4 to 8 inches (10 to 20 centimeters) long are coarsely dentate with the tip of each tooth drawn to a fine point and curved inward. The leaf can be confused with those of chestnut oak and chinquapin oak, but on the oaks the teeth are rounded. Few chestnuts reach sexual maturity; saplings probably rise from old root systems, even though no vestige of the parent tree is visible.

Mockernut Hickory, *Carya tomentosa.* The mockernut is a common hickory, perhaps the most common, on mature mesic slopes throughout the Piedmont. It apparently tolerates a considerable range of moistness within the mesic scale and slightly outside it on both ends; *C. tomentosa* can be seen in alluvial soils and on dry hilltops. Sexual maturity of the oaks generally coincides with the arrival of mockernut saplings. The connection may be that squirrels regard the presence of acorns as a signal to begin burying nuts in

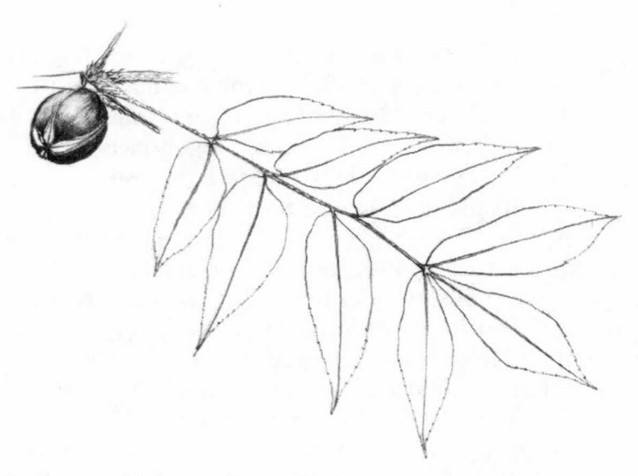

47. Mockernut hickory, *Carya tomentosa*

general, and hickories are almost certainly dependent upon squirrels for distribution. The bark is smooth on saplings, becoming shallowly braded with age. The 12 inch (30 centimeter) leaf is odd-pinnately compound with five to nine leaflets, each elliptical, serrate, and tomentose below. The leaflets graduate in size distally. The one-inch nut is slightly flattened laterally and has five distinct ridges matching the segments of the husk. The husk is ¼ inch (6 millimeters) thick, and the entire nut with husk is larger than a golf ball.

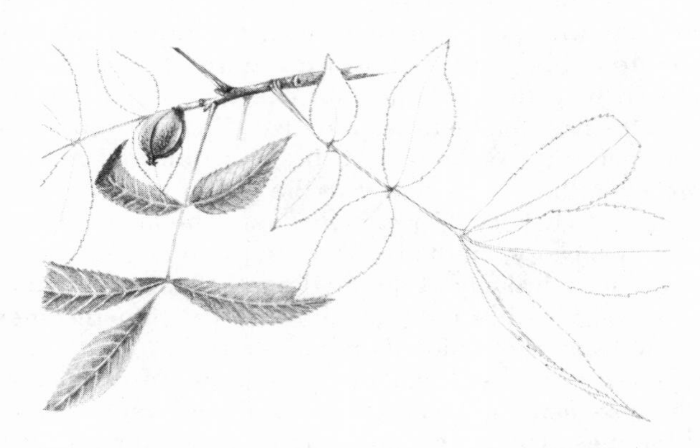

48. Pignut hickory, *Carya glabra*

Pignut Hickory, *Carya glabra.* The pignut endures drier soils than any other hickory in the Piedmont and it is therefore the first of its genus to arrive at some locations. The species' range encompasses the entire province, with densest distributions in the south. The odd-pinnately compound leaf is about 10 inches (25 centimeters) long, the five to seven leaflets are narrowly elliptic and pointed. Bark is slightly shreddy. The nut is smooth, lacking the ridges of the other hickories. The husk is only 1/10 inch (2 millimeters) thick.

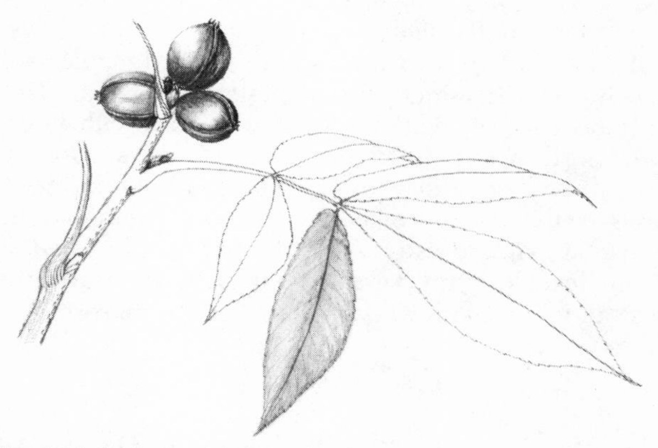

49. Shagbark hickory, *Carya ovata*

Shagbark Hickory, *Carya ovata.* The shagbark prefers moist soils but can often be found on advanced sites, particularly those of low acidity and correspondingly high availability of nutrients. Oaks must first be established. Very old individuals may exceed 36 inches (90 centimeters) dbh. Shagbarks are usually in the canopy and sexually mature before achieving 12 inches (30 centimeters) dbh. There are usually five leaflets, broadly elliptic, on the 12 to 14 inch (30 to 35 centimeter) leaf. Proximal leaflets are smallest. The nut is four-ribbed and, with the husk on, may be nearly as large as a baseball. The bark on the mature trees is the principal identifier: it is exceedingly shaggy, having scales to 24 inches (60 centimeters) long curved outward at the ends.

Tulip Tree, *Liriodendron tulipifera.* See page 124. The tulip tree may be one of the earliest woody plants to become established on a mesic site, and it may persist, and replace itself, well into the mature phase of seral advancement. Due to its age and rate of growth, the tulip may have a larger dbh than any other species in the mature forest. The period photograph of the Farnsworth site at Gettysburg shows a tulip tree prominently in the background; today, the same individual is approximately 48 inches (120 centimeters) in diameter.

Red Maple, *Acer rubrum.* See page 117. In the mature mesic forest, red maple is a stunted member of the understory. Many individuals seem unable to rise above the shrub layer, where it is common to see red maple saplings adjacent to their mimic, the mapleleaf viburnum. The red maple occasionally achieves canopy stature with beech on lower, north-facing slopes. Reference to the section on hydric soils will show that this species, while important in all phases of the mesosere, performs more vigorously in the mature alluvial forests. Perhaps red maple's most significant feature is the versatility it displays in adapting to life at all levels of all woody phases of the mesosere.

Trees of the Understory

Black Gum, *Nyssa sylvatica.* See page 160. The black gum, or black tupelo, is the largest tree that is restricted to the mature understory. It is found throughout the province, but is most significant in the northern and western sections. Look for black gum upslope from an understory dominated by beeches.

American Hornbeam, *Carpinus caroliniana.* This small, moisture-loving tree grows beneath the mature canopy up the mesic slopes from the bottomlands, where it is most common. The musclewood, as it is also called, is often encountered far from water, however, where the slopes are mature and face northward. The trunk is fluted and covered in smooth gray bark suggesting taut skin over rippling muscles. The leaf is elliptic with a rounded base, pointed, doubly serrate, 2 to 4 inches (5 to 10 centimeters) long and half as wide. Seeds are borne in clusters of foliate bracts to 6 inches (15 centimeters) long.

Flowering Dogwood, *Cornus florida.* See page 123. Arriving late or early in the woody sere, dogwood is almost always present in the understory at maturity throughout the Piedmont. Old trees may be no more than 30 feet (9 meters) tall and 8 inches (20 centimeters) in dbh, but the bark, deeply cleft into cubes, signifies age. Seedling and sapling dogwoods are also present in the mature forest, the seedlings recognizable by the candelabralike arrangement of three pairs of opposite leaves. The clustered red berries are eaten and distributed by birds and squirrels.

Redbud, *Cercis canadensis.* See page 149. On some sites, particularly north of the Potomac, redbud diminishes as the location matures; on others, the Judas tree, as it is also known, continues as a prominent member of the understory.

Sourwood, *Oxydendron arboreum.* See page 148. From the Potomac southward the sourwood is common in the mature understory, typically having become established during the growing forest phase. Its presence decreases with advancing maturity of the site. Wherever found, it is an important source of nectar for honeybees.

Eastern Hop Hornbeam, *Ostrya virginiana.* Although generally associated with xeric soils, the hop hornbeam is quite common on mesic sites at maturity. In stature the species seems genetically limited to 30 feet (9 meters) in height and 6 inches (15 centimeters) dbh. The bark is light brown and finely shreddy. The leaves, like those of its close relative the American hornbeam, are doubly toothed. The seeds are clustered at the tips of the twigs in nested bracts, the entire fruiting structure vaguely resembling a small pine cone, but green. Hop hornbeam is commonly attacked by a fungus which causes the growth of thick clusters of fine twigs, at a glance suggesting birds' nests, a disorder known colloquially as witch's broom. The condition is so common as to be mistaken by some for a normal characteristic of the tree.

Shrubs

Mapleleaf Viburnum, *Viburnum acerifolium.* This chest-high shrub is distinctive of the mature forest. The opposing leaves have three pointed lobes and serrate margins. The plant has several main stems sprouting from a system of stolons. Flowering stems have one pair of leaves. The inflorescence is a flattened cyme; the fruits are black drupes compressed laterally, ½ inch (12 centimeters) long.

Downy Arrowwood, *Viburnum rafinesquianum.* See page 150. This shrub is more characteristic of the growing forest than of the mature stands, but on some sites it persists indefinitely and may be found alongside mapleleaf viburnum. On some mature sites the species may be stunted.

Wild Azalea, *Rhododendron nudiflorum.* "Pinxter" is another common name of this strikingly-flowered deciduous shrub which thinly colors the mature mesic slopes in early spring. The 2 inch (5 centimeter), trumpetlike pink corollas are arranged in turrets of four, the red-tipped sexual parts extending outward and upward. Leaves are alternate, elliptic to oblong, smooth and green above, lighter and fuzzy beneath. Seed capsules persist through the growing season. Pinxter is common in the mature mesic habitats of the Piedmont and is easily the most colorful in bloom.

Blueberry, *Vaccinium* spp. See page 150. The blueberries, or huckleberries, are clearly associated with drier soils and they tend to become less numerous as the sere progresses. Any of the species described in the material on shrubs of the growing forest may be present.

Strawberry Bush or **"Hearts-a-bustin'-with-love,"** *Euonymus americanus.* With its thin green stems, this deciduous shrub can be mistaken for an herb. The flowers are green and inconspicuous, but the fruiting structure is something to behold! The pink, spiny globe opens in late summer into four or five spherical segments, each bearing an orange berry at the tip, the entire structure being about 1 inch (25 millimeters) across. The plant is associated with mesic soils in the Piedmont and is scattered throughout the growing and mature forests.

Shadbush, *Amelanchier arborea.* See page 152. Shadbush is affiliated with the growing and mature forests of the mesic Piedmont principally north of the Roanoke River.

Herbs

Jack-in-the-pulpit, *Arisaema triphyllum.* Vigorous individuals of this woodland herb grow to waist height; knee height is more common in the mesic Piedmont. There are usually two leaves, each pinnately compound with three leaflets suggestive of those of poison ivy. The inflorescence consists of a vertical spathe 3 to 5 inches (8 to 13 centimeters) long—the "pulpit"—enclosing a spadix—"Jack"—which bears the flowers. At maturity the spathe withers and a conical cluster of red berries is revealed on the spadix. Jack-in-the-pulpit prefers moist conditions, but is a frequent member of the herb community beneath mature mesic canopies.

Mayapple, *Podophyllum peltatum.* The species shares the habitat and moisture preferences of Jack-in-the-pulpit, and the two often grow side by side. Rising to one foot above the leaf litter in midspring, a single stalk of this rhizomatous perennial bears two peltate (the stem attaches inside the leaf margin) leaves and a single, nodding, waxy, white flower with yellow sexual parts. Some stems bear a single leaf and no flower. The fruit is a yellowish "apple" 1½ inches (4 centimeters) in diameter.

Skunk Cabbage, *Symplocarpus foetidus.* Blooming in late winter or early spring, skunk cabbage thrusts its rusty, fleshy spathe above the moist soils of alluvial bottoms and seeps on mesic slopes. The strong odor attracts flies and other pollinators. As the weather warms, the ovate leaves unfold to about 18 inches (45 centimeters) in length. Their color is a brilliant green against the leaf litter.

Catesby's Trillium, *Trillium catesbaei.* The Piedmont's most common trillium by far, Catesby's is identified by a single nodding flower with three pinkish, recurved petals. Trilliums have three of everything; three leaves in a horizontal whorl at the top of the stem, three petals, three sepals, three stygmas in the flower. Catesby's trillium grows on the upper slopes on mesic soils, generally upslope from the majority of woodland spring wildflowers.

Bloodroot, *Sanguinaria canadensis.* Toxic crimson sap flows from the thick rhyzome of this perennial when it is dug. The single reniform (kidney-shaped) leaf is palmately cleft into seven lobes. As the plant emerges in very early spring, the leaf embraces the floral bud growing on a separate, parallel stem. The floral stem advances above the leaf, blooms and develops a pointed seed capsule. The flower is a 4 inch (10 centimeter) "wagon wheel" of ten white spokes and a hub of yellow pistils and stamens. The leaf persists well into summer at 6 inches (15 centimeters) above the leaf litter.

Cranefly Orchid, *Tipularia discolor.* In midsummer a single stem rises to 12 inches (30 centimeters) and bears a series of half-inch, brownish flowers with purple accents. The floral display is small, delicate, and easily overlooked. Seeds develop in pods at the tips of the peduncles. The stalk hardens and stands for several months, perhaps into winter. After the leaves fall from the trees, a single 2 inch (5 centimeter), elliptical leaf grows at the base of the plant and spends the winter prostrate on the litter. Venation is parallel, as is characteristic of orchids.

Lily-leaved Twayblade, *Liparis lilifolia.* This orchid presents its leaves and flowers simultaneously in midspring. There are two leaves: basal, glossy, ovate, rigid, sessile (without stems), and semierect. A dozen or more flowers bloom in distal sequence on the single vertical raceme.

Partridge Berry, *Mitchella repens.* Named, like the highest mountain in eastern North America, for the pioneering botanist Elisha Mitchell, partridge berry is a creeping perennial herb that grows in thick mats on moist mesic sites in the mature forest. The flowers are composed of white, four-lobed, trumpetlike corollas which grow in pairs and result in a double-seeded berry, shiny and red. Leaves are ½ inch (12 millimeters) long, broadly ovate, smooth, and leathery. They are arranged in opposing pairs on the viney stems.

Indian Cucumber Root, *Medeola virginiana.* This lily is not common, but it is structurally unique and a delightful find on a springtime hike. The plant consists of a single delicate stem approximately 12 inches (30 centimeters) tall with two stacked whorls of foliage, the lower radiating five to ten sessile, elliptic leaves and the upper bearing three, much smaller. At the tip of the stem, the single greenish flower

nods to the level of the upper whorl on a 1 inch (25 millimeter) pedicel. In late summer the upper whorl reddens at the center and dark berries grow on separate red pedicels.
False Solomon's Seal, *Smilacina racemosa.* Another lily, this perennial grows a single, arching, zigzag stem to 36 inches (90 centimeters) long, from which the parallel-veined leaves alternate. The inflorescence is a 3 inch (8 centimeter) raceme of faintly fragrant white florets at the top of the stem, ripening in late summer to produce a tight cluster of red berries not unlike those of jack-in-the-pulpit.
Windflower, *Thalictrum thalictroides.* This delicate perennial emerges in very early spring, with the ephemerals, to display ½ inch (12 millimeter) white blossoms, usually three, at the tips of 2 inch (5 centimeter) wire-like pedicels. The upper leaves are in a whorl of five; basal leaves are in groups of three. The plant is limited to 6 inches (15 centimeters) in height and is characterized by a springy rigidity which causes it to vibrate in the prevernal winds. The foliage persists into summer.
Hepatica, *Hepatica americana.* The thick, waxy leaves with three rounded lobes grow on lax, individual stems which let the leaves rest on the litter. The leaves, like those of the cranefly orchid, are purplish below. In late winter the flowers rise on 2 inch (5 centimeter) stems, hardly clearing the litter, which provides shelter from the icy winds. The petals are usually lavender but range in color from white to blue.
Trout Lily, *Erythronium americanum.* Trout lily is the most common spring ephemeral in the Piedmont. Its two fleshy, pointed, elliptical leaves are waxy, green, and mottled with purple. The folk name may result from the coloration or from the belief that the plant appears in early spring as the trout begin to bite. The nodding flower rises on a 6 inch (15 centimeter) stem, the lower portion of which is sheathed in the leaves. The petals and sepals (three each) are yellow within and purple on the back. An underground bulb stores starches after the leaves wither in midspring.
Spring Beauty, *Claytonia virginica.* Each plant of this ephemeral bears two grass-like leaves and a cluster of sequentially blooming floral buds arranged in a "fiddlehead." Each flower lasts one day. There are five petals, white with pink and lavender accents. An underground organ stores

starches for the 48 to 50 weeks of the year during which no part of the plant is above the litter.

Toothwort, *Dentaria laciniata.* One foot (30 centimeters) in height and topped by a cluster of ¾ inch (20 millimeter), white, four-petaled flowers, toothwort occupies positions generally upslope from the two ephemerals above. There are usually three leaves on each plant, each suspended upon a 2 inch (5 centimeter), erect petiole. The leaves are three- to seven-lobed, deeply cleft and dentate. They resemble the leaves of marijuana.

Dutchman's Breeches, *Dicentra cucullaria.* The foliage, to 12 inches (30 centimeters) in length, is delicate, lacy, and fernlike. It precedes the floral stems by a fortnight, garnering sunlight beneath the still-open canopy. The corollas are white with two basal spines suggesting the plant's common name. The flanged opening of the flower is rimmed in yellow. Dutchman's breeches grows on the slopes of the Piedmont's oldest and richest woods.

The Mature Forest—State by State

NEW JERSEY

White oak and red oak are the dominant trees of the canopy, with assistance from mockernut and shagbark hickories. At the Hutcheson Memorial Forest (see Part IV, Special Places), sweet pignut, *Carya ovalis,* is the principal hickory; its bark is shaggy but substantially less so than that of shagbark. Red maple and cherry birch, *Betula lenta*, are masters of the understory. White birch, *Betula populifolia*, is an understory occasional and saplings (to 6 inches/15 centimeters dbh) of American chestnut can be found. Hemlock and white pine, where present, are commonly restricted to the understory; they are presumably, but not certainly, transgressives headed for the canopy. Black oak and beech are common in the mature canopy. Flowering dogwood over spicebush and mapleleaf viburnum account for much of the two lowest layers of woody vegetation.

Toward the southern part of New Jersey's Piedmont, in the vicinity of Princeton, sweet gum and tulip trees may persist in the mature canopy alongside oaks larger than the sweet gums. In the Herrontown Woods (see p. 384) black oak is at least as common as red oak, and in many respects the mature forests of the region are indistinct from those of more southerly latitudes of, say, Virginia and North Carolina.

Featured in the herb layer in New Jersey's mature tracts are the spring ephemerals: mayapple, Solomon's seal, wood anemone, *Anemone* spp., partridge berry, New York fern, *Thelypteris noveboracensis*, and Christmas fern.

PENNSYLVANIA

In Pennsylvania the main axis of the Piedmont turns westward and the province moves away from the sea and its moderating influence upon the weather. Due to the somewhat cooler climate, and probably also to edaphic changes, the black oak, sweet gum and the pioneering pine species common in New Jersey are absent from comparable latitudes in Pennsylvania. In consequence, the northerly and westerly parts of the Piedmont in Pennsylvania are the most northerly in vegetative aspect of the entire province, even though parts of New Jersey are higher in latitude.

Red oak is chief among the pointed-leaved oaks and in many mature woodlots is more numerous than white oak. Scarlet oak is a common canopy tree, as is tulip tree. Mockernut and shagbark hickories are common but numerically inferior to the oaks. Basswood is scattered in Pennsylvania's mature mesic woodlots, particularly on the lower slopes. White pine and Canadian hemlock are widely but thinly distributed in the mature canopies and in the understories as transgressives. Black oak and sweet birch are similarly scattered. Red maple occasionally reaches the canopy, but is more commonly found in the understory.

Black tupelo is the red maple's most common companion in the subcanopy. Beneath them thrive the flowering dogwood, sourwood, witch hazel and redbud. Transgressives of hickory, oak, white pine, and hemlock weigh heavily in the understory populations. Perhaps anomalously, small American elms are frequent in the understory of many oak–hickory stands. It is doubtful that they will reach the canopy.

Shadbush and mapleleaf viburnum are the most common

shrubs. The shrub layer is typically thinner under the mature canopy than in the growing forest, perhaps because it is the nature of spicebush, chief shrub of the penultimate phase of the sere, to grow in thick profusion. Some spicebush persists on mature sites but at restrained densities.

Mayapple, jack-in-the-pulpit, true and false Solomon's seal, windflower, and the spring ephemerals are to be expected in the displays of spring wildflowers.

It may be fairly observed about woodlots in Pennsylvania and in other parts of the northern Piedmont that a high percentage of them are mature. Farmland is not being abandoned at a significant rate nor has it been, probably since the region was initially settled. Farmlands are then generally stable, and many contain blocks of acreage which have been kept as woodlots for two centuries or more to be used on a sustained-yield basis for timber and firewood. These woodlots tend to be established on the locations least desirable for farming, such as hill crests, and many are therefore more xeric than mesic. The mature vegetation will distinguish the two soil types at a glance.

MARYLAND

The mature forests of Maryland are closely similar to those of Pennsylvania, but nuances of "southerliness" differentiate them. Occurrences of hemlock and white pine are widely scattered, red oak is slightly less complete in its dominance, scarlet oak is reduced to casual status, and black oak is more in evidence. White oak plays a somewhat more aggressive role than to the north and it is on many sites the most numerous oak. Mockernut is the most common hickory. American beech, red maple, sourwood, dogwood, redbud, musclewood and witch hazel are all important in the understory. The shrub and herb layers do not differ materially from those of Pennsylvania.

VIRGINIA

The history of agricultural land use in the northern half of Virginia's Piedmont is a continuation of that pertaining to the province in Maryland and Pennsylvania. The farms are well cared for and recent abandonment is minimal to moderate. The landscape is generally stable and many of the woodlots are mature. This condition changes at the James River, south of which there has been considerable abandonment in the twentieth century. Open fields are fewer, and the forests tend to be in continuous tracts checkered by recent cuttings. There are more woodlots but fewer mature forests.

Virginia is also a zone of transition in the composition of the forest at maturity and at earlier seral stages. In the nearly one-third of the Piedmont's length which lies within Virginia's borders, the mature forest shifts in complexion from northern to southern. The rate of change is greatest between the Potomac and the James, for it is in that region that the red oak loses its grip on the canopy, while black oak makes proportionate gains. Spanish oak, most strongly identified with the southern Piedmont, becomes established in the same transitional range. White oak is a constant; it is the most numerous single species in the mature mesic canopies of Virginia, as it is in perhaps most parts of the Piedmont.

The pignut hickory also shows a southern orientation in the Piedmont, and it gains in prominence between the James and the Roanoke Rivers, particularly on the better-drained slopes. Mockernut is the most common of Virginia's hickories; on old moist slopes shagbark joins the stand.

Black tupelo ranges consistently in the understory throughout Virginia in company with red maple. Dogwood, redbud, and sourwood dominate the next lower level. Replacement transgressives of oak, hickory, and tulip tree complete the understory, with the occasional participation of witch hazel, musclewood, and hop hornbeam. Shadbush, mapleleaf viburnum, and wild azalea are the prominent shrubs. Strawberry bush is common. Black cohosh, *Cimicifuga racemosa*, with its spire-like (but foul-smelling) white racemes, is the tallest herb. Other notables of the herb community include Solomon's seal, false Solomon's seal, perfoliate bellwort, *Uvularia perfoliata*, *Phacelia* spp., jack-in-the-pulpit, mayapple, and the spring ephemerals. Orchids of the mature forest include the rattlesnake orchid, *Goodyeara pubescens*, lily-leaved twayblade, and cranefly orchid.

NORTH CAROLINA

Elements of the northern Piedmont forest, if present, are vestigal. Basswood and red oak are uncommon, the latter having been effectively replaced by the black oak. Spanish oak is prominent, making the distribution of oaks about evenly divided between Spanish, black, and white. Tulip tree is consistently present. Ascending the slopes one encounters, in order, shagbark, mockernut, and pignut hickories. Beech is common on the lower slopes, particularly in the transition zones just uphill from the alluvial soils and in sheltered coves.

Red maple dominates the understory, with less competition, on many sites, from black tupelo than from sourwood. Flowering dogwood is everywhere. Red mulberry, *Morus rubra*, is common but scattered on the lower slopes.

Amelanchier Shadbush is restricted to the western part of North Carolina's Piedmont. Elsewhere, mapleleaf viburnum, strawberry bush, and wild azalea control the shrub layer. Downy arrowwood and several species of *Vaccinium* are usually present. Notably absent from the shrub layer of the mature forests is the escaped ornamental, silverberry, *Elaeagnus umbellata*, which completely usurps that stratum under some growing forests.

The spring ephemerals stage some of their grandest displays in the mesic Piedmont of North Carolina. Look for spring beauty, trout lily, and toothwort in early April, just after the flowers of hepatica and trailing arbutus are spent. Bloodroot and Catesby's trillium are common to profuse in the mature woodlots, downslope and upslope, respectively. Mayapple and jack-in-the-pulpit are present, but only in the seeps and other moist spots; this contrasts with their vigor on purely mesic sites to the north. Later in the season, the Solomon's seal, false Solomon's seal, butter-and-eggs, *Chrysognomum virginianum*, and false foxglove, *Aureolaria virginica*, a parasite on the roots of black oak, are in bloom. Beechdrops bloom around the roots of beech trees. Indian pipes and pinesap are common but scattered, rising through the leaf litter briefly at any time during summer.

SOUTH CAROLINA

Mature canopies in South Carolina differ little from

those in North Carolina. White oak, black oak, and Spanish oak dominate. Post oak is scattered but is more common than to the north. Mockernut, pignut, and shagbark hickories are common in proportion to the age of the site. Tulip tree is faithfully present.

In the understory on mature sites, red maple and black gum receive considerable competition from winged elm, presumably persistent from earlier in the sere. Dogwood, redbud, and sourwood occupy the lower subcanopy. In the shrub layer, French mulberry, *Callicarpa americana*, and fringe tree, *Chionanthus virginicus*, join the Viburnums, (*V. acerifolium*, *V. rafinesquianum*, and *V. prunifolium*) as well as strawberry bush and occasional vacciniums. Hawthornes, *Cretagus* spp., generally xeric in their associations, occur on many mesic sites in South Carolina.

Locations of sufficient maturity to support spring ephemerals must be sought out in South Carolina. Such locations may also sustain Catesby's trillium, bloodroot, windflower, and hepatica. Secluded coves and seeps may contain black cohosh, mayapple and jack-in-the-pulpit. Solomon's seal and false Solomon's seal are common; Indian pipes and pinesap are scattered and short-lived in the old woods of South Carolina (as they are throughout the province) and may be seen at any time from midsummer to frost.

GEORGIA

Spanish oak and post oak reach their maximum importance as Piedmont canopy dominants in Georgia. Water oak, generally associated with the bottomlands, moves onto the mesic slopes. On selected sites any of the foregoing may be the most numerous oak. Black oak and white oak are also important, and tulip tree is regularly found in the canopy, often as the largest member. Shagbark is less common than to the north; mockernut and pignut are the more common hickories. Old sweet gums and black cherries may persist from the growing forest.

The subcanopy in Georgia's mature woods is populated by red maple (and sugar maple on the very oldest sites), sourwood, dogwood, redbud, and white mulberry. Winged elm is common as a growing forest remnant, though seedlings and saplings are in evidence. On the very oldest sites, cucumber tree, *Magnolia acuminata*, rises into the understory. Shrubs include strawberry bush, wild azalea, French mulberry, *Viburnum nudum*, (no common name), dwarf azalea (*Rhododendron atlanticum*, a Coastal Plain shrub), and horse sugar (*Symplocos tinctoria*). The dwarf azalea is smaller (3 feet/1 meter or less) than the wild azalea and is distinguished by its hairy twigs. Horse sugar is a shrub or small tree, witch hazel-size, with alternate, entire sweet-tasting leaves which are elliptic, 5 inches (13 centimeters) long and half as wide.

Ferns can be prominent in the herb layer and may include Christmas, bracken, cinnamon, and netted chain ferns, as well as ebony spleenwort. Excellent displays of the spring ephemerals and their associates can be found at selected locations on old slopes. Cranefly orchid, wild yam, wild ginger, *Hexastylis* spp., pipsissewa and hepatica are also common.

On the Dahlonega Plateau, 1000 feet (300 meters) higher than the rest of the Georgia Piedmont, we expect and find that the mature woodland communities are of generally more northern composition. The southern oaks (Spanish, water, and post) are fewer; black and, occasionally, red and scarlet oaks are present. The plateau's abrupt elevation difference invites comparative botanizing.

ALABAMA

The crowns of Alabama's mature mesic forests are under the control of Spanish, white, and post oaks. Water oak is commonly involved in the mesic canopy, although the species prefers bottomlands. Mockernut is the principal hickory. Tulip trees are occasional. The pecan, *Carya illinoensis*, is native to the southern Mississippi basin, but it has become established in Alabama's mesic forests as an escapee from cultivation. Winged elm, redbud, dogwood, and white mulberry hold a majority of the secondary layer. French mulberry, wild and dwarf azaleas, *Viburnum nudum*, horse sugar, and strawberry bush are principals in the shrub layer. The forest herbaceous community is comparable to that of Georgia. Prime mature sites are uncommon, and must be "staked out" whenever found, for exploration in early spring.

Animals of the Mature Forest

Insects, Arachnids and Other Arthropods

The majority of the forbs in the mature forest bloom in early spring or before. We know that any plant which produces a showy inflorescence attracts insects which will, in the course of foraging, transfer pollen from the male sexual parts of the flower to the female, thus fertilizing the ovum (or ova) so that seeds may develop. It seems logical that an intense and simultaneous flowering of many species, involving large numbers of individuals from each, attracts a larger and more varied group of pollinators than would thin flowerings uncoordinated in time. We are not surprised to find the prevernal forest abuzz with pollinating insects, for that is the time of the great flowerings on the forest floor. In observing the spring ephemerals and their contemporaries, we are therefore obliged to consider a series of interlocking relationships between the plants and animals of the mature forest. The ephemerals, clearly, have made evolutionary adjustments to fit their schedules to those of the canopy trees—that is, they reproduce and photosynthesize before the trees take the sunlight. The pollinating insects must observe the same schedule to take advantage of the nectars (quick-energy carbohydrates) and pollens (sources of protein). The non-ephemerals, such as hepatica, jack-in-the-pulpit, bloodroot, and mayapple, continue to photosynthesize in the weak light available to them in summer. They bloom with the ephemerals and avail themselves of the pollinators which attend the ephemerals. Thus the trees control the schedules not only of the ephemerals but of the non-ephemerals as well, and of the insect pollinators of both.

The solitary bees (those which nest singly) and certain flies are the principal pollinators of the prevernal forest. Many of the bees are from the family Andrenidae and Halictidae, and they tend to share common habits. The males emerge earliest in the season and visit flowers promiscuously in quest of nectar. The females fix upon a single species of flower from which to gather pollen, thus maximizing their effectiveness as pollinators. The females collect wads of pollen on the bristles of their hindmost legs, then prepare balls of the pollen to be stuffed into the burrows in the forest soil in which eggs have been laid. The larval bees, one per pollen ball, eat the pollen, grow, pupate for most of the year, then emerge as adults at the right prevernal moment of the following year. Numerous additional species of bees, wasps, and flies complete their reproductive cycles by parasitizing the nests of the pollinating bees, laying an egg on the unguarded pollen ball to hatch before that of the host and commandeer the food. Bee flies (family Bombyliidae, squat and fuzzy flies that resemble bees) and syrphid flies (family Syrphidae) are among the prominent dipteran pollinators; there are, of course, numerous other families involved.

The litter on the forest floor churns with an essential biotic process that accounts for half the forest cycle of food and nutrients. The process of decomposition involves numberless arthropods. Above the leaf litter, growth occurs; that is, atmospheric carbon is fixed in plant tissue by the process of photosynthesis. In the litter (the collection of fallen leaves, twigs, and woodpecker chips) the forest materials are desynthesized and the carbon, nitrogen, and minerals are released. The forest plants are composed largely of cellulose and lignin, complex compounds in which carbon is the key element. These compounds are chemically decomposed by sequential invasions of fungi and bacteria which return the carbon to the atmosphere in the form of carbon dioxide gas (CO_2).

Many (perhaps thousands) of species of mites too small to be seen with the unaided eye graze upon the fungi and bacteria in the litter; their populations are estimated in the millions per square meter. Insects in the order Collembola (the springtails) are legion in the litter, exploiting a variety of scavenging and predatory feeding opportunities. Other arthropods important in the litter are the centipedes (class Chilopoda), with their flattened bodies and many body segments, each with a single pair of legs; centipedes are predators on other arthropods and are somewhat dangerous to handle. The millipedes are a distinct class (Diplopoda) with two pairs of legs per body segment, the members of which gain their living by grazing on plant and fungal detritus. *Sigmoria aberrans*, identified by a 2 inch (5 centimeter) black body with orange trim, is a common millipede in the mature woods of the southern Piedmont. The group also includes the pill millipedes, 1/3 inch (8 millimeters) in length and capable of rolling into a protective ball when disturbed, but it does not embrace the very similar, woodlice which are land crustaceans living in the moist litter and feeding upon plant detritus. The pill bugs (family Armadillidiidae) and the sow bugs (family Oniscidae) are among the more commonly seen woodlice. Craneflies (family Tipulidae) are associated with the moist soil litter where their larvae eat decaying vegetation. The adults resemble huge mosquitoes, but they do not bite. Daddy longlegs (family Phalangidae) are arachnids with extremely long legs and single-unit, oval bodies. They scavenge for dead insects.

Sprouting from the leaf-strewn floor of almost every growing and mature forest in the Piedmont are the leathery, heart-shaped leaves of wild ginger, *Hexastylis* spp., plants with particularly intimate relationships with the insects and other arthropods of the litter. The bulbous, fleshy flowers are concealed beneath the litter and they depend for pollination upon the visits of springtails and beetles. When the seeds mature, each bears a fat body (small nodule of lipids) highly favored in the diet of ants. The ants store the fat bodies in their galleries in the soil, thereby distributing the attached seeds.

Chewing audibly at the standing deadwood are the long-horned borers (beetle family Cerambycidae) and the horned beetles, *Passalus cornutus*. Flatheaded borers (family Buprestidae) are often found in oak and hickory logs that died standing and were later split for firewood; they are prominent among those insects which channel deep into the deadwood, opening galleries of ingress for other insects and fungi.

The towering oaks attract a variety of leaf-eating lepidopterans (see page 175), such as the buckmoth, oakworms (north of the Potomac), red-humped oakworm, and the rough, prominent, and variable oakleaf caterpillars. The hickory horned devil, the hickory tiger moth and the luna moth feed heavily on the leaves of hickory. Acorn and nut weevils (subfamily Curculioninae) eat the fruits of both these canopy dominants. The red-spotted purple, the comma, the question mark, and the mourning cloak are common as adults in the mature forest, flickering through the shadows to dab at tree sap. Katydids (Pseudophyllinae) and walkingsticks (Phasmatidae), the twiglike relatives of the mantises, eat the leaves of oak. Cynipid wasps (see page 175) produce the papery oak apple galls common in every mature woods.

Reptiles and Amphibians

The reptiles and amphibians of the Piedmont's forests at maturity do not differ materially from the populations of the growing forests (see pages 178–79). Bear in mind that as forests mature, the soils become more moist and therefore more amenable to the needs of many amphibians, particularly the salamanders. Herpetological populations are more dense in the mature forests than in the previous seral stage.

Birds

Much of the material regarding birds of the previous seral stage (see pages 180–82) applies to those of the mature forest. Absent or reduced are those birds whose principal orientation is toward conifers, notably the pine warbler, pine siskin, brown-headed nuthatch, golden-crowned kinglet and red-breasted nuthatch. The white-breasted nuthatch, largest of the clan, is plentiful in all seasons and is able somehow to extract grubs from the bark even in the dead of winter. Woodpeckers abound. The downy, hairy, red-bellied, red-headed, and pileated woodpeckers and the flicker (also, of course, a woodpecker) give the impression of occupying the larger mature tracts at saturation densities; that is, at something like the maximum populations the habitat can support given the species' territorial needs. In winter the yellow-bellied sapsucker inflicts its minor phlebotomies on oaks, hickories, and maples, opening tree-sap carbohydrates to the rest of the avifauna, insects, and mammals.

The woodpeckers provide nesting cavities for the great crested flycatcher, the black-capped and Carolina chickadees, the tufted titmouse and the eastern bluebird (at the ecotones), and for flying squirrels. The excavations of the large pileated woodpecker—sometimes hammered directly into living oak—serve admirably for gray squirrels.

The black, quarter-inch (6 millimeters) berries of black tupelo, dominant in the mature understory, attract large flocks of robins in winter, as well as rusty blackbirds and cedar waxwings. Dogwood berries feed the same birds, in addition to cardinals, bluejays, and evening grosbeaks. Goldfinches and purple finches take the samaras of maple and tulip trees. Fox sparrows scratch in the litter.

Nesting birds of the habitat include the scarlet and summer tanagers, red-eyed and yellow-throated vireos, eastern wood peewee, wood thrush, veery (Potomac northward), black and white warbler, worm-eating warbler, parula warbler, black-throated blue warbler (Potomac northward), ovenbird, Kentucky warbler, hooded warbler, redstart, northern oriole, cardinal, rufous-sided towhee, blue-gray gnatcatcher, blue jay, common crow, fish crow, yellow-billed cuckoo, and wild turkey.

Spring and fall migrations are particularly intense in the mature forests, especially in the western Piedmont. The spring passage affords splendid opportunities to see birds in their breeding plumages and singing their territorial songs; the mature mesic hills are prime habitat for certain of the migrants. Look for the Swainson's thrush and the veery. Expect the blue-winged, Cape May, bay-breasted, blackpoll, Tennessee, black-throated blue, black-throated green, Blackburnian, chestnut-sided, and yellow-rumped warblers and the rose-breasted grosbeak. Their calls are essential aids in locating and identifying these jewels of the forest canopy. Even the experienced naturalist must relearn some—if not most—of the calls each spring as the warblers pass through the Piedmont, but an ever- deepening intimacy with the climax forest rewards the effort.

Among birds of prey active in the mature forest, the Cooper's hawk, though not common, is a formidable presence, preying on smaller birds. The broad-winged hawk is no doubt the most numerous diurnal raptor of the habitat in the nesting season and during spring and autumn migrations. The sharp-shinned hawk does not spend much time in the mature forest but flits along its edges to dine upon the many smaller birds who also use the ecotones. In the southern Piedmont the barred owl is associated with alluvial forests, but north of the James it hunts in the mesic woodlands as well. The great horned and screech owls are ubiquitous in the mature forest habitat.

Mammals

The mammalian species present in the mature forest are, generally, those which populate the previous stage (see pages 182–84). The greater degree of maturity of the oaks and the arrival of the hickories probably results in higher carrying capacities for the eastern gray squirrel and the eastern fox squirrel. Associated with advanced maturity is additional deadwood shelter for birds and mammals. Deciduous trees die piecemeal and tend to leave deadwood standing for longer periods than do pines, and woodpecker excavations proliferate. Also, deciduous trees develop hollow centers as they age, much to the benefit of bobcats, gray foxes, gray squirrels, flying squirrels, fox squirrels, opossums, skunks, and raccoons.

CHAPTER SIX

The Xerosere, Dry Soil Habitats

THE AMOUNT OF MOISTURE in the soil and its availability to plants is the single most important factor in selecting the vegetative mix which composes the sere at any site. The pH of the soil (see Glossary) is significant, though secondarily, because some plants tolerate a narrow range of pH and because pH regulates the availability of nutrients in the soil. Nutrients, even when present, are less usable by plants in acidic soils than in circumneutral and basic soils. This explains, for example, why the spicebush is able to grow on mesic sites such as the Farnsworth field at Gettysburg and the slopes of Bull Run Mountain although at most other places the plant is restricted to bottomlands. Gettysburg and Bull Run Mountain are underlain by diabase rock and the resulting soils are basic in pH, therefore rich in available nutrients.

Soils of low moisture content are termed xeric (the Greek *xeros* means 'dry'). In the Piedmont, dryness of the soil is the result of its inability to retain moisture, rather than any pronounced lack of precipitation, for the province is blessed with generous, consistent, and evenly distributed rainfall. Sites at the tops of hills, for example, tend to be xeric, or at least drier than the lower slopes, because water runs downhill. Hilltop soils are also thinnest and least able to retain moisture because the erosive forces of wind and water act most forcefully there. They carry the finer soil particles downslope, leaving at the summit the coarser, most recently weathered chips, which retain water poorly. Finally, it is obvious, but particularly significant in the context of soil moisture, that hills are hills because they are underlain by rock more obdurate than that beneath the adjacent lowlands, that is, rock better able to resist weathering. And weathering, the chemical decomposition of rock beneath the surface, is one of the initial steps in building soils.

Dryness is associated with "poorness," and in many xeric contexts in the Piedmont the terms are synonymous. Dry soils tend to be thin, coarse, leached of minerals, and lacking in the decomposed organic material, fungi, and bacteria which are the essential components of humus. Soils may be thin, poor, and dry because of natural features of the terrain or because of manipulation by humans. Man-made xeric areas are found on badly eroded abandoned fields where the A horizon (topsoil) has been washed nearly or completely away, and on graded lawns, roadsides, and sites where the soils have been similarly ravaged by earth-moving equipment.

Dryness has two distinctive effects on vegetation; it slows the progress of the sere and it tends to select plant species that are least demanding of moisture and nutrients. For example, of two sites similar except for a moderate difference in the soils' ability to retain moisture, the vegetative mix will be similar but the individuals will grow more slowly and the progress of the sere will be retarded on the drier site. If the difference in moisture is great, the mixture of plants, at all seral stages, will be different. In the Piedmont, the differences are more often subtle than pronounced, and the observer judges a site's position on the drainage scale by shifts in the portion of the canopy occupied by mid-mesic plants as opposed to those at the dry end of the mesic range. However, extremes do exist. True xeric plants are distinctive, and the very dry soils on which they grow are easily recognized, with experience, by their characteristic vegetation.

Dryness is expressed by the vegetation, at seral maturity, when certain oaks appear in the dry–mesic mix, namely, the post oak, the black oak and the Spanish oak. Truly xeric conditions are revealed by the presence of blackjack oak and, later, by chestnut oak and the associated shrub layer of heaths (family Ericaceae). The chinquapin and dwarf chinquapin oaks are regionally affiliated with the chestnut oak–heath portion of the xeric regime, and because such forests are typically situated on the tops of ridges in the Piedmont and in the adjacent montane provinces, they are called ridgetop communities. Where conditions are still more severe due to thin soils on high ridges closely underlain by rock, the mature canopy may be stunted and scattered and may include pitch pine and table mountain pine as well as chestnut oak. The shrublike bear oak may be present. On the most harshly xeric rock outcrops the trees may be severely stunted, prostrate, or absent altogether.

The primary succession process can be observed on some granitic outcrops in the Carolinas and Georgia. The arrival of pioneer life forms and the initial process of soil-building on the exposed rocks involves some rare plants endemic to the Piedmont's granite domes. The granitic outcrops are discussed in Part IV, Special Places.

The Herbaceous Phase

The traditional Clementian view of edaphic succession (the building of soils), in which all sites tend toward a mesic edaphic climax, is perhaps most valid where applied to abandoned fields that have been rendered xeric through agricultural abuse. Such sites were probably mesic before the topsoil was lost to erosion, and due to their slopes and other physical characteristics they will again seek a mesic equilibrium as the natural process of secondary plant succession works to restore the soil and enhance its ability to retain moisture.

At the outset, however, eroded abandoned fields offer a glimpse of the herbaceous phase of the xerosere. The shoulders and banks of graded roads offer additional opportunities to observe xeric herbs, though we must bear in mind that the progress of the roadside sere is repeatedly arrested by mowing, a practice that kills young pines, stunts the other woody plants, and leaves the tall herbs in command indefinitely. Also, roadsides are often sown with herbs introduced to stabilize the embankments.

Where the drainage rating is naturally xeric, abandoned fields are rare simply because such sites are rarely cultivated. A strict comparison with abandoned mesic fields is therefore hard to draw—the true herbaceous xerosere is a rare condition. Even after a forest is burned, the seeds and root systems of woody plants are still quite viable and vigorous; immediate woody growth usually abbreviates the herbaceous phase. On the occasional sites which afford an observation of the herbaceous xerosere (most of which will likely be erosion gullies in otherwise mesic abandoned fields) the following herbs may be present:

Table 5: Xeric Grasses and Forbs

Grasses	*Forbs*
Poverty grass, *Aristida dichotoma* and *Aristida oligantha*	Peppergrass, *Lepidium virginicum*
Naked beard grass, *Gymnopogon ambiguus*	Bracted plantain, *Plantago aristata*
Needle-leaved panic grass, *Panicum aciculare*	Lance-leaved plantain, *Plantago lanceolata*
Broomsedge (dry-soil variety), *Andropogon tenarius*	Creeping buttonweed, *Diodia teres*
	Pinweed, *Lechea leggettii*
	Bitterweed, *Helenium amarum*
	Low ragweed, *Ambrosia artemisiifolia*
	Sheep sorrel, *Rumex acetosella*
	Wild onion, *Alium vineale*
	Blazing star, *Liatris graminifolia*
	Butterfly weed, *Asclepias tuberosa tuberosa*
	Pineweed, *Hypericum gentianoides*
	various low, compact lichens

The sequence of herbs on xeric lands is less rigidly predictable than on mesic sites, and on moderately dry soils in general, herbs of the mesic regime begin to take over from the characteristically xeric herbs within a few years. Thereafter, the vegetative procession may approximate the mesosere, though at a slower pace. For example, thin stands of poverty grass may unite with buttonweed, bracted plantain, and certain tall lichens (such as British soldiers, *Cladonia cristatella*) to form an initial mat of soil-holding vegetation. Taller xeric herbs may arrive within a few years—the dry-soil broomsedge, ragweed, sheep sorrel, and pineweed. They may be joined, depending on the degree of dryness, by plants of comparable seral function from the mesic ranks, or, in truly xeric situations, the dry-soil herbs may hold exclusive tenure. In later years of the herbaceous xerosere, the strikingly-flowered bitterweed, butterfly weed, and blazing star may bloom on the dry soils.

Xeric Woody Succession

Pines are noted for being tolerant of dryness and undemanding of nutrients. These traits place the pines in a position to exploit the xeric sites and to dominate the early woody xerosere even in localities normally pioneered by deciduous trees. Among the Piedmont's pines, Virginia pine, *Pinus virginiana*, appears to be the most tolerant of dryness and poorness of soil. This species is the principal xeric woody pioneer of the province, a function it shares with pitch pine, *P. rigida*, in some localities. In the broad zone in the central Piedmont where the ranges of the loblolly, shortleaf, and Virginia pine overlap, the tracts occupied by *P. virginiana* have the poorer and drier soils. In the xeric Piedmont, pioneer Virginia pines are often stunted, scattered, and slow-growing. The species is a xeric hallmark, particularly in the central and southern Piedmont, where its presence on mesic soils is unlikely (though it is common on mesic soils north of the Potomac and on the Dahlonega Plateau in Georgia).

The Growing Forest

Given that it is less common to find a totally xerophytic community than one in which xeric and mesic plant types are mixed, we can recognize the xeric growing forest by its association of Virginia pine, xeric oak transgressives, and heath shrubs (family Ericaceae). As the pines mature—which may take decades longer on a xeric site than under mesic conditions—they are joined by saplings of blackjack oak, *Quercus marilandica*, and chestnut oak, *Quercus prinus*, and by sourwood, *Oxydendron arboreum*, red maple, *Acer rubrum*, and flowering dogwood, *Cornus florida*.

50. *Vaccinium tenellum*

As the growing forest progresses, heath shrubs appear beneath the pines and transgressives. *Gaylussacia* is an ericaceous genus closely associated with xeric woodlands, most notably *G. frondosa*, *G. baceata*, and *G. dumosa*. The gaylussacias are easily confused with the smaller species of *Vaccinium* such as *V. tenellum* and *V. vacillans*, which in all probability will be present in a representative xeric growing forest alongside the gaylussacias. *Vaccinium vacillans* has perhaps the highest tolerance for dryness of any member of the genus in the Piedmont.

The Mature Forest

The chestnut oak is characteristic of the ridgetop forests in the Piedmont as well as in the mountain provinces to the west. The blackjack oak occupies sites drier still, particularly those reduced to xeric condition through agricultural abuse. Consequently, stands of blackjack oak are commonly succeeded by chestnut oak as the soils develop. The chestnut oak signals dryness in the Piedmont just as reliably as do the cacti in the western deserts. Though the chestnut oak forests, even at maturity, lack the cathedral grandeur of the great mesic forests—heights above 60 feet (20 meters) are uncommon on dry hilltops—these gnarled and angular patriarchs of old ridgetop forests take on the nobility of the venerable. These forests are rarely timbered, for the sawmill cannot use misshapen logs, but the slow-growing wood is densely packed with combustible cellulose and makes excellent firewood. Wind-thrown trees are common in the ridgetop woodlands because the roots have but scant purchase in the thin soils closely underlain by rock.

Lack of diversity is a quality of the ridgetop forests that quickly captures the observer's attention. The American chestnut shared the dryland canopies until its extirpation earlier in this century; now it is common to find chestnut oak occupying more than 90 percent of the xeric canopy. The subcanopy is also simplified; sourwood tends to control that stratum with help from dogwood, red maple, and sassafras. Black tupelo is casually present in the understory. Saplings of pignut and mockernut hickories are common transgressives, reaching the canopies on very mature sites.

The shrub layer can be somewhat more diverse. Mountain laurel, *Kalmia latifolia*, is a montane ericad (member of the family Ericaceae) present in the Piedmont under some chesnut oak canopies, generally on north-facing slopes somewhat below the ridges. Species of *Vaccinium* and *Gaylussacia* may persist from the growing forest phase. Wild azalea, *Rhododendron nudiflorum*, tolerant of a wide range of moisture conditions, is common under the ridgetop canopies. These genera, as well as *Lyonia* and *Gaultheria*, all members of the ericaceous heath family, are associated with the Piedmont's dry forests and with acidic soils. Non-ericads on the ridgetops and dry slopes include occasionals of mountain ash, *Sorbus arbutifolia*, and witch alder, *Fothergilla major*. The hawthorns, *Cretagus* spp., show a preference for dry hilltops, and they are casually present in xeric woodlots throughout the Piedmont. Trailing arbutus, *Epigaea repens*, a prostrate shrub with alternate, leathery leaves, and spotted wintergreen, *Chimaphila maculata*, indicate the presence of a certain shoelace-high microhabitat exploitable by very low shrubs (the wintergreen is sometimes called a subshrub).

The association of the ericads with the dry forests is common; the family's own true herbaceous genus, *Monotropa,* is frequently seen in the habitat's rather scanty herb community. Pinesap, *Monotropa hypopithys,* may find its saprophytic (see Glossary) lifestyle well-suited to living in the dry forest. The plant has no chlorophyll and does not photosynthesize; it therefore probably does not experience the loss of moisture associated with that process. The bean family (Fabaceae) contributes goat's rue, *Tephrasia virginiana,* and, in the clearings, butterfly pea, *Centrocema virginianum,* which add nitrogen to the soil of the chestnut oak–heath woods. The coriaceous (leathery), somewhat succulent leaves of *Hexastylis virginica* make it the form of wild ginger probably best suited to the dry woods. Wild yam, *Dioscorea villosa villosa*, is a common vine of the habitat, being differentiated only racially from *D. villosa hirticaulis*, whose adaptation at the other end of the drainage scale locates it in the alluvial woods of the eastern Piedmont and Coastal Plain. Bracken fern, *Pteridium aquilanum*, is a dry-land fern found in woods and open spaces.

Trees

Pitch pine, *Pinus rigida.* See page 114. This pine is identified with dry soils at ridgetops and outcrops in the western Piedmont, although in Pennsylvania and New Jersey the species participates vigorously in the mesosere. Under xeric conditions the tree is not large. Cones are squatty, almost as wide as long. The needles are rigid, in clusters of three.

Table Mountain Pine, *Pinus pungens.* In the Piedmont, this pine occurs only at the borders of outcrops and other truly xeric conditions. It is rare and localized. The needles, 1 to 2 inches (25 to 50 millimeters) long are twisted slightly and grow in clusters of two. The cones are 2 to 3 inches (50 to 75 millimeters) long, spiny, and painful to handle. Twigs are supple and nearly impossible to break.

51. Table mountain pine, *Pinus pungens*

Virginia Pine, *Pinus virginiana.* See page 112. Virginia pine is important in mesic succession north of the James River. In more southerly parts of the Piedmont, except for the Dahlonega Plateau, it is associated with poor, dry sites. It is likely to precede the chestnut oak–heath forests on ridgetops throughout the Piedmont.

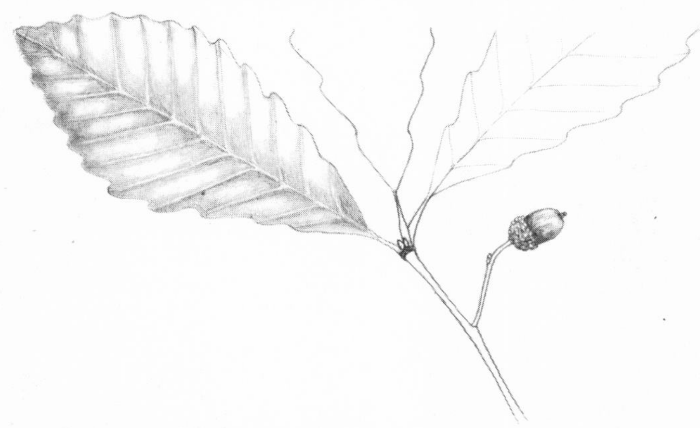

52. Chestnut oak, *Quercus prinus*

Chestnut Oak, *Quercus prinus.* This is a medium to large tree, depending on soil depth and microclimate. The leaf is elliptic to obovate, widest past the midpoint, with rounded dentate margins (the teeth of the leaves of the true chestnut are pointed). The acorn is large, 1 to 1½ inches (25 to 40 millimeters) long; a funnel-shaped cup encloses the first third. The bark is dark and deeply furrowed. The soils on which the chestnut oak grows are usually thin and acidic, with the result that many individuals are gnarled and stunted. It could no doubt grow in better soils but is probably excluded by competition from the mesic oaks.

Chinquapin Oak, *Quercus muehlenbergii.* The chinquapin oak's leaves are similar to those of the chestnut oak but are conspicuously more narrow and the teeth are somewhat sharper, though not so sharp as those of the true chestnut, for which "chinquapin" is a folk name. The acorn is about ¾ inch (18 millimeters) long with a rounded cup covering nearly half. The chinquapin oak is a southerly species in the Piedmont, though it ranges to the Great Lakes west of the province. It is not common, but is recorded in the dry, rocky soils of the western half of the Piedmont in the Carolinas and Georgia and in all Piedmont counties in Alabama. Northward, the species spills over locally from the mountain provinces to the western Piedmont.

53. Chinquapin oak, *Quercus muehlenbergii*

Dwarf Chinquapin Oak, *Quercus prinoides.* This oak rarely attains tree status; it is found mostly in the shrub layer. It is more frequent in the northern Piedmont than in the south, though it is generally uncommon. The leaves are 2 to 4 inches (5 to 10 centimeters) long and elliptic, with short rounded teeth. They are smaller and have fewer teeth than the leaves of chestnut oak but are otherwise similar. Acorns are nearly spherical, ½ inch (12 millimeters) in diameter, and half enclosed in a thin cup.

Post Oak, Scarlet Oak, Spanish Oak, and **Black Oak.** See pages 154–59. These oaks are capable of vigorous growth on the dry end of the mesic scale. Technically termed "xerophylophilous," these species are commonly found mixed with the true xerics, chestnut oak and blackjack oak, on marginally dry soils and on slopes where the xeric and mesic regimes meet.

Blackjack Oak, *Quercus marilandica.* This small ragged tree occupies dry, rocky, poor sites, sometimes preceding the chestnut oak in succession and sometimes contributing to the xeric climax. The bark is dark, coarse, and chunky. The branches droop. The leaves are obovate and three-lobed, fan-shaped at the ends. There is a bristle at the tip of

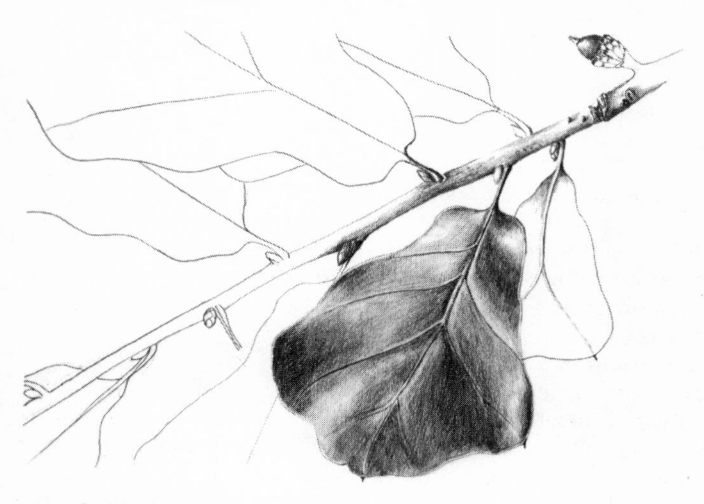

54. Blackjack oak, *Quercus marilandica*

each lobe. The acorns are ¾ inch (18 millimeters) long, oblong, and tipped with an elongated nipple. The coarsely scaled cups cover half the acorn.

Bear Oak, *Quercus ilicifolia.* This is a shrub, found only on selected xeric sites—in the Piedmont, generally the prominent monadnocks and outcrops. The leaves are oblong, 2 to 5 inches (5 to 13 centimeters) long and half as wide, with five to seven pointed lobes separated by shallow sinuses. The acorns are ½ inch (12 millimeters) long and half covered in a thick cup. The bark is nearly black and the twigs are covered with short, soft, brown fuzz.

55. Bear oak, *Quercus ilicifolia*

Mockernut Hickory and **Pignut Hickory.** See page 193. These are the Piedmont hickories that show the greatest tolerance for dry conditions. Transgressives, particularly of mockernut, are common in the otherwise greatly simplified xeric forests of solid chestnut oak over sourwood and dogwood.

Sourwood, Flowering Dogwood, Red Maple, and **Sassafras.** See pages 117, 123. These trees can thrive in a wide range of moisture conditions. They are common species in the chestnut oak–heath understory throughout the Piedmont and in other habitats as well.

Sweet Birch, or **Cherry Birch,** *Betula lenta.* See page 160. Sweet birch grows well on ridgetops, as well as on mesic slopes, but is limited to the northern Piedmont and the prominent monadnocks.

Shrubs

Gaylussaccia spp. This prominent genus of xeric shrubs is offered as a group because they are very similar in appearance to the vacciniums, and differentiation in the field between the genera is all that can be hoped for without considerable experience and a detailed taxonomic manual. The *Gaylussacia* (gay-lu-*say*-sha) are low shrubs growing from rhizomes, generally less than 3 feet (1 meter) in height. Leaves, commonly not uniform in size, are usually of rounded elliptic shape. The feature which distinguishes the *Gaylussacia* is the presence of minute resinous glands on the undersides of the leaves, easily visible through a watch glass and, with experience, to the naked eye. The fruits have ten seeds. *Gaylussacia frondosa*, *G. baceata*, and *G. dumosa* are among the more common Piedmont species. All *Gaylussacia* are grouped under the folk name "huckleberry."

56. *Gaylusaccia frondosa*

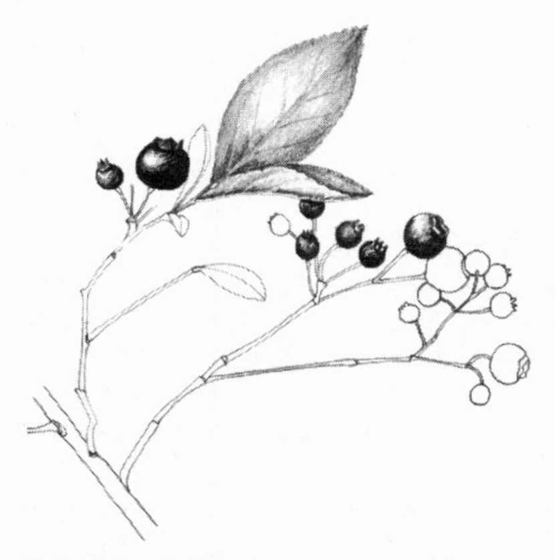

57. Black highbush blueberry, *Vaccinium atrococcum*

Black Highbush Blueberry, *Vaccinium atrococcum.* This shrub grows to 16 feet (5 meters) height. The alternate leaves are pubescent along the midrib on the underside, 2 to 3 inches (5 to 8 centimeters) long, elliptic to oblanceolate (see Glossary). Throughout the Piedmont this blueberry is restricted to dry woods, usually in the pine stage.
Sparkleberry, *Vaccinium arboreum.* At maturity, this vaccinium is not mistakeable for any other, for it grows to more than 30 feet (10 meters), the size of a healthy dogwood. It is found on mesic slopes but does well on and commonly occupies xeric sites. The leaves are varied in shape, entire,

58. Sparkleberry, *Vaccinium arboreum*

generally elliptic, and leathery and tend to be retained late into autumn.

Vaccinium tenellum and **V. vacillans.** See page 220. are common on mesic sites as well as xeric. They tend to be associated with the pine stage on mesic soils and to remain into the climax on many xeric sites.

Mountain Laurel, *Kalmia latifolia,* is an evergreen shrub with thick leathery leaves 3 to 6 inches (8 to 15 centimeters) long, shiny, and alternate. The fruit capsule is a five-segmented globe about ¼ inch (6 millimeters) across, from which a persistent stygma stands erect. Each of the florets

59. Mountain laurel, *Kalmia latifolia*

in the inflorescence, a compound corymb, shows a white corolla one inch or less across with pink to purple trim. The pollination mechanism features spring-loaded anthers tucked into the pockets in the corolla; they release when touched by an insect, often a bumblebee or a honey bee, dusting the visitor with pollen. Mountain laurel is, as the name suggests, principally a montane plant, but it is present beneath the Piedmont's chestnut oak canopies on ridgetops and monadnocks. It is one of the principal identifiers of the xeric forest.

Wild Azalea, *Rhododendron nudiflorum.* See page 198. This ericad thrives on soils wet or dry, provided they are acidic. Wild azalea grows beneath the streamside sycamores and at rocky summits in the Piedmont.

Wintergreen, or **Checkerberry,** *Gaultheria procumberns.* This rhizomatous shrub grows to only a few inches height, its evergreen leaves clustered at the stem tips. The leaves are elliptic. The solitary white flowers develop into ¼-inch (6-millimeter) red berries. Colonies of wintergreen grow on xeric sites in the western Piedmont.

Witch Alder, *Fothergilla major,* is another of the true xeric shrubs. It is not common because its habitat is limited and localized. It is a colonial shrub about 3 feet (1 meter) in height, mistakable for its close relative of lower slopes, the witch hazel. Witch alder's leaves are approximately round with edges entire (lacking witch hazel's dentition) and about 5 inches (13 centimeters) across. In spring the brushy white inflorescence is conspicuous in the dry woods, even though the flowers lack petals, for the filaments are numerous and densely packed.

Hawthorn, *Crataegus* spp. A rosaceous shrub closely related to the mountain ash, the hawthorn is identified with dry poor soils in the Piedmont. There are numerous species and they are very difficult to differentiate in the field. Most species are about 10 feet (3 meters) or less in height, have long, spurlike thorns on the twigs and leaves with serrate edges. Leaf shape varies with species; most are elliptic to spatulate and 1 to 2 inches (25 to 50 millimeters) in length. The flowers are typical of the family and suggestive of, but smaller than, the domestic apple. Fruits grow singly or in clusters.

Trailing Arbutus, *Epigaea repens,* is a prostrate shrub with evergreen leaves 1 to 3 inches (25 to 75 millimeters) long, leathery, wavy, growing from a system of stems laced into the leaf litter on dry slopes. The leaves lie directly on the litter. In late winter the white to pale blue flowers grow in clusters close to the stem, sheltered by the litter.

Viburnum cassinoides is a montane shrub present on some of the rocky slopes of the western Piedmont. Growing from an otherwise barren, vertical crevice on Moore's Wall in Hanging Rock State Park, North Carolina, one *cassinoides* plant serves as a much needed protection (from falling) on an exposed climbing route. The leaves are opposite, entire, and rounded to elliptic in shape. The flowers occur on stems which have two pairs of leaves. The fruits are drupes which change in color from pink to deep blue when ripe.

Herbs

Poverty Grass, *Aristida dichotoma* and *A. oligantha.* The grasses of the genus *Aristida* are called the "three awn" grasses because of the three awns (bristles) projecting from the seed sheaths. *A. oligantha's* are exceptionally elongated, to 1 inch (12 millimeters), and all three are of even length. In *A. dichotoma*, the more common of the two, only one awn is visible, and it is perpendicular to the seed sheath. Both are tufted annuals.

Bracted Plantain, *Plantago aristata,* is a grasslike winter annual with a taproot helpful in seeking deep moisture in poor soils and withstanding continued erosion. The leaves are basal and vertical, to 3 inches (8 centimeters) in height. Floral stems rise vertically another three inches above the leaves, each bearing a wheatlike floral head at the tip.

Creeping Buttonweed, *Diodia teres,* is not really a creeping plant; it is an erect and spreading, often profusely branched annual rising perhaps to knee height. Leaves are linear to lanceolate, 1 to 2 inches (25 to 50 millimeters) long, opposite and sessile (see Glossary). Flowers have white corollas ¼ inch (6 millimeters) long with four lobes; they occur singly at the leaf nodes.

Low Ragweed, *Ambrosia artemisiifolia,* can become a tall herb, reaching head height. The plant is freely branched. The leaves have a compound appearance, being deeply cleft into ragged leaflets, the proximal pairs being opposite, the distal alternate and irregularly lobed. Flowering stems grow as erect spikes from the leaf sheaths, the individual florets being green and indistinct. Ragweed is associated with poor, eroded soils.

Blazing Star, *Liatris graminifolia.* This asteraceous perennial grows to waist height on poor soils on steep slopes and ridgetops where there are openings in the canopy. As the name implies, the lower leaves resemble blades of grass. Small clusters of frilly florets alternate along the rachis. The corollas are purple.

Butterfly Weed, *Asclepias tuberosa tuberosa,* is identified by alternate leaves about 2 inches (5 centimeters) long, widest past the middle with entire, flat, unwrinkled edges. Plants are usually less than 24 inches (60 centimeters) high, spreading. The inflorescence is a 3-inch (5-centimeter) umbel of red to yellow florets with reflexed corollas.

Pinesap, *Monotropa hypopithys*, is a perennial herb in the heath family. It is saprophytic on decaying plant matter, therefore is without chlorophyll and is yellow with amber highlights. The translucent stems, about 6 inches (15 centimeters) high, support no real foliage but are studded with bracts pressed against the stem. Several florets, composed of bracts and sepals similar to the rest of the plant, nod at the top of the stem.

Stonecrop, *Sedum* spp. Several species may be found among the rocks on monadnocks and outcrops. The leaves are simple, often entire but toothed in some species, and succulent. The plants are low and spreading. Flowers of most are star-shaped, about ½ inch (12 millimeters) across, with four or five petals and an equal number of sepals.

The Xeric Habitats—State by State

We observe that the xeric life systems of the Piedmont are not so richly diverse as are the habitats that supply more moisture, and that relatively few plants comprise the vegetative mix at any point in the xerosere. The mature hilltops in Passaic County, New Jersey, may have the same chestnut oak–heath forests as do prominences on the Dahlonega Plateau in Georgia. An overgrazed, south-facing, rocky hillside abandoned in New Jersey will probably develop a community of poverty grass, ragweed, and creeping buttonweed, before moving into a dry-soil variety of broomsedge, thence to Virginia pine and to chestnut oak. This successional pattern closely parallels that of a sister tract in Georgia.

Truly xeric sites tend to be cloaked in mature vegetation because they are too unproductive for farming or timber-raising. The successional process is therefore not easily observed on xeric sites, and the drawing of regional comparisons is correspondingly perilous. The regional differences, and there are some, are based on geology and climate. The blackjack oak appears to be excluded from the glaciated portion of the Piedmont. Also, in the extreme northern sections of the province, climate somewhat restricts the conifers and forces the truly xeric places into a woody succession more deciduous than they might otherwise experience. Southward, monadnocks, granitic outcrops, and serpentine barrens enrich the province in xeric habitats. The serpentine barrens and granitic outcrops are described in detail in Part IV, Special Places.

NEW JERSEY

The hills and monadnocks adjacent to the Ramapo Scarp in upper Morris and Passaic Counties support ridgetop chestnut oak–heath communities, although in most instances the terrain is not severe enough to create strongly xeric conditions. Scarlet oak, red oak, sweet birch, tulip tree, and ash may share the canopy on north slopes. Black oak, white oak, and hickory are commonly present with the chestnut oaks on southern exposures over subcanopies of dogwood and sassafras. Shrub layers are richest in the shelters of the north-facing slopes; they include species of *Kalmia, Gaylussacia,* and *Vaccinium.* Among the dry-forest herbs of New Jersey are wild ginger, sasparilla, snakeroot, and columbine. Grasses and sedges may be present on south slopes, particularly where the canopies are least complete—and sparse canopies are characteristic of the chestnut oak stands.

In Sommerset County, the Watchungs also support some ridgetop dry woods similar to those near the New England Province but generally are less montane in vegetative aspect. Post oak is present, more or less in lieu of sweet birch. Pitch pine persists even into edaphic maturity around some rocky places. South of the Raritan River (i.e., south of the glaciated Piedmont) there are few dry woodlands in New Jersey's Piedmont.

PENNSYLVANIA

Many of the hilltops in French Creek State Park and other locations in Pennsylvania's northern Piedmont grow the stunted canopies of chestnut oak over gnarled red maple and dwarf chinquapin oak, *Quercus prinoides,* characteristic of the northern xeric climax. The shrub layer can be somewhat weak on the hilltops, strengthening downslope, particularly on the north side. At the summit of Big Roundtop, at Gettysburg, a few *Paulownia* trees grow, showing the affinity of this weedy alien for places as dry and stony as a battlefield hilltop or a crack in an urban sidewalk. Downslope on Big Roundtop grows a spotty canopy of chestnut oak mixed with a few stunted mockernut hickories and red maples. A short distance farther downslope, red oak, black oak, and other mesic species resume control.

MARYLAND

The xeric sites of Maryland are equivalent to those of Pennsylvania and northern Virginia; chestnut oak over red maple, sourwood, dogwood, and dwarf chinquapin oak. Shrub layers of *Kalmia*, *Vaccinium*, *Gaylussacia*, and other heaths tend to be stronger downslope, with the shrub layer at the summits sparse to open. Parrs Ridge, running northeast to southwest just west of Westminster, supports some of the Maryland Piedmont's most clearly defined ridgetop woods, the Catoctins of Maryland being considered part of the Blue Ridge Province.

VIRGINIA

The inner (western) Piedmont in Virginia is geologically part of the Catoctin Mountains, which in Maryland and Pennsylvania are said to be portions of the Blue Ridge Province. This places miles of ridgetops in the Catoctin and Bull Run mountains within the Piedmont. In addition, the great belt of monadnocks which occupies the western 20 miles (32 kilometers) of the Piedmont in Virginia south of Culpeper adds thousands of acres of ridgetop chestnut oak–heath forests, giving Virginia probably a larger measure of Piedmont xeric forest than any other state. The aspect of the canopies is broken, partially open, and composed of twisted chestnut oak, occasional blackjack oak, and residual Virginia and pitch pine over dogwood, sourwood, and sassafras. Ericads compose the shrub layer, which is richest on north-facing slopes.

Bull Run Mountain and some of the monadnocks are capped with quartzite and other very resistant rocks. Where these monoliths lie at or very near the surface, a rock-outcrop plant community exists. At some sites, such as the summit of Bull Run Mountain north of Thoroughfare Gap (U.S. Route 55), conditions are locally too harsh even for the chestnut oak. In some spots the canopy is composed of table mountain pine, *Pinus pungens*. Pitch pine is also present. Chestnut oaks on the site are twisted and deformed by ice and wind; some individuals approach 24 inches (60 centimeters) dbh without exceeding 35 feet (11 meters) in height. Sourwood is present in the understory around the outcrops over the shrubby bear oak, *Quercus ilicifolia*, which is known at relatively few locations in the Piedmont, all of which are on xeric high ridges. Somewhat downslope from the quartzite cap of Bull Run Mountain and similar sites in Virginia, the chestnut oaks grow with greater strength over understories of dogwood, sourwood, and sassafras and vigorous shrub layers of *Kalmia*, *Gaylussacia*, and *Vaccinium*. Farther downslope the traveler encounters Bull Run Mountain's unique mesic realm, described on page 353, with regal tulip trees over dense spicebush.

NORTH CAROLINA

The monadnock belt continues across the Virginia border into North Carolina. At Pilot Mountain, Hanging Rock State Park (Moores Wall), Stone Mountain in Wilkes County, Rocky Face Dome in Alexander County and at King's Mountain–Crowders Mountain, southwest of Gastonia, the formations reach their most spectacular expressions. In keeping with the geology, the ridgetop forests and outcrop vegetation on these monuments carved from earlier erosion cycles and resting like vast pyramids on the present peneplain offer their unique forms of botanic splendor. The table mountain pine, pitch pine, and bear oak are present on the higher monadnocks, and the vegetation on some shows strong montane influence. Mountain shrubs present as disjuncts on Carolina's monadnocks include *Vaccinium constablaei*, *Rhododendron catawbiense*, *Viburnum cassinoides*, *Clethra accuminata*, and *Ilex ambigua*. Among the herbs of the high monadnocks are the stonecrops, *Sedum* spp., sandworts, *Arenaria* spp., saxafrages, such as *Saxifraga michauxii*, and *Silphium compositum*, with its kidney-shaped leaves 1 foot (30 centimeters) across. Rock cap fern, *Polypodium virginianum*, is common on the outcrops.

There are many less severe but still quite xeric sites on the ridgetops of the lesser monadnocks and in the Brushy Mountains in Alexander and Wilkes counties and in the Uwharrie Mountains in Montgomery and Randolph counties and at other scattered sites. Occoneeche Mountain, near Hillsborough, is one of the easternmost xeric sites in North Carolina, with gnarled chestnut oak canopies over *Kalmia* and *Gaylussacia*.

SOUTH CAROLINA

The Piedmont of South Carolina is nearly devoid of hills high and dry enough to support the Piedmont's characteristic ridgetop communities of chestnut oak and ericads. The most spectacular examples of xeric and outcrop forests accessible to the Greenville–Spartanburg metropolitan areas are at Kings Mountain–Crowders Mountain just across the state line in North Carolina. Paris Mountain, just north of Greenville, supports mixed xeric–mesic forests at its summit. On other sites where chestnut oak does occur, it will likely be mixed with black, scarlet, post, and white oak. Chinquapin oak, *Q. muehlenbergii*, with leaves similar to those of chestnut oak but conspicuously more narrow, is present but uncommon. Hawthorn and french mulberry, *Callicarpa americana*, are scattered in the shrub layers, with *Vaccinium* and occasional *Gaylussacia* also present.

GEORGIA

Most of Georgia's xeric woods are found on the Dahlonega Plateau. The altitude of the plateau itself varies from 1500 to 1800 feet (450 to 550 meters), with peaks rising to well over 2000 feet (600 meters). At such prominences, the chestnut oak–heath compositions may predictably be found, some touched by montane influence (see description of the North Carolina monadnocks, above). Chinquapin oak may be locally present.

Off the plateau, the driest sites are likely to be too mesic to support chestnut oak–heath communities. Some will resemble the dry ridge above Pumpkinvine Creek north of Atlanta, a notably mature bottomland site with cove hardwood aspects (associations of maple, buck, basswood, and tulip seen in the sheltered coves in the Blue Ridge). Here the canopy is of stunted post, white, and black oaks over several species of hawthorn (*Cretagus* spp.) and sparkleberry, *Vaccinium arboreum*.

ALABAMA

My field notes from the Alabama Piedmont record few sightings of chestnut oak and none of a clearly developed chestnut oak–heath community. Blackjack oak is seen on poorer sites. It will be a challenge to the Alabama naturalist to find this habitat in the Piedmont of that state and to enjoy, in advance, an understanding of its unique adaptations and strategies for success where soils are poor and thin. The Talladega National Forest along the Alabama Piedmont's northern limits, particularly in the vicinity of Cheaha, could be a fruitful region to search.

Animals of the Xerosere

Insects, Arachnids and Other Arthropods

The reduction in plant diversity characteristic of xeric life systems results in a corresponding simplification of the fauna. For example, the scarcity of showy-flowered forbs (the substitution of, say, a xeric composite like ragweed for goldenrod or daisy fleabane) reduces the number of flying insects in the habitat, which in turn eliminates most of the orb-weavers and the crab spiders. All insects whose life cycles require aquatic or moist conditions are absent. Because of the dryness, few species can be said to proliferate.

Yet any stressful environment offers opportunities for some specialists to adapt and to enjoy the inestimable boon of living where many of his competitors cannot. One such creature is a very small homopteran for whom I find no common name but whose habits—leaping by the hundreds in unison but at random directions on the face of an exposed boulder—leads me to call them "rock hoppers." They are members of the family Delphacidae, which includes the smallest of the plant hoppers.

Among the arachnids, those who fare best are the dwarf

spiders (Linyphiidae) living by the thousands in the leaf litter, the funnel-weavers (Agelenidae), whose funnel-shaped webs nested in the rock crevices catch jumping insects, and the harvestmen or daddy longlegs (Phalangiidae), which live by predation and scavenging. Mites, woodlice (land crustaceans) and pill millipedes are also present.

Because the larvae of many species of moths and butterflies specialize in eating mesic and hydric plants, only a restricted cadre of lepidoptera are present in the xeric life systems. The oakworm moths, the variable oakleaf caterpillar, and certain geometrids are present to feed on the leaves of chestnut oak in the ridgetop forests, usually in the spring soon after the foliage opens. Luckily, for the trees, this is about the time the spring warblers are passing through. Some adult anglewing butterflies (the comma and the question mark) are commonly present to dab at the sap of oak and maple.

Reptiles and Amphibians

The herpetology of the Piedmont shifts with dryness away from the amphibians and in favor of the reptiles. Most amphibians require aquatic or damp conditions for their eggs and larvae and a moist environment for the adults. Life in xeric habitats is therefore difficult for all except perhaps the toads, which do frequent ridgetop forests and other dry habitats.

Dryness excludes many reptiles, too, but some thrive in xeric woods and rocky places. The northern fence swift, the broad-headed skink, and the southeastern five-lined skink hunt insects amid the rubble of ridgetop boulders. While rock-climbing at Stone Mountain and Hanging Rock in North Carolina, I have watched skinks scamper blithely across the same vertical rock faces to which I clung so tenuously hundreds of feet above the treetops. The ringneck snake is predictably present in the dry rocky woods and on the outcrops, where it preys on small snakes and lizards. The ubiquitous black rat snake pursues the rodents of the dry habitats, and the hog-nosed snake is present wherever there are toads. The timber rattlesnake may be encountered on xeric slopes and ridges. Close approach sometimes evokes a quick withdrawal into a striking coil and a conspicuous rattling display. Move very slowly if you're within five feet; retreat, give the old boy a wide berth, and press on.

Birds

Birdlife, too, is less varied than in more moist circumstances, but it is still rich; more, it is complemented by the presence of the bird of birds in the Piedmont, the common raven. It is difficult for an easterner to think of this magnificent, soaring corvid as "common," regardless of its name, for there is only a small enclave of breeding ravens east of the Mississippi, a population once thought restricted to the highest Appalachians. However, ravens can reliably be seen in the monadnocks of the western Piedmont. They shuttle between Pilot Mountain and Hanging Rock State Park in North Carolina, and I have also seen them at Stone Mountain and at Kings and Crowders mountains. My most surprising encounter with this mystic bird was at Bull Run Mountain, just west of Washington, D.C., one misty autumn morning in 1977 when a pair glided silently past at the rocky summit. After the birds disappeared into the ragged scud draped over the rocks, a brief exchange of throaty croaks confirmed the sighting. It is surely one of the notable rewards given the seeker of the Piedmont's wild places to be hailed by a passing raven while climbing a monadnock's vertical granite.

In summer the broadwinged hawk is the chief diurnal raptor of the Piedmont's hilltop forests. Here competition from the formidable red-tailed hawk is minimal and the broadwing is free to monopolize the rodents abroad in the rocky woods by day. At dusk, the horned and screech owls take over.

In spring, the northbound wood warblers coordinate their arrival with the hatching of the early broods of caterpillars in the chestnut oaks; in a morning's observation of this intense predation, one wonders if the trees would survive a spring without the warblers. Other passerines of the ridgetop forests include, in their seasons, the wood thrush, the red-breasted and white-breasted nuthatches, the brown creeper, the common crow, the black-capped and Carolina chickadees, the tufted titmouse, the blue jay, the blue-gray gnatcatcher, and the ruby-crowned and golden crowned kinglets.

The windswept ridges bear the brunt of winter's assault on the Piedmont, and in consequence the ridgetop forests sustain considerable breakage and deadwood. This is of particular interest to the woodpeckers, and every Piedmont species may be found in the xeric woods—the downy, the hairy, the red-bellied, the red-headed, the flicker, and the pileated. In this habitat, the pileated shows a behavior I have not seen it exhibit in moister forests; it drills into the living wood of chestnut oak, Spanish oak, and *Paulownia*, apparently to gain access to the beetle larvae living in the dead heartwood of these trees. The benefit to the xeric life system is undeniable; squirrels make their homes in the vacant cavities, as exposed nests of twigs and leaves might fail in the winds. Screech owls also find in these cavities the needed accommodations to roost and nest in the dry forest.

Mammals

The eastern gray squirrel is numerous in the ridgetop forests, feeding on the acorns of the chestnut oak. Due to the sparseness of the canopy in some areas, squirrel populations in xeric forests often do not reach the abundance of populations in mesic stands. A compensating factor is that squirrels in the ridgetop forests sometimes have the benefit of pileated woodpecker excavations in the living wood of chestnut oaks for use as dens (see above and page 213).

The eastern chipmunk may be the only mammal whose numbers respond favorably to the dryness of this habitat. Chipmunks appear to be most plentiful in mature rocky ridgetop forests where crevices and spaces between the rocks provide den space and shelter from predators. The presence of the chipmunks, and possibly the short-tailed shrew, attracts the long-tailed weasel, a carnivore aptly suited to hunting in the granite's crannies. The gray fox may take up residence in larger crevices.

CHAPTER SEVEN

The Hydrosere, Wet Soil Habitats

THE SUCCESSIONAL PROCESS on wet soils is called the hydrosere. In the Piedmont, most hydric habitats occur in the alluvial soils adjacent to flowing water.

The hydric habitats are the Piedmont's most intense and diverse life systems. Many of the province's wet and moist habitats have been inaccessible for development and consequently have not sustained agriculture or human occupancy. In addition to being the richest of our natural communities, hydric habitats are also our wildest. The islands in the Potomac River and the banks along its sides where it flows into Washington, D.C., for example, support primal hydric life systems probably akin to the pre-Columbian wilderness. Immense bottomland trees are adorned with ospreys, pileated woodpeckers, wood ducks, red-shouldered hawks and great blue herons in a setting where only the incessant whine of jet engines overhead reminds the canoeist that this is the twentieth century.

There is a sense of mystery about the bottomlands; they are places where primitive creatures dwell and they seem to hold a certain hostility toward humans. We have traditionally viewed the bottomlands as primeval and vaguely sinister and as niduses of disease; to a part of the human psyche these places are forbidden and forgotten. Benignly neglected, the Piedmont's hydric life systems have flourished.

Unfortunately, the qualities which have made the wet places unappealing (until recently) to largescale human enterprise have fostered the view that they are worthless. What good are wildlife or scenic resources, asks a certain logic, if they are inaccessible? The public, knowing little about the bottomlands and floodplains is hardly prepared to protest their destruction when they are flooded behind a dam or filled-in for commercial development. Today these places are being destroyed at an alarming rate.

The Piedmont's hydric communities are not simply lowlands on which live the same organisms which occupy the adjacent slopes. The hydric places are entirely distinct from drier communities in their vegetation and in their fauna. Bearing in mind that mesic systems account for the overwhelming majority of the Piedmont's surface—perhaps 90 percent or more—and that the remaining fraction is divided about equally between the true xeric and the true hydric habitats, we see that any further reduction in the bottomland acreage materially threatens the balance of the province's life systems.

The Piedmont's hydric habitats range in moistness from those soaked in standing or running water to those that intergrade with the mesic slopes. The wet places are the low places, generally the lower the wetter. Those habitats with visible water at the surface are said to be aquatic. They are separable into *lotic* (from the Latin *lotus*, 'a washing') for habitats with flowing water, and *lentic* (Latin *lentus*, 'slow') for ponds, lakes, marshes, and other habitats of standing water. We will examine the aquatic habitats, then move shoreward to follow the hydrosere in the alluvial forests and the swamp forests.

Flowing Water: The Lotic Habitats

Only a few plants are capable of living in rapidly flowing water because the current uproots all but the most tenacious. Duckweed, *Lemna perpusilla*, said to be North America's smallest flowering plant, avoids this problem by growing no roots. These tiny (1 to 3 mm) plants with rounded, obovate fronds float in dense colonies in quiet pools and eddies on the surface. Riverweed, *Podostemum ceratophyllum*, attacks the problem directly by clinging to the rocks in rapids and resisting the current with short, pliant, woody stems. Water cress, *Nasturtium offininale*, a tasty member of the turnip family (Brassicaceae), grows stems to several

60. Knotweed, *Polygonum persicaria*

feet in length from which roots grow at intervals anchoring the plant as it trails with the current. One of the knotweeds, *Polygonum persicaria,* specializes in life in the shallow rapids. With exceptionally strong roots but pliant stems this plant forms dense stands which appear grasslike from a distance, cloaking the rocky islands and gravel bars in the channel where sunlight is unobstructed. Bur reed, *Sparganium americanum,* waves its 36-inch (90-centimeter) sheathed, grasslike foliage from shoals in the flowing water. Its seeds are contained in marble-size, spiny spherical achenes. In the streams' calmer waters, water nymph (*Najas* spp.) extends its submerged, slender stems with sparse foliage into the lazy currents. Where currents are still slower, the elliptic, basal leaves of water plantain, *Alisma subcordatum,* rise toward the surface on erect petioles up to 12 inches (30 centimeters) long. All the foregoing species must be capable of completing part or all of their life cycles beneath the surface of flowing water including growth in reduced sunlight. It is not surprising that vegetative reproduction is widely employed among plants of lotic habitats, but sexual reproduction by crosspollination is necessary at intervals to retain genetic vigor.

There is abundant animal life in the Piedmont's rivers and streams, much more than could be sustained by the

61. Bur reed, *Sparganium americanum*

primary production of the comparatively sparse lotic flora. The food available to the aquatic animals is augmented by algae such as *Spirogyra* spp., whose slimy green threads lace the backwaters. The more current-resistant *Cladophora* spp. cause some of the rocks in the water to appear hairy. Certain mosses and liverworts may also be present. However, the majority of plant detritus suspended in the current and available to a stream's fauna comes not from aquatic plants but from land plants. Every tributary bears detritus picked up as water seeps and flows over the land, minute fragments of leaf litter and animal remains, any organic matter not yet entirely decomposed. Thus the entire watershed contributes to the lotic larder.

Numerous invertebrates feed on the current-borne detritus. Among the simplest are the rotifers, the one-celled protozoans, the bryozoans, and the fresh-water sponges. Such organisms are generally too small to be seen individually with the unaided eye. The latter two groups are colonial, occurring, respectively, as brainlike masses attached to vegetation in back-waters and as encrustations on the rocks in rapids.

A photograph taken shortly after the battle of Gettysburg (above) shows where Union General E. J. Farnsworth fell while leading a cavalry charge across an open field. Today the spot is cloaked in a growing mesic forest in transition to maturity (below), an example of undisturbed old-field succession in the northern Piedmont.

PLATE I

Broomsedge, then pine, takes over abandoned fields within a few years after cultivation ceases.

Deciduous successors germinate under a pine canopy. Pines and other coniferous pioneers do not replace themselves but yield dominance in a few decades to broad-leafed trees.

PLATE II

On a few north-facing slopes of diabase or other basic rock, Dutchman's breeches grow in the herb layer (above) along with the more common trout lily and toothwort (below).

A green heron fishes from water lilies in a farm pond.

A crayfish devours a small fish in a Piedmont stream glistening with particles of suspended detritus. Small black dots in the skin of the fish are the encysted larvae of fresh-water mussels.

PLATE IV

A low mound of mud and sticks on the banks of the Rocky River in North Carolina is home to a family of beavers. Tunnels in alluvial embankments are a more common shelter. Beavers build dams only on small creeks in the Piedmont.

Typical Piedmont farmland, particularly in the northern part of the province, includes pastures, cropland, woodlots and farm buildings.

A raccoon feasts on persimmons in a Piedmont hedgerow.

North America's oldest mammal, the opossum, is abundant throughout the Piedmont.

A red-tailed hawk feeds its young a morsel of meadow vole in a nest high in a loblolly pine. This is the Piedmont's largest soaring hawk and is the most numerous bird of prey over open country.

Young red-tailed hawks in the nest.

The lepidoptera are of immense importance in the flow of energy from plants to higher animals in the Piedmont. There are thousands of species. Above is an unidentified moth larva.

On the tip of a thistle leaf an arrowhead micrathena spider waits for prey to touch her web.

The Alluvial Forest

Friction against the banks and bottom slows the flowing waters, leaving the greatest velocity at the midstream surface. During times of flooding, the friction causes the waters lapping over the streams' banks to slow, eddy, and deposit some of their silt burden, largest particles first, causing natural levees of sand and pebbles to accumulate. As a result, the soil elevation immediately adjacent to the banks may be greater than that a few yards inshore. The "high ground" soils are notably less moist than those of only slightly lower elevation which may contain wet-weather sloughs and seasonal pools, and significant differences in the sere(s) may result. Elevations which are free of standing water are said at maturity to support alluvial forests. In contrast, swamp forests grow in the depressions where water stands during wet weather and where conditions generally are more moist than on the dikes. Each plant's tolerance for standing water is the main factor selecting the vegetation of these two distinct types of bottomland forest.

Although generally free of standing water, the alluvial forests are subjected to the mechanical forces of flooding when a stream overflows its banks. Tempted by the uncanopied sunlight, many streamside trees lean riverward, often to their peril, for they will have more of their trunks immersed in fast-moving water in time of flooding. Attrition among riverbank trees is high. The stand of alluvial forest adjacent to the channel therefore tends to remain at an arrested successional stage, perhaps never reaching the climax which soil conditions would permit in the absence of mechanical flood damage. Inland from the channel there are elevated terraces of fluvial deposit which offer immunity both from standing water and from high-energy flooding. Growing in rich, sand–silt soils of consistent high moistness, the alluvial forest here may realize its edaphic potential.

Succession begins in the alluvial soils very quickly upon abandonment, and in the case of pasturage the cultivated grasses and legumes may fail to compete with the wild hydrophilic herbs during cultivation. These wild annuals and perennials are usually in place at the time of abandonment and are in a position to seize dominance at the first opportunity. Prostrate mats of a low grass, *Microstegium vimineum*, and the much taller purpletop, *Tridens flavus*, are prominent hydric graminoids (grasses). Asters appear in the alluvial meadows immediately upon abandonment; they are unsuppressable even during cultivation. The daisy fleabane, *Erigeron annus*, and the goldenrod, *Solidago* spp., are familiars from the mesosere, though some of the goldenrods found in the bottomlands, such a *S. gigantea*, prefer this habitat. Ragweed, *Ambrosia trifida*, stands twice a human's height and is anchored against flooding by a deep taproot; it branches freely, spreading trident-shaped leaves 10 inches (25 centimeters) across. The tall, yellow-flowered *Coreopsis tripteris* dominates locally. Several species of *Eupatorium* are common: boneset, *E. perfoliatum*, bears white, tufted florets in broad corymbs which are inconspicuous against the smoky green, almost frosted, foliage. Another, mist flower, *E. coelestinium*, shows bold violet corollas. The most conspicuous aster of the bottomland swales and one which clearly identifies alluvial soils in the Piedmont is the purple-flowering ironweed, *Vernonia noveboracensis*. Blooming at a height of 10 feet (3 meters) in late summer, ironweed drops a tincture of fuchsia into the low meadows.

Certain species of milkweed are firmly associated with the open bottomlands. *Asclepias incarnata*, called swamp milkweed, holds its elliptical follicles (seed cases) erect, in contrast with the drooping posture of the beanlike follicles of *A. perennis*. A rich array of blue-flowered lobelias, *Lobelia* spp.—perhaps a half-dozen species—thrive in the herbaceous hydrosere. They are known by their erect racemes, each floret showing a more or less tubular corolla with distinctive upper and lower "lips" at the opening. Spotted touch-me-not, *Impatiens capensis*, bears pendant yellow hornlike corollas with a curled spur at the rear. Varying amounts of red speckle the translucent yellow. Seeds mature in spring-loaded pods which snap open when touched—a startling experience that accounts for the common name—to eject and distribute the seeds.

Sedges (family Cyperaceae) and rushes (family Juncaceae) are grass-like plants strongly associated with the bottomlands. The stems of sedges are triangular in cross–

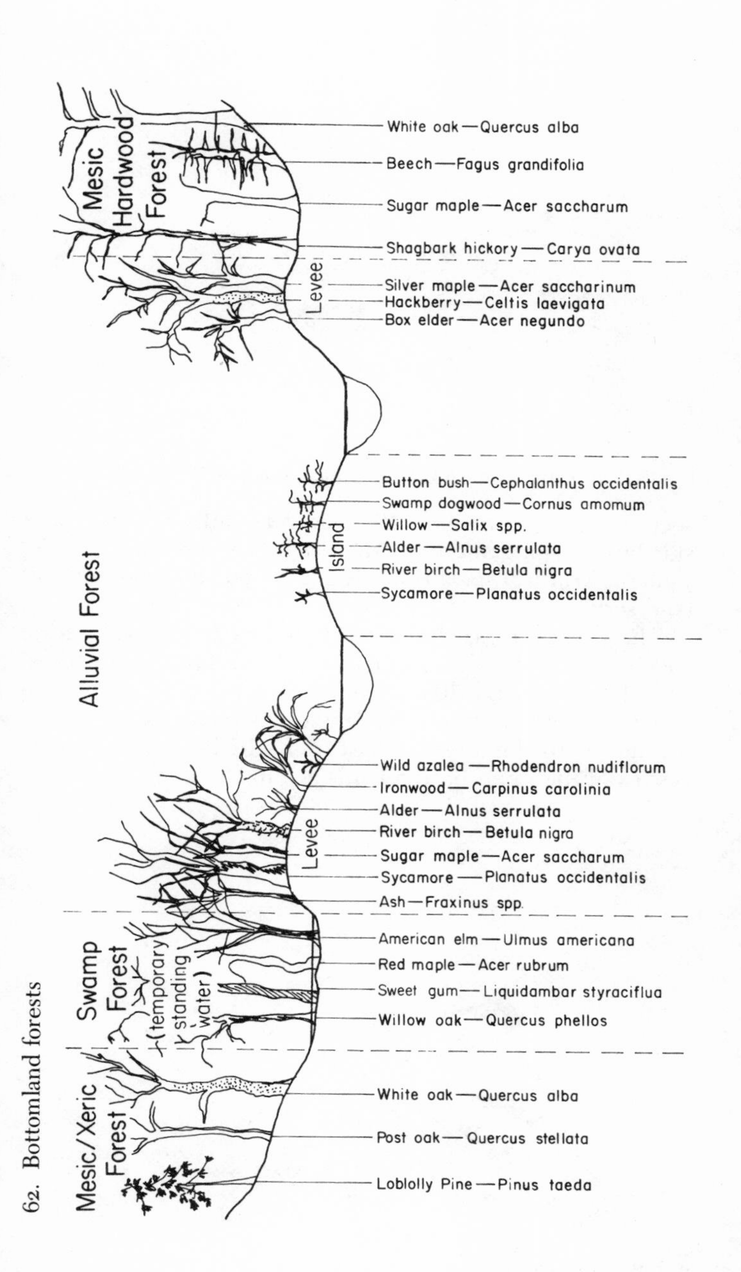

62. Bottomland forests

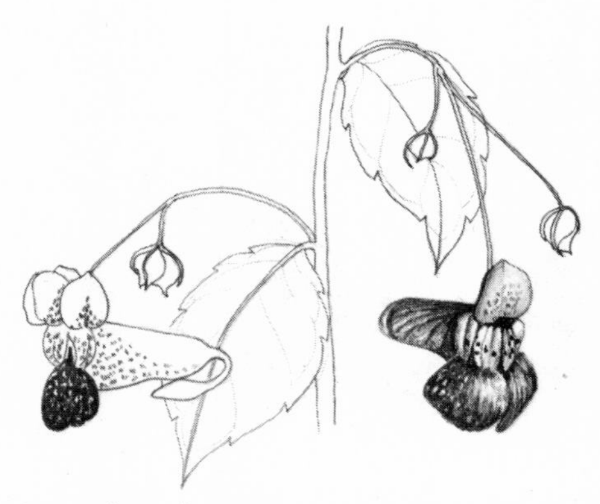

63. Spotted touch-me-not, *Impatiens capensis*

section. Many forms bear their flowers and seeds in ornate star-bust arrangements. Most species are perennials. *Cyperus erythrorhizos* is an example of a common bottomland sedge.

The rushes are divided into two genera: *Juncus,* whose species have rounded stems resembling knitting needles, and the very grass-like *Luzula,* or wood rushes, whose stems are flat. As the name suggests, wood rushes are associated with the forested part of the hydrosere, and most rushes of the open bottomlands are in the genus *Juncus.*

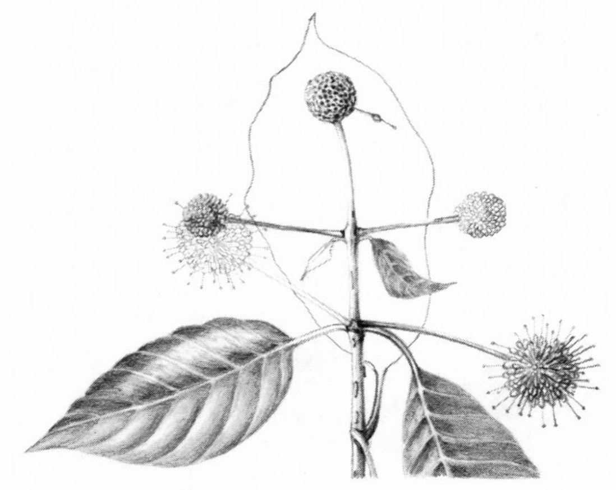

64. Buttonbush, *Cephalanthus occidentalis*

Woody succession on alluvial lands not subjected to standing water commonly begins with a shrub community of buttonbush, *Cephalanthus occidentalis,* swamp dogwood, *Cornus amomum,* and *Viburnum dentatum* (no common name). Buttonbush grows to waist height and bears its reproductive structures in a terminal sphere somewhat like the fruits of sycamore. Its leaves are broadly elliptic and opposite. The swamp dogwood and the wetlands viburnum *(V. dentatum)* are head-high shrubs easily confused due to the similar size and shape of their elliptical leaves which are arranged in opposing pairs on the twigs and because of the similar open cymose arrangement of the flowers and dark

65. Swamp dogwood, *Cornus amomum*

drupes, raisin-size but smooth. A penknife will show that the pith of the dogwood twigs is dark. The margins of the swamp dogwood leaves are smooth; those of the viburnum are dentate. Saplings of black willow, *Salix nigra,* and tag alder, *Alnus serrulata,* soon appear in the wettest parts of the bottomlands, the willow being identified by its erect twigs and narrowly elliptic 3-inch (8-centimeter) leaves and the alder by its finely serrate, ovate leaves and persistent pistillate (female) catkins resembling tiny pine cones 1 inch (2 centimeters) or less in length. The alder is a shrub, identified as such by its multiple stems, but it may achieve 15 feet (5 meters) in height. Box elder, *Acer negundo,* a tree, is

66. Black willow, *Salix nigra*

67. Tag alder, *Alnus serrulata*

a trifoliate maple whose samaras hang in dried clusters throughout winter. The alder, willow, and box elder are among the woody pioneers, and as the sere advances they overtop the shrubby buttonbush and swamp dogwood, establishing a layered community a few years into the successional sequence. Within perhaps five years the alders reach their maximum height and are passed by the box elder and willow.

68. Box elder, *Acer negundo*

69. Silver maple, *Acer saccharinum*

Maples, genus *Acer*, are important at all stages of alluvial succession. Red maple occupies the alluvial terraces, germinating among the first woody plants. From the James northward, silver maple, *Acer saccharinum*, may dominate the riversides and terraces early in the sere and, where flood damage arrests the successional advance, may persist indefinitely. Silver maple is identified by its 6-inch, (15-centimeter) deeply cleft leaves with coarsely serrate margins;

the silvery underside is easily seen in wind. Sugar maple, *Acer saccharum*, is similarly distributed. The tree is identified by the smooth margins of its 3 to 4-inch leaves. Young sugar maples may retain their dried, papery leaves through the winter. This is an upland tree in its range to the north of the Piedmont, but in our province it is largely restricted to the bottomlands, particularly south of the Potomac. Growing alluvial forests are likely to contain the red, sugar, and silver maples as well as the box elder.

70. Sugar maple, *Acer saccharum*

71. River birch, *Betula nigra*

72. Sycamore, *Platanus occidentalis*

River birch, *Betula nigra,* is among the riverside principals at midsere. The papery, peeling, light-brown bark is distinctive. This is a true birch, and it may attain considerable girth (2 feet/60 centimeters or more) while remaining beneath the stream-bank dominant, the sycamore, *Platanus occidentalis.* The bark on the upper trunks and branches of maturing sycamores is smooth and chalky white, but on the lower trunk there are large irregular brownish scales. The leaf is large, to 8 inches (20 centimeters) across and rather maple-like in shape with three to five shallow but sharply pointed lobes. The fruit is a 1-inch (25-millimeter) globe of densely packed, pappus-bearing seeds dangling on a 3-inch (75-millimeter) stem. It is comparable to the seed globes of the sweet gum but lacks the spines. The courses of the Piedmont's streams are visibly traced in winter by the zones of white-crowned sycamores lining the banks. As the river birches and sycamores mature, they are joined by an understory of hornbeam, *Carpinus caroliniana,* also called ironwood or musclewood, whose smooth, gray bark is stretched over a muscular, textured trunk to 8 inches (20 centimeters) in diameter. Tag alder, *Alnus* spp., may persist well into the woody sere at the stream banks, particularly where there has been recent flood damage. Shrubs of wild azalea, *Rhododendron nudiflorum,* are frequent on the banks. Cardinal flower, *Lobelia cardinalis,* a 24-inch (60-centimeter) unbranched herb with spire-like racemes of crimson flowers with deep, conical corollas, spotted

73. Hornbeam (ironwood), *Carpinus caroliniana*

touch-me-not, *Impatiens capensis,* and other herbs take advantage of temporary gaps in the streamside canopy. On circumneutral (pH 6.5 to 7.5) or alkaline soils, the pale touch-me-not, *Impatiens palida,* is more likely than the spotted touch-me-not.

It is at the point of dominance by sycamores and river birches, sometimes in league with the green ash, *Fraxinus pennsylvanica,* that the wrenching forces of high water commonly arrest the streamside sere. Huge sycamores may be swept away, leaving an opening to be usurped by willows and alders, thus restarting the woody sere. It is only behind the levees on the fluvial terraces that the alluvial forest reaches its seral destiny. Growing in rich, sandy soils carried from the slopes miles upstream, the river birches and sycamores grow to maturity and yield to another generation of forest giants. The green ash may arrive toward the end of the box elder–river birch–sycamore tenure or it may progress with these pioneer trees. As this canopy matures they shade the transgressives of the climax alluvial forest—red maple, sugar maple, hackberry, *Celtis lavaegata,* American elm, shagbark hickory, *Carya ovata,* water oak, *Quercus nigra,* and willow oak, *Quercus phellos.* Sweet gum and tulip tree are occasionals in the growing and mature alluvial forests. Black walnut and red mulberry, *Morus rubra*, are scattered in the streamside woods, particularly in the Southern Piedmont.

74. Hackberry, *Celtis lavaegata*

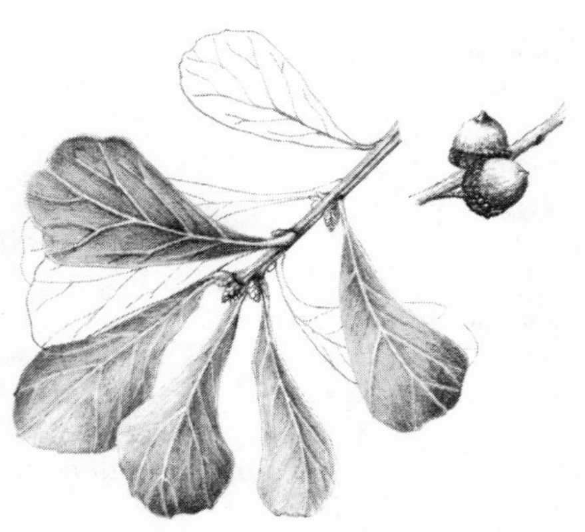

75. Water oak, *Quercus nigra*

It is possible to find mature alluvial forest tracts of great antiquity—usually small parcels under steep slopes which have made the little pockets inaccessible for farming or timbering. There the trees just mentioned may gain considerable size and may fashion beneath them a true alluvial forest community. Wild grape vines, *Vitis* spp., entwine in the tree tops and dangle from the upper branches. Vines of poison ivy, *Rhus radicans*, or possibly poison oak, *R. Toxicodendron* (the difference may be limited to location and mode of growth), cling to the trunks of the alluvial forest trees with small tentacles so thick they resemble fur. Both vines bear fruits which are important food for birds and mammals. Beneath the canopy, old, usually decadent, red mulberry, *Morus rubra*, may be scattered. American holly, *Ilex opaca*, is in many localities the understory dominant in the mature alluvial forest; its fruits, too, are edible. Beech, *Fagus grandifolia*, is common on the terraces and on the transitional slopes where the hydric and mesic regimes meet; it may achieve canopy status or may remain a subdominant. Ironwood and flowering dogwood are scattered in the alluvial forest subcanopy, the transition from *Cornus amomum* to *C. florida* having taken place while the ashes, sycamores, and river birches held the canopy. Hazel nut, *Corylus americana*, a close relative of the tag alder, is an occasional in the alluvial forest; its female flowers are only a few millimeters across yet conspicuous as rubies backlit by the low, late-winter sun. Witch hazel, *Hamamelis virginiana*, is another tall shrub of this habitat, particularly in the southern Piedmont. South of the James River it is largely restricted to the bottomlands, as is spicebush, *Lindera benzoin*, though both are common on the uplands from the James northward. Black haw, *Viburnum prunifolium*, deciduous holly, *Ilex decidua*, and strawberry bush, *Euonymous americanus*, are old friends from mesic sites who grow vigorously in the bottomlands. The naked-flower rhododendron is present. Buckeye, *Aesculus sylvatica*, is a common, sometimes dominant small tree of the alluvial forests, its palmate leaves of five elliptic leaflets and its orange–yellow flowers setting the tone of the shrub layer under streamside north-facing slopes in early spring. A wild, native hydrangea, *Hydrangea arborescens*, stands 6 feet (2 meters) tall at the edges of the bottomlands; the shreddy bark, large, opposite, ovate, coarsely serrate leaves and broad cymes of white florets show distinct kinship with the cultivated varieties.

The rich, moist alluvial soils sustain a vibrant community of forest herbs, including the Piedmont's most notable displays of spring ephemeral wildflowers. Many of the herbs grow best where the alluvial terraces are sheltered beneath steep north-facing slopes that protect against the drying effects of the sun. The ephemerals of the habitat include golden corydalis, *Corydalis flava,* wearing delicate fernlike foliage and 1/4-inch (6-millimeter) spurred yellow flowers; spring beauty, *Claytonia virginica,* with bladed grasslike leaves and floral buds nested in fiddleheads for sequential blooming, one each day during the prevernal period; and trout lily, *Erythronium americanum* (see page 201). Toward the edges of the terraces and on the adjacent lower slopes grow toothwort, *Dentaria* spp., and on selected, moist, lower north-facing slopes, Dutchman's breeches, *Dicentra culcullaria,* dangles its pendant white and yellow, pantaloonlike flowers against delicate frilly foliage, the whole plant perhaps 10 inches (25 centimeters) high. On the skirts of the very most favored slopes and adjacent fluvial terraces—there are just a few such sites in the entire Piedmont—grows the false rue anemone, *Isopyrum biternatum,* its frail, 1/2-inch (12-millimeter), three-lobed leaves resembling the foliage of sassafras in miniature.

Non-ephemeral wildflowers—those whose foliage persist after the forest canopy closes in spring—also abound in the bottomlands, some blooming with the ephemerals but most waiting until after the canopy closes. Jack-in-the-pulpit, *Arisaema triphyllum,* known also from the mesic seeps in mature forests (see page 199), and mayapple, *Podophyllum peltatum* (see page 199) favor the alluvial terraces above all other Piedmont habitats. Chickweed, *Stellaria media,* forms mats on the forest floor, each of the five petals on its wheel–like white flowers being divided into two halves, giving the effect of ten equally spaced spokes. Several species of violets may be present in the shady dampness. In the more open alluvial woods, tall asteraceous herbs find advantage. The head-high stems of *Coreposis* spp. and ragweed, *Ambrosia trifida,* may usurp the shrub layer where sunlight is adequate, particularly in the zones nearest the water, where flood damage is apt to keep the understory sparse. Wood nettle, *Laportea canadensis,* is a hazard in summer, for it can inflict a painful, stinging itch if the trichomes on the stems and petioles are touched.

Grasses are important in binding the alluvial soils at all

seral stages. Beneath the alluvial forest canopy, one is likely to find cane, *Arundinaria gigantea,* growing to 30 feet (10 meters) tall, its blade-like, linear leaves arranged in vertically aligned fans from the woody perennial stems. *Uniola latifolia,* a relative of the famed sea oats of the Atlantic coastal dunes, is a rhizomatous, colonial perennial which grows to 3 feet (1 meter) and dangles its laterally compressed seed heads, oatlike, on one-inch peduncles. The *Microstegium vimineum,* a grass with no common name, which grows in mats on the alluvial soils during the herbaceous phase, may persist indefinitely.

The low woods abound in ferns, particularly on the terraces and at the edges of the alluvial soils under north-facing slopes. The waist-high fronds of cinnamon fern, *Osmunda cinnamomea,* and the separate spore-bearing leaves lend a tropical luxuriance to the shaded seeps and stream banks in spring and summer. Alluvial woods with gaps in the canopy are more suited to the needs of royal fern, *Osmunda regalis,* known by the tiny petioles which attach the smooth-margined subleaflets to the rachis of the leaflet. The subleaflets of the cinnamon fern, in contrast, are joined near the rachis of the leaflet. The cut-leaf grape fern, *Botrychium dissectum,* produces a single, triangular (but compound and dissected), fleshy leaf about 4 inches (10 centimeters) long. The leaf is highly variable in color and shape. The sporophyll (spore-bearing leaf) is born on a single stalk about a foot (30 centimeters) tall and overtopping the leaf stalk. The horseshoe-shaped fronds of maidenhair fern, *Adiantum pedatum,* line the seeps in limestone soils. Christmas fern is common. Fancy fern, *Dryopteris intermedia*—the fern of the florist—grows in erect clusters in circumneutral alluvial soils. Where streams have cut into rock and left the stony surfaces exposed at streamside, they may be festooned with rock cap fern, *Polypodium virginianum,* or with resurrection fern, *P. polypodioides.* Both grow from creeping rhyzomes from which proliferate leaves cut once; that is, the leaves are divided into leaflets, but the leaflets are not divided into subleaflets. Resurrection fern is known for the way it shrivels during drought then freshens when water is available. Both ferns favor the acidic conditions which may be present in the microhabitats on certain rocks, regardless of the pH of the adjacent soils, due to the activities of lichens or to the chemistry of the rocks themselves. The leaves of rock cap fern may be 7 inches (18 centimeters) long; those of resurrection fern are limited to about 4 inches (10 centimeters). On the mossy rocks of a few north-facing, stream-cut outcrops, the slender, extended leaves of walking fern, *Camptosorus rhyzophyllus*, arch away from a central parent plant and touch down a foot or so away to sprout satellite plantlets. In its favored limestone microhabitats, the walking fern is generally found in association with saxifrage, *Saxifraga* spp., a flowering herb which sends its floral stem 3 to 4 inches (8 to 10 centimeters) up from a compact rosette of obovate basal leaves.

Along certain stream banks and river's edges grow the horsetails, *Equisetum* spp., probably the most primitive of the Piedmont's vascular plants. These relics of the coal age spout vertical, terete stalks, 1/2 inch (12 millimeters) or so in diameter, some species branching radially at intervals, others producing ringed, solitary stems. The genus *Equisetum*, with its 25 members worldwide, is the sole survivor of the order *Equisetales*, which with the club mosses dominated the earth soon after it was colonized by land plants.

The Swamp Forest

Standing water, rather than roaring torrents, is the main selecting force in the swamp forest. No woody plant or herb can live in the swamp forest unless it can endure sustained periods when its roots and lower trunk are immersed in water. Whereas the distinction in elevation between those parts of the bottomlands that sustain swamp forest and those occupied by alluvial forests may appear slight, the number of sequential days during which the two types are likely to be immersed varies considerably. Rivers rise out of their channels only after heavy rains, and after a few days—rarely as much as a week in the Piedmont—subside to normal flow. Trees on the levees and terraces experience flooding for two or three days, perhaps, while those growing in the sloughs, ditches, and depressions may remain soaked for weeks.

Each species of tree has a more or less consistent limit to its tolerance to flooding. The tulip tree, for example, can endure only thirty days of immersion and is therefore not a regular component of the swamp forest. Sweet gum, on the other hand, is a major presence in the swamp forest because it can endure up to eighty days of innundation during the growing season. Most species common in the swamp forests are capable of surviving fifty days or more of continuous flooding.

The swamp forests of the Piedmont constitute our closest approximations of true wilderness. Because the soils are too wet to farm, and often too wet for profitable timbering, some of the swampy tracts are largely intact. Here may be found the wildest, richest, and most diverse habitats available to the Piedmont naturalist. Most are inaccessible by foot during the wetter seasons—winter and spring—but they may be explored by canoe at any time of the year. The chances are good that each of us lives within striking range of one of these wet forests. If they are not visible where the roads we commonly travel cross rivers and streams, the chances are good that a canoe trip downstream between two road crossings will yield a visit to a swamp forest.

Conditions favorable to the development of swamp forests are found in the floodplains of streams of relatively flat gradients and shallow entrenchment. Where intrusions of nonresistant rock, such as gabbro, have weathered beneath the surface and shrunk in volume, broad linear depressions at the surface may result, and these depressions become the avenues of egress for waters leaving the Piedmont en route to the Atlantic. Since they are comparatively flat, the waters flow slowly and do not entrench themselves deeply, and the banks are easily overflowed in times of heavy rainfall. Much or all of the depressed area may remain under standing or sluggishly flowing water during the wet season, creating, in effect, a river perhaps hundreds of yards wide. Being poorly drained, these floodplains retain pockets of water even during dry weather.

Because the soils that are likely to sustain swamp forests are unlikely to be farmed, such areas are infrequently observed in a purely herbaceous phase. Herbaceous dominance may occur where a swamp forest was long ago cleared, drained by ditching, then put into pasture, subsequently to be abandoned with the eventual filling or clogging of the drainage ditches. In one such instance I am aware of in the eastern Piedmont of North Carolina, beavers have dammed a network of old drainage ditches, effectively restoring the water level to that of a swamp forest, which, I predict, the tract will become in a few decades.

Rushes and sedges will be more prominent in the herbaceous phase of swamp forest soils than in soils which produce alluvial forests. Only the most hydrophytic asters, *Coreopsis* spp. and *Ambrosia trifida,* and ironweed, *Vernonia noveboracensis,* are present. The spotted touch-me-not, *Impatiens capensis,* thrives except where the alluvial soils are rich in transported particles of limestone, in which case the pale touch-me-not, *I. pallida,* is more likely.

Woody succession is largely similar to that on alluvial forest soils in the initial phases. Buttonbush, *Cephalanthus occidentalis,* swamp dogwood, *Cornus amomum,* and *Viburnum dentatum* rise with the tall herbs and are quickly overtopped by black willow, *Salix nigra,* and tag alder, *Alnus serrulata.* Saplings of red maple, sugar maple, and box elder—*Acer rubrum, A. saccharum,* and *A. negundo*—are also among the woody pioneers, along with sweet gum, *Liquidambar styraciflua.* Zonation is commonly evident between areas occupied by alder, willow, and river birch, suggesting that they follow one another in that sequence. Such a sequence also suggests a gradual buildup of soils and a slight rise in elevation with seral progress, with later arrivals taking advantage of progressively drier (though still very wet) circumstances. In the sequence just mentioned, sycamore is the likely inheritor of a canopy once dominated by river birch.

The canopy of the growing swamp forest may be indistinct from that of the alluvial forest at the same stage—sweet gum, sycamore, box elder, river birch (usually decadent), the maples, green ash. In the understory, however, a distinctive group of transgressives outlines the forest community to come. Willow oak and water oak, *Quercus phellos* and *Q. nigra,* are present as transgressives in both habitats. Other quercine transgressives give the swamp forest its character. The overcup oak, *Q. lyrata,* for example, grows exclusively in the swamp forests. It is named for the large, flattened acorns that are enclosed, except for the

76. Overcup oak, *Quercus lyrata*

very tip, in a shaggy cup. The leaves are divided by wide sinuses into irregular lobes rounded at the tip but drawn nearly to a point. The leaves of swamp chestnut oak, *Q. michauxii*, are distinguishable from those of the upland namesake by being slightly broader relative to the length and by having fewer lobes. The acorns are oblong, up to 1 1/2 inches (4 centimeters) long, and the cup, covering roughly one-third of the acorn, is covered with spiny scales in contrast with the irregular, fused scales of the chestnut oak, *Q. prinus*. In all, the swamp chestnut oak is more readily identified by its habitat than its morphology—it is a

77. Swamp chestnut oak, *Quercus michauxii*

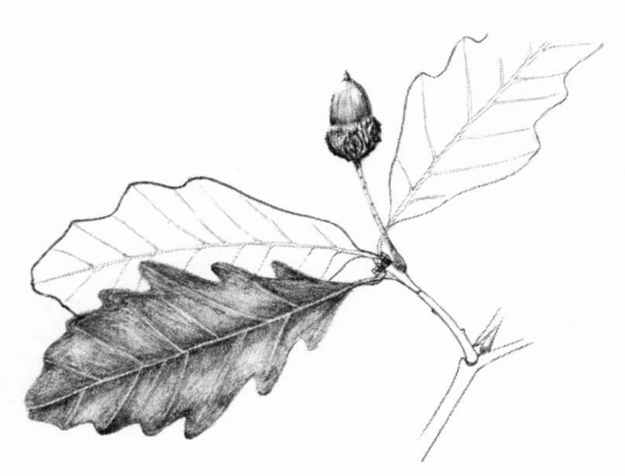

78. Swamp white oak, *Quercus bicolor*

predictable component of the swamp forests south of the Potomac. The swamp white oak, *Q. bicolor,* is scattered in the swamp forests throughout the Piedmont, with more frequent occurrences in the northern half of the province. Its leaf is suggestive of that of the swamp chestnut oak but is smaller (to 6 inches/15 centimeters) and has larger and fewer lobes—perhaps a half-dozen on each margin. The one-inch acorns occur in pairs. The pin oak, *Q. palustris,* is similarly distributed, and in the northern Piedmont it is likely to occur in the mesosere as well as in the bottomlands (see page 265). The shingle oak, *Q. imbricata,* is common in

79. Pin oak, *Quercus palustris*

80. Shingle oak, *Quercus imbricata*

the swamp forests north of the James. It is known by elliptic leaves to 6 inches (15 centimeters) long, waxy and dark green above, pale and tomentose beneath. The acorns are rather small (1/2 inch/12 millimeters across) and squatty.

In the swamp forests of the southern Piedmont grows a variety of the Spanish oak, *Q. falcata pagodaefolia,* commonly called the swamp Spanish oak, which differs from the typical species in having a pointed base and more regular lobing to the leaves. As with the swamp chestnut oak, the swamp Spanish oak is more readily told from its upland counterpart by habitat than by shape. At scattered locations, mostly south of the James, one may encounter the Shumard or swamp red oak, *Q. shumardii.* Its leaves are deeply cleft and the lobes are regular, symmetrical and pointed—generally not unlike those of the scarlet oak (see page 159) but broader, shorter, and with fewer lobes.

When these oaks mature, they claim the canopy more or less indefinitely, considering the rather plastic nature of the bottomland soils and the likelihood that man-made disruptions to the soils upstream could result in rapid changes in the rate of siltation and soil buildup. It is likely that in the mature swamp forest the hydrophytic oaks will share the canopy with sweet gum, American elm, and hackberry (in the north, *Celtis occidentalis,* in the south, *C. laevigata*). Additionally, there are hydrophytic hickories of importance in the habitat. The shagbark, *Carya ovata,* looms like a

81. Swamp red oak, *Quercus shumardii*

Pleistocene beast in the morning mists. The thick husks split into four segments revealing a ridged nut 1 1/2 inches (4 centimeters) in diameter. The shagbark is distinguished from the bitternut hickory, *C. cordiformis,* by having five leaflets on its odd-pinnately compound leaves as opposed to the bitternut's eleven. The bitternut's bark is either smooth or shallowly furrowed. On all hickories the twigs and leaves are alternate in arrangement, and this provides a key distinction between the smoother-barked hickories, such as the bitternut and the mockernut, and the ashes, the leaves and twigs of which are opposite.

The mature swamp forest canopies in the northern part of the Piedmont are likely to include silver maple, red maple, and sugar maple; those of the south are likely to contain only the latter two. Box elder may be found in selected canopies, particularly in the vicinity of openings, but it is more often confined to the subcanopy. The eastern cottonwood, *Populus deltoides,* is an occasional in the mature hydric canopies throughout the Piedmont, though it occurs more frequently along northern streams. It is a large tree with leaves roughly triangular, twenty or so teeth on each side, whose seeds drift through the spring air on cottony pappuses. Unlike the yellow poplar, or tulip tree, *P. deltoides* is a true poplar. In the low woods as far south as the Yadkin River in North Carolina, the pin oak, *Quercus palustris,* joins the canopy, its lower branches tangled and

82. Bitternut hickory, *Carya cordiformis*

drooping. It is a strong presence in the hydric forests of the northern Piedmont, occupying positions of dominance along the Millstone and Raritan Rivers in New Jersey, for example. There are regional substitutions along the length of the province, but the similarities seem to override the peculiarities. The constants of the bottomland forests include the green ash, river birch, sycamore, red maple, sweet gum, black willow, box elder, the hydrophytic oaks, and the shagbark hickory.

83. Pawpaw, *Asimina triloba*

84. Buckeye, *Aesculus sylvatica*

The shrubs beneath the swamp forest canopy are, by and large, peculiar to that habitat. This shrub layer is the richest in the Piedmont. It is only in the swamp forest shrub community that we find the pawpaw, *Asimina triloba,* the hazel nut, *Corylus americana,* the painted buckeye, *Aesculus sylvatica,* the bladdernut, *Staphylea trifolia,* the parsley hawthorn, *Crataegus marshallii,* and the winterberry, *Ilex verticillata.* The community also includes familiars from the mesic realm such as possum haw, *Viburnum prunifloium,*

85. Parsley hawthorn, *Crataegus marshallii*

86. Winterberry, *Ilex verticillata*

which bears a striking resemblance to the winterberry and to that shrub's upland and swamp-forest congener, the deciduous holly, *Ilex decidua;* all have short, sharpened twigs and smooth, light-gray bark. They are distinguished by the hollys' alternate and the viburnums' opposite arrangement of leaves and twigs. The wild azalea, *Rhododendron nudiflorum,* abounds under the swamp forests. Spicebush, *Lindera benzoin,* and witch hazel, *Hamamelis virginiana,* are present in the alluvial soils of circumneutral or basic pH. The chemistry of alluvial soils is a function of upstream geology and not of the rocks immediately beneath them, so it is possible to find these and other shrubs of calcareous (basic) affinities on the bottomlands in close juxtaposition with granite slopes clad in acid-loving plants of the family Ericaceae such as mountain laurel, *Kalmia latifolia.* Also characteristic of the mature bottomland forests (though thinly distributed and perhaps uncommon or rare) are leatherwood, *Dirca palustris,* with its smooth-margined elliptic to obovate, alternate, irregular-size leaves and blood-red ellipsoid drupes, and the richly flowered storax, *Styrax grandifolia,* known by its thin, pale-green, smooth-margined obovate leaves to 7 inches (18 centimeters) long and its profuse cymes of white-petalled florets, each one an inch (25 millimeters) across.

The herb communities under the mature swamp forest canopies are somewhat less colorful. The spring beauty, *Claytonia* spp., and mayapple, *Podophyllum peltatum*, are consistently present and may be considered chief among those forest herbs which can withstand both shading and prolonged flooding. The true and false Solomon's seals, *Polygonatum pubescens* and *Smilacina racemosa*, may endure, and in some cases thrive, in these conditions, but they are generally found in association with spicebush, signifying a requirement for nonacid soils. Jack-in-the-pulpit is similarly distributed. A type of toothwort, *Cardemine bulbosa*, is scattered in certain very wet woodlands. One herb closely identified with the habitat in the Piedmont (north of the Roanoke River) is the skunk cabbage, *Symplocarpus foetidus*, a perennial whose purple spathe precedes the green, cabbagelike foliage in very early spring. Numerous species of mints (family Lamiaceae) asters, and buttercups (family Ranunculaceae) may be locally abundant. The sloughs, some wet on a semipermanent basis, are likely to contain such aquatics as arrowhead, *Sagittaria latifolia*, the leaves of which are aptly described by the name, and golden club, *Orontium aquaticum*, with its elliptic, parallel-veined, erect leaves and yellow spadixes. The touch-me-nots, *Impatiens* spp., and cardinal flower, *Lobelia cardinalis*, may take advantage of gaps in the canopy.

The Bottomlands, State by State

NEW JERSEY

In the alluvial forests of the northern anchor of the Piedmont, silver maple is commonly substituted for sycamore as the streamside dominant. Pin oak dominates landward, especially where the bottomlands are of open character owing to frequent flooding. Green ash is prominent. For example, if you scout the low woods in the (unnamed) state park north of State Road 514 where it crosses the Millstone River (between the river and the canal tow path) you'll find an ash over 4 feet (120 centimeters) dbh a quarter mile or so upstream of the bridge. The same woods, and others like it in the New Jersey Piedmont, contains shagbarks to 3 feet (90 centimeters) dbh, and red maple, honey locust, and black walnut to 2 feet (60 centimeters) or more. The swamp white oak is a dominant at seral maturity in some of the state's wetter woodlands. Wild grape, poison ivy, and honeysuckle interweave the canopies.

Examples of earlier phases of the bottomland hydrosere are common in New Jersey's Triassic lowlands, the sandstone-based depressions which account for much of the Piedmont in New Jersey east of the Hunterdon Plateau (i.e., western Hunterdon County) and outside the highlands enclosed by the Watchung Mountains in Somerset County. The bottomlands along the Lamington River—where it is crossed by Interstate 87, for example—and along Interstate 287 in the vicinity of Basking Ridge and New Vernon exhibit early woody succession dominated by red maple overtopping swamp dogwood, alder, and *Viburnum dentatum*. Rich displays of ferns are common here. They are likely to include marsh fern, *Thelypteris palustris*, with its 18-inch (45-centimeter) fronds bearing twelve or more leaflets cut not quite to the midvein and untoothed, the third pair of leaflets from the bottom usually being the longest; and sensitive fern, *Onoclea sensibilis*, whose 24-inch (60-centimeter) leathery leaves are cut once to leaflets with wavy margins and netted veins. Cinnamon fern, *Osmunda cinnamomea* (see page 258), is plentiful. In some abandoned wet meadows the silver maple is the initial woody dominant; it may retain its primacy late into the sere, particularly near stream edges where flood damage is likely.

PENNSYLVANIA

In separating Pennsylvania from New Jersey, the Delaware River cuts across the long axis of the Piedmont and lays bare the region's geologic bones. One is hard-pressed to name a part of the Piedmont whose natural beauty and gentle human stewardship yield greater scenic wealth. The picturesque drive along Route 263 from New Hope to Upper Black Eddy offers splendid insights into the natural and cultural histories of the northern Piedmont and draws the traveler into the beauty that binds the people of the Piedmont to their province. It should not be missed.

The magnificent Delaware is deeply entrenched—in places 30 feet (9 meters) or more—into its own alluvium, and as a result its floodplain is quite well drained. In some sections nearly a mile wide, the Delaware's Pennsylvania bottomlands are ideal croplands and they have been farmed for almost three centuries. Abandonment is rare, and early secondary succession in this part of the Piedmont is not easily observed. It is imputed, however, from the growth along the immediate banks where the powerful Delaware must thin the canopy at frequent intervals. Here sycamore and green ash dominate in company with silver maple and, reflecting the near-mesic character of the drainage, red oak. Black walnut and honey locust are prominent over red mulberry. Within the throw of minor floods, periodic seral reversals from mechanical damage are expressed in the presence of black willow.

Smaller watercourses with wetter bottomlands, such as Octoraro and Brandywine creeks in Chester County, exhibit the expected transition of ash, sycamore, black walnut, and box elder overtaking decadent willows.

In French Creek State Park near Hopewell Village there are wet woodlands more akin to swamp forests. The mature canopy is shared by red maple and swamp Spanish oak over a rich shrub stratum of hazel-nut, pawpaw, spicebush, and wild azalea. Herbs include jack-in-the-pulpit, windflower, fancy fern, and a mint, *Lycopus virgincus*, very common in this wet forest habitat throughout the Piedmont.

MARYLAND

The bottomlands along the Monocacy River, a tributary of the Potomac, in the vicinity of Frederick support mature alluvial forests representative of this habitat in the Maryland Piedmont. Crossing the river on Route 26 near Ceresville, there is an unpaved rural road that follows the river downstream on the east bank into an open bottomland forest dominated largely by silver maple. Where local terrain offers protection from flood damage there are large sycamores, American elms, and hackberries, as well as walnut. Box elder is important here. The appearance of the elm and hackberry in significant numbers differentiates the Maryland bottomlands from those of the extreme northern Piedmont. Their proliferation in the vast central and southern segments of the Piedmont indicates seral maturity. The pin oak, dominant in kindred bottomlands in New Jersey, is only an occasional here. This silver maple–sycamore–hackberry–American elm forest characterizes the Potomac's drainage along the tributaries and on the banks of the Potomac itself as it bisects the Piedmont at its narrowest neck. A bird's eye look at this forest's crown is available from the bridge across the Potomac at Point of Rocks, the only crossing of this great river in the Piedmont upstream of the system of bridges in the vicinity of Washington (except for White's Ferry near Poolesville without question the most picturesque crossing of any of the Piedmont's major rivers).

To see the stunning riverside forests of the north-central Piedmont, journey by foot, bicycle, horseback, or canoe along the historic C&O canal on the Maryland bank of the Potomac. The canal's towpath is maintained as a hiking trail and the public is welcome. Much of this forest, and particularly that in the Great Falls National Park upstream from Washington, is at its edaphic maturity and features the silver maple–American elm dominance interspersed with sycamore, black walnut, tulip tree, and occasional eastern cottonwood and pin oak. It is interesting that the green ash occurring in these mature tracts tend to be notably smaller (about 12 inches/30 centimeters dbh) than the dominants, suggesting that they are opportunists filling in canopy gaps in an otherwise mature forest. The subcanopy is occupied by box elder, red maple, red mulberry, and musclewood. Pawpaw, spicebush, *Viburnum prunifolium* and sweetshrub, *Calycanthus floridanus,* with its deliciously scented flowers are prominent in the shrub layer. Spring herbs include robust jack-in-the-pulpit, pale touch-me-not (signifying nonacid soil), blue phlox, *Phlox divaricata* (identified by its light lavendar corollas), wood nettle, *Laportea canadensis,* and garlic mustard, *Alliaria petiolata,* with crenate-margined, heart-shaped leaves 2 inches (5 centimeters) across.

The role of the black walnut on alluvial soils of circumneutral to basic pH cannot be overemphasized. Early in the woody sere, the species may dominate where soil porosities and pH are suitable, overtopping willows, alders, *Viburnum dentatum,* and red maple to establish apical primacy (dominance) which may last until edaphic maturity is approached.

The Japanese honeysuckle, wild grapes, and poison ivy common in the bottomland canopies further to the north are joined in Maryland by cross vine, *Anisostichus capreolata,* a pencil-thin (or thinner) woody vine that reaches into the subcanopy and is identified by its paired leaflets separated by a terminal tendril. The leaves are actually pinnately compound, arising at opposite positions on the vine. They are semievergreen, glabrous (hairless and somewhat shiny), elliptical, and smooth-margined.

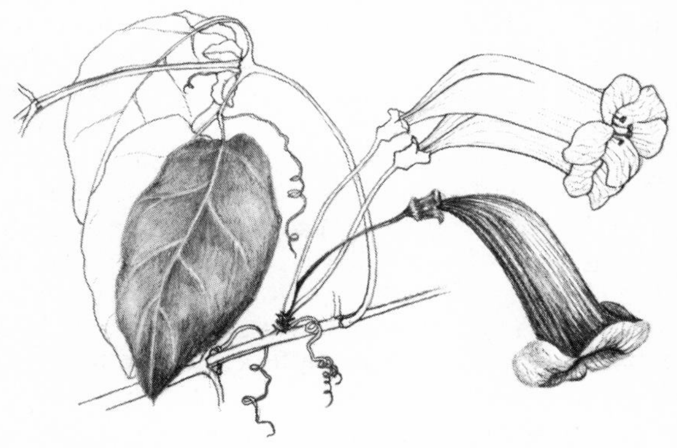

87. Cross vine, *Anisostichus capreolata*

VIRGINIA

Important vegetative transitions occur in the bottomlands of Virginia's Piedmont, particularly among the hydric oaks. The pin oak is present but not frequent. The swamp white oak, dominant in the swamp forests of the northern Piedmont, occurs only in scattered localities south of the Potomac. Instead, the swamp chestnut oak penetrates from the south into the hydric forests of Virginia's outer Piedmont. The willow oak is prominent on alluvial soils, arriving early in the woody hydrosere and remaining to seral maturity. The Shumard oak is principally a tree of the southern bottomlands but it straddles the Potomac in a disjunct range covering the northern third of Virginia's Piedmont. Red oak is found along some streams in the inner Piedmont.

It is also in Virginia that the silver maple yields its early woody dominance and stream-bank eminence to red maple, a transition which is completed between the Potomac and the James. Bottomland pastures that in New Jersey are invaded by saplings of silver maple as woody pioneers would in Virginia experience red maple or sycamore. Moreover, the immediate streamsides in Virginia are knit together against the forces of rushing water much more often by sycamore than by the silver maples characteristic of more northerly streams. Poison ivy, cross vine, wild grape, and Japanese honeysuckle are prominent vines.

Accotink Creek Bottomland Woods in the Virginia suburbs of Washington, D.C. (see page 379) handsomely demonstrates these transitions in bottomland vegetation. Here the canopy in the swamp forest sections is shared by the northerly pin and swamp white oaks and by the Shumard, swamp chestnut, and willow oaks of southerly affinities. The Accotink forest is also representative of the region's alluvial forests; its better-drained sections are populated by bitternut hickory, green ash, sweet gum, American and winged elms, box elder, and red maple. The stream banks are lined with sycamore, silver maple, eastern cottonwood, and river birch, with ironwood in the understory.

Herbaceous plants of the bottomlands in Virginia include jack-in-the-pulpit, spring beauty, mist flower, cardinal flower, touch-me-not, wood nettle, and bottle brush grass, *Hystrix patula*, to name a few. Skunk cabbage extends into a few of North Carolina's northern Piedmont counties, but effectively finds its southern limits in the alluvial soils of Virginia.

NORTH CAROLINA

The southward-moving naturalist finds that North Carolina's alluvial soils support the northernmost of the truly southern swamp forests. Several isolated ranges of the northerly swamp white oak add diversity in central North Carolina, but the species is one of the last links with the bottomlands of the northern Piedmont. Enter the overcup oak, the swamp Spanish oak, and the water oak, all characteristic of the southern Piedmont's mature hydric forests. They share the canopy with swamp chestnut oak, the Shumard oak, and the willow oak, which are also found scattered in the bottomlands of Virginia's Piedmont. Silver maple is uncommon to rare, its few occurrences south of the Roanoke being in the western part of the province. One must remember that the Piedmont widens from 25 miles (40 kilometers) where it is cut by the Potomac to nearly 150 miles (240 kilometers) in some parts of North Carolina, so the state, with its considerable range of Piedmont elevations, shows substantive variations in its vegetation in all moisture regimes across the province. This is another way of saying that the inner Piedmont of North Carolina may resemble, in some aspects, the northern parts of the province.

Sycamore is North Carolina's chief riverside dominant where it towers over river birch, box elder, red maple, and ironwood. Behind the levees grow green ash, sweet gum, tulip tree, bitternut hickory, hackberry, and occasional loblolly pine over American holly, deciduous holly, and winterberry. The latter are all in the genus *Ilex* and show preference for decreasing soil acidity in the order presented. Parsley hawthorn is prominent in the shrub layer under mature canopies. Spicebush and witch hazel signify circumneutral to basic soils. Pawpaw, bladdernut, *Staphylea trifolia,* and storax, are scattered in North Carolina's mature moist forests.

The early phases of the hydrosere feature the pan-Piedmont sequence of woody shrubs—buttonbush, swamp dogwood, black haw, *Viburnum dentatum,* and tag alder, *Alnus serrulata*—yielding to red maple, black willow, box elder, and river birch. Moisture zonations in the North Carolina bottomlands early in the woody sere commonly show bands of alder, willow, and river birch, suggesting that the latter have succeeded the former.

The prominence of lianas in these bottomland forests can astonish. Tall oaks and hackberries festooned with wild grape vines dangling free of any attachment save tendrils interlaced in the canopy perhaps 100 feet (30 meters) overhead always raise the engineering question of how the vines made the leap to the treetops. One suggestion that comes to mind is that they grew with the forest, in which case some of the grape vines may be of great age. A vine 6 inches (15 centimeters) in diameter and hundreds of feet in length may be presumed venerable. The furry stems of poison ivy, which may also achieve a six-inch diameter, do not hang free but cling to the trunks with numberless sucker-footed tendrils. Both are enormously productive of drupe-like fruit eaten by birds and animals.

SOUTH CAROLINA

Basswood, *Tilia heterophylla*, generally associated in the northern Piedmont with rich, sheltered slopes, is scattered in the bottomlands of South Carolina, particularly toward the inner Piedmont. Shumard, willow, water, swamp chestnut, and swamp Spanish oaks are its companions in the mature hydric canopies, together with shagbark and bitternut hickories, hackberry, American elm, tulip tree, and sweet gum. On the stream banks sycamores guard river birch, black walnut, green ash, and box elder. Similar associations mark seral disruption behind the levees and in other sheltered alluvial soils. Red maple may be present in any part of a bottomland, but it might not reach the canopy. It may share the subdominant stratum with walnut, box elder, winged elm, ironwood, honey locust, and occasional catalpa. Sassafras, growing tall and straight and perhaps even reaching the canopy on younger sites (in contrast to its remaining a twisted resident of the understory or shrub layer on mesic sites), is a major presence on some alluvial soils in the Carolinas. Flowering dogwood is also common in the hydric understory. Umbrella tree, *Magnolia tripetala*, is an understory occasional in the inner Piedmont of the Carolinas; its 18-inch (45-centimeter), obovate, smooth-margined, deciduous leaves are clustered at the twig tips.

Rich shrub layers diversify the southern swamp forests. Mature canopies over alluvial soils not subject to frequent flood damage in South Carolina are likely to support spicebush, pawpaw, painted buckeye, bladdernut, parsley hawthorn, black haw, deciduous holly, and winterberry. Cane, although actually a grass, is a tall (to 30 feet/10 meters) woody perennial more likely to be thought of as a shrub than as an herb. It is common on open, flood-swept bottomlands in the Carolinas together with the pan-Piedmont herb of the same habitat, *Ambrosia trifida*. Strawberry bush, *Euonymus americanus*, is another shrub–herb of these bottoms. Plants unmistakably herbaceous are Solomon's seal, pipsissewa, spring beauty, touch-me-not, and jack-in-the-pulpit. Golden club, *Orontium aquaticum*, and duck potato, *Sagittaria latifolia*, are common in vernal pools and sloughs in the swamp forests. Ferns include grape fern, *Botrychium* spp., Christmas fern, maidenhair fern, and cinnamon fern.

Lianas become increasingly important in the alluvial vegetation in the southern Piedmont. In the early woody phases, virgin's bower, *Clematis virginiana*, may lace into the branches of willow and alder along with Japanese honeysuckle. Cross vine, trumpet creeper, *Campsis radicans* (a relative of both cross vine and catalpa tree, and known more from mesic sites than from hydric), wild grape, and poison ivy ascend the growing and mature forests.

GEORGIA

Willow oak, swamp chestnut oak, swamp Spanish oak, water oak, overcup oak and Shumard oak are the quercine dominants of the mature swamp forests of Georgia's Piedmont. We must bear in mind that South Carolina is the last of the Piedmont states which are stacked north-to-south; the Piedmont in Georgia and Alabama are therefore more to the west than to the south of South Carolina, and they are not necessarily more austral in vegetative aspect. However, distinctions exist. The Oglethorpe oak, *Q. oglethorpensis*, a medium-sized tree endemic to the Piedmont of South Carolina and Georgia, reaches its southern limit in the Monticello Bottomland Woods in Jasper County, Georgia. It is identified by its light-gray, scaly bark and 6-inch (15-centimeter) elliptical leaves with entire margins. American elm, shagbark hickory, bitternut hickory, southern sugar maple, *Acer saccharum floridanum*, sweet gum and loblolly pine are to be found in the hydric forests of this state, often over a subcanopy of winged elm, box elder, ironwood, and red mulberry. Hackberry, black walnut, and tulip tree are common. Sycamore, river birch, and green ash occupy the riverbanks and other sites characterized by mechanical seral disruption. Basswood is an occasional.

As we would anticipate, the shrub layers of the Georgia Piedmont's swamp forests are particularly exciting. At Pumpkinvince Creek, northwest of Atlanta, for example, leatherwood, *Dirca palustris*, spicebush, and pawpaw grow in zonated profusion. Black haw and *Viburnum nudum* (no common name) are locally important, and parsley hawthorn is widespread. Painted buckeye, *Aesculus sylvatica*, deciduous holly, and winterberry are present in significant growths. Coastal Plain influence places dwarf palmetto, *Sabal palmetto*, in the shrub layer of some swamp forests. Rattan vine, *Berchemia scandens* (also a Coastal Plain invader), cross vine, wild grape, and poison ivy are important lianas.

Several anomalies in Georgia's bottomland habitats deserve mention, some to be examined more closely in Part IV, Special Places. The Alcovy River swamp forests near Covington, 70 miles (110 kilometers) or more into the Piedmont, contain stands of coastal bottomland hardwoods dominated by water tupelo and swamp red maple with buttressed bases. The banks and slopes of Cooler Branch (see page 369), west of Manchester, combine such decidedly montane assemblages as chestnut oak, mountain laurel, and azalea with Coastal Plain communities of water tupelo, sweet bay, and *Vaccinium atrococcum.* Panther Creek (see page 371), in Stevens County, provides an exceedingly rich example of a mountain cove hardwood community. The bottomland forests are the most luxuriant and diverse of the Piedmont's major habitats, and they are nowhere more worthy of investigation than in Georgia.

ALABAMA

The hydric forests of the lower end of the Piedmont are populated by the same wetland oaks that control the habitat south of the Roanoke River—the willow oak, water oak, Shumard oak, overcup oak, swamp Spanish oak, and swamp chestnut oak. The laurel oak, *Quercus laurifolia*, is principally a tree of Coastal Plain affinities but is also common in Alabama's outer Piedmont. Live oak, *Q. virginiana*, is decidedly a Coastal Plain species, and occurrences in the Piedmont are probably the result of plantings by humans as an ornamental. Non-quercines wetland dominants include sweet gum, red maple, loblolly pine, hackberry, American elm, and tulip tree. The understory features box elder, ironwood, and alder. Streamsides are lined with sycamore over, or with, river birch, green ash, and box elder. Indeed, the very common early woody succession association of tag alder yielding to black willow, followed by river birch and sycamore, is as prevalent in Alabama as in any part of the Piedmont.

In the outer and central parts of the province in Georgia and Alabama, the vegetation seems to be somewhat more susceptible to influence from the Coastal Plain than are sections of the province to the north. Fingers or pockets of Coastal Plain vegetation may be found well inland of the fall zone in Alabama. Plants from the two provinces mix at many sites. A Piedmont site in Lee County, which I was shown only after swearing to guard the secret of its exact location, contains a species of palm (needle palm, *Rhapidophyllum hystrix*), touching our province not only with Coastal Plain influence but with a taste of the semitropical.

Animals of the Lotic Habitats

Mollusks

The mussels, or freshwater clams, spend their larval stage, called *glochidia*, embedded in the skin and fins of fish. The cysts in which the glochidia enclose themselves erupt and the mussels sink to the stream bottom as miniature adults. There they burrow into the sand to filter plant and animal detritus from the water by taking in water through a syphon, passing it over mucous-coated gills, and expelling the filtered water, along with their own body wastes, through another opening. The parasitism of the glochidia serves to distribute the mussels upstream, sometimes into surprisingly small headwaters. Abrasion sometimes peels away the dark outer covering to the shells of the adults, revealing a pearly white subsurface. Such shells are commonly seen along the banks and sandbars of cleaner streams in the Piedmont, discarded by raccoons, mink, muskrats, otters, and turtles.

Eastern North America is the world's richest center for freshwater mussels. Many are endemics, localized to specific rivers and drainage systems. Although there are many species in the Piedmont's water courses—perhaps one hundred or more—each may tolerate water within only a narrow range of physical and chemical properties. Those found in flowing water cannot exist in lakes, for the damming of a river causes the bottom—once rocky and scoured by well-aerated currents—to be smothered in sediments. The decomposition of organic matter in the bottom sediments which settle out of the stilled waters behind an impoundment depletes the water's dissolved oxygen and replaces it with carbon dioxide, simultaneously lowering the pH (raising the acidity). Obstructing the flow with dams separates the glochidia from their host fish, a critical deprivation because each unionid—all freshwater bivalve mollusks belong to the family Unionidae—parasitizes a single species of fish. An additional threat, the introduction of the Chinese clam, *Corbicula malinensis*, into the Atlantic drainage, may eliminate Piedmont unionids; it is suspected to have done so in the lower Mississippi drainage.

Identification of unionids is a matter for experts. The genera *Elliptio*, *Anodonta*, and *Alasmidonta* are well represented in the Piedmont's streams. The naiads (the term means "fresh-water bivalve" as well as "the aquatic larvae of certain orders of insects") are quite important in the Piedmont's benthic biota. As filter feeders, they help clean the water and fix the filtered material in the bottom sediment where it can be available to other benthic life forms. They are an important food source for higher organisms; their list of predators in the nineteenth century includes *Homo sapiens*, who, prior to altering his feeding habits in favor of fast foods, took commercial advantage of this natural delicacy. Being long-lived, the unionids record in their shells the chemistry of the water over many decades, a property which makes these animals useful in monitoring water pollution. The loss of these unobtrusive creatures would be a tragedy. Dr. Rowland M. Shelley, Curator of Invertebrates at the North Carolina Museum of Natural History, points out that given the present rate of habitat destruction many species of naiads "will become extinct before they can even be discovered."

The freshwater snails, another group of mollusks, are common in the Piedmont's streams. They are in a separate class (Gastropoda) from the bivalves with whom they share the fresh-water portion of the phylum Mollusca, the soft-bodied animals whose bodies (excepting slugs, squids, and octopuses) are enclosed in shells. The aquatic snails are, of course, close relatives of the land snails. They are divided into those who breathe through gills and those with internal lunglike organs. The gilled snails can retire into their shells and seal behind them a bony trap door called the operculum. Pulmonate snails have no operculum. There are two extensile, sensory tentacles with an eye at the base (not the tip) of each. Snails build their shells from lime secreted by the mantle of tissue surrounding the body, and they therefore have somewhat rigid requirements for the chemical properties of the water in which they dwell; calcium is necessary for the manufacture of lime, and the pH must not be acidic or the shells would dissolve. It might be possible to estimate crudely the pH and the calcium content of soils drained by a given stream by observing the presence of snails in that stream.

Where conditions are suitable, as they are in many Piedmont watercourses, snails in abundance cling to rocks and vegetation, gleaning detritus from the thin layers of sediments usually present on most stationary objects in a stream bed. Some can be seen in quiet waters attached to the underside of the surface film feeding on material trapped in the surface tension.

Crustaceans

The distribution of crayfish in the Piedmont's streams is comparable to that of the mussels; there are numerous species (over 200 recorded in North America) and some are limited in range to one or a few drainages. With the lobsters, shrimps, and crabs, the crayfish are in the order Decapoda, signifying that there are ten legs used in locomotion, feeding, defense, or some combination thereof. Actually, the crayfish sense and manipulate their surroundings with 19 pairs of appendages—two vertically aligned pairs of small antennae called antennules, the main antennae, which may equal the length of the body, five pairs of bailers for handling food and moving water through the gills, the big claws (chelae) used for fighting and grabbing prey, four pairs of walking legs, the first two of which are clawed, and five pairs of swimmerets under the abdominal segments which fan water fore-to-aft to aid in breathing in still water. Finally, the tail is divided into two pairs of uropods and a telson (see Glossary), together forming a fanlike paddle to propel the crayfish in quick rearward escape. Males have larger claws and narrower bodies than females. Many of the Piedmont's crayfish are in the family Cambaridae, within which the genera *Cambarus*, *Procambarus* and *Orconectes* are well represented.

Crayfish exploit diverse feeding opportunities, sifting detritus from the current, scavenging animal remains, and preying directly on other benthic creatures. In turn, they are eaten by the streams' mammals, birds, and reptiles. They are chiefly nocturnal, spending much of the daylight hidden beneath rocks and debris or in the holes they dig in the streambanks an inch or so in diameter. Some species excavate into the alluvial soils beside stream channels, creating small vertical chimneys at the openings. Crayfish grow in instars—three in most species—and shed their shells periodically to grow. The second joint of each limb is designed to break off easily, as in combat, and thereafter to regenerate. To conserve minerals, crayfish often eat the lost limbs and carapaces they shed.

There are numerous other decapods in Piedmont streams, although many are small and difficult to observe; most occupy the quiet sloughs and backwaters. In addition, there are fairy shrimp (order Anostraca, which includes the brine shrimp fed to aquarium fish) which swim on their back and grow to perhaps 1 inch (25 millimeters) in length. Water fleas (order Cladocera) are abundant algae eaters, somewhere near the visible threshhold in size, and are important in the diets of small fishes. Copepods (order Copepoda) may also be present in the stream's plankton; many are large enough to be seen in a test tube or vial held to the light.

Fish

Feeding on the streams' teeming insect populations are larger predators and scavengers. Small fish are important in this part of the predation scale. The colloquial term "minnow" embraces a variety of finger-length fishes including the cut-lipped minnow, *Exoglossum maxillingua,* identified by the lateral indentations in the lower lip, the golden shiner, *Notemigonus crysoleucas,* a deep-bodied fish the mature males of which are gilded during the breeding season though females and nonbreeding males are silvery, and the very common red-bellied dace, brown above black median stripe and silvery below except in breeding males whose colors name the species. These fish, particularly the dace, occupy the farthest headwaters of virtually every tributary. In fact, they prefer streams small enough for a human to stand astride, for such waters are usually free of the next larger tier of fish, the creek chub, *Semotelus atromaculatus*, a monster minnow to one foot long marked with a dark spot on the dorsal fin, and the grass pickerel, *Esox americanus*. When the pickerel is alarmed, which it usually is in the presence of humans, its escape is too quick for the human eye to follow, a swiftness which may account for the paucity of sightings, for the species is common in Piedmont streams. Larger chubs and pickerel must seek wider waters, but surprisingly large individuals of both these predators hunt smaller fish wherever there is flowing water.

There are two groups of fishes who have adapted to life in the current by resting on the bottom or grasping with pectoral fins while facing upstream. They are the darters and the sculpins. The finger-length Johnny darter, *Etheostoma nigrum*, is abundant in Piedmont streams, resting on the bottom and darting forward for aquatic insects or simply to change position. The sculpins, or muddlers, of which more than one species in the genus *Cottus* may be present in Piedmont streams, grasps rocks with its fanlike pectoral fins to hold a position against the current. Sculpins grow to 8 inches (20 centimeters). Both groups are camouflaged to resemble the substrate, are negatively bouyant so that they sink quickly when not swimming, and have eyes positioned near the top of the head for better vision upward.

Human activity has rendered the water quality less than pristine in some of the Piedmont's streams. Catfish and eels have a greater tolerance for pollution than most other fish, and they may dominate the piscine ranks in murky waters. They are also present in clean streams. Madtoms, *Noturus* spp., are the smallest at 3 to 4 inches (8 to 10 centimeters) of the catfish clan. They are followed in size by the stonecat, *Noturus flavus*, identified by the connection of the tail and the rear of the dorsal fin, and the bullheads, *Ictalurus* spp., who grow to 12 inches (30 centimeters). The magnificent channel catfish, *I. punctatus,* sleek and silvery with dark speckles and weighing up to 20 pounds (45 kilograms), haunts the Piedmont's major rivers. Catfish are scavengers on vegetable and animal matter in the streams and they take live insects and smaller fish. They are without scales, have sensory barbels around the mouth, which they probably use in finding food in murky waters, and have spines in the dorsal and pectoral fins that can inflict painful injury on the careless handler.

"Anadromous" means "going upstream to spawn" and applies to such fish as the salmon that battle the currents in western rivers. The herring family are the most notably anadromous fish plying the rivers of the Piedmont; they are characterized by laterally compressed bodies, deeply forked tails, and spineless fins. In spring the alewife, *Alosa pseudoharengus,* and the American shad, *A. sapidissima,* leap over the fall line in the major rivers, particularly the Potomac and the Delaware, and provide considerable sport and commercial fishing. The adults return to marine or estuarine waters, but the fingerlings spend the summer in the creeks and rivers, moving seaward in autumn.

The American eel, *Anguilla rostrata,* spends its youth and adulthood in the Piedmont's streams and rivers, then swims downstream, out the bays and into the Atlantic Ocean south of Bermuda. There in the deep waters of the Sargasso Sea the females lay millions of eggs, then die. The blade-like larval eels then find their way back to the rivers, creeks, and remote streams where they feed and grow to adulthood over several years. Eels are common and important fish in the flowing waters of the Piedmont—they link the farthest tributaries (of undammed rivers) with the sea.

Ranking highest with Piedmont anglers, and often highest on the piscine food ladder, are the sunfishes. This group includes the pumpkinseed, the bluegill, and others commonly understood to be sunfish, or *Lepomis* spp., as well as the largemouth and smallmouth bass, *Miropterus salmoides* and *M. dolomieui.* In addition to being the province's capital native game fish, the sunfish are important predators on other fish and aquatic animals. The sunfish have a single dorsal fin. They build and defend circular nests in quiet waters, but they have no aversion to the current.

Insects

Many insect species of the Piedmont's rivers and streams are airborne or terrestrial as adults, but pass their larval stages in the water. This phenomenon provides many examples of the life cycles of modern species reenacting their evolutionary history; that is, their phylogenies included life cycles which were at first entirely aquatic but with passing millenia the adult phases "emerged" and became terrestrial. Some of the larval insects of the streams are among the planet's oldest living forms. The dragonflies, mayflies, stoneflies and caddis flies were present in the coal age of 200 million years ago, and their larvae, called naiads, are alive and well in today's streams and ponds.

The mayflies (order Ephemeroptera) of which 1500 species have been identified worldwide, are first in the feeding hierarchy. They graze on plankton, algae, and suspended detritus principally, though some forms are carnivorous. Mayflies are identified by their three caudal (tail) filaments. Frilly gills containing trachea (air ducts) undulate from the abdominal segments. The legs are shaped so that the current will press the naiad to the rocks rather than sweep it away. The larvae range in size from the barely visible early instars to about 1 inch (25 millimeters) in length. The Ephemeropterans are the only insects which fly before reaching adulthood. The subimago (the stage after pupation but before adulthood) lasts only an hour or two in most mayfly species before undergoing a final molt into adulthood, the imago. Adults may live for only a day or two, the males of some species flying in nuptial swarms. Females reenter the water and attach their eggs to current-swept stones on the stream bottom.

The plecoptera, or stone flies, are similar in appearance as larvae to the mayflies, having broad heads and flattened bodies, but they are distinguishable by their two caudal filaments and by the location of the trachial gills on the thorax. In the Perlidae, a prominent family of stone flies in the Piedmont, there are gill tufts on all three thoracic segments. Stone fly naiads prey on the naiads of mayflies and on other aquatic larvae such as those of the midges (the dipterous family Chironomidae). As adults, the stone flies retain vestiges, called cerci, of the two caudal filaments.

The wings are folded along the back, wasp-style, and when extended the hindwings are seen to be broader than the forewings. The larval stage lasts one to four years; adulthood is brief—only a few days in those species which are seen in summer, for they do not eat. Species whose adults emerge in autumn and winter have mouth parts adapted to feeding on algae and aquatic vascular plants above the surface, and they probably live for a few weeks. Stone fly eggs are laid and hatch on the surface film; the naiads then swim to the bottom.

The odonata are members of an ancient and primitive order composed of the dragonflies (suborder Anisoptera, meaning "unequal wings" because the hindwings are broader at the base than the forward pair) and the damselflies (Zygoptera, from the Greek *zygon*, 'yoke' or 'pair'). Both are familiar and conspicuous flying forms along the Piedmont's streams and ponds, the dragonflies being generally the larger and holding their wings horizontally and perpendicular to the body at rest. The forms of damselflies found in the Piedmont fold their wings together above the body at rest. This is considered an important evolutionary advancement, for holding the wings horizontally prevents walking, burrowing and other important pedestrian activities of which the higher insects are capable.

As naiads, the dragonflies are among the more capable insect predators in the Piedmont's streams. They grow in instars to 2 inches (5 centimeters) or more over an aquatic adolescence of perhaps five years. They capture prey, often naiads of stone flies and mayflies, and the aquatic larvae of beetles, with a hinged, extensible lower lip. Dragonfly naiads are recognized by their broad abdomens tapering to a sharp point. There are no external gills, the transfer of oxygen from water to bloodstream being accomplished by drawing water into the rectum, where the gills are located. These creatures normally move with deliberation but when frightened can expel the water from the gills and jet away at great speed.

The naiads of damselflies are also active predators of the rocky bottoms of Piedmont streams, and like the dragonfly naiads capture other aquatic insects with a hinged labium. Their bodies are elongated (up to 1 1/2 inches/4 centimeters) and slender. The three external, leaf-like gill plates at the tip of the abdomen are diagnostic.

Predation is a way of life for the odonates in adulthood as in adolescence. Along the Piedmont's streams, rivers, and ponds the dragonflies and damselflies patrol rigidly demarcated territories, holding their bristly legs in a netlike scoop to capture other flying insects. There are five thousand species worldwide in this order. All are aquatic as naiads and most spend their adult lives near water, though some dragonflies venture over the Piedmont's fields and forests.

Several common forms of damselflies are in a group known as the narrow-winged damselflies (family Coenagrionidae) which includes the bluets, *Enallagna* spp., with clear wings, the violet dancer, *Argia violacea*, and several species in the genus *Amphiagrion*, identified by their black bodies and red abdomens. Along riverbanks and streams we see the stately, waltzing flight of the black prince, a broad-winged damselfly of the family Calopterygidae. The male has an iridescent green body and glossy black wings; the female's body is duller and the wings are gray with white stigmata (small spots, one on the leading edge of each wing near the tip). Male damselflies are generally more colorful than the females. Pairs frequently fly coupled, the male holding the female behind the head with claspers at the tip of his abdomen. During copulation, an activity to which damselflies devote considerable time, the female arches her abdomen ventrally, touching the male's genitals at the lower rear of his thorax.

The darners (family Aeshnidae) and skimmers (families Macromiidae and Libellulidae) account for most of the dragonflies seen along running water in the Piedmont. The darners are largest, their long (to 4 inches/10 centimeters) slender bodies usually green or brownish and their wings clear. The libellulids (common skimmers) typically have stubby bodies and bands in the wings. River skimmers (macromiids) are similarly shaped but the wings are clear.

Among the beetles (order Coleoptera) there are many whose larvae are aquatic and live in the Piedmont's flowing waters. The water penny beetles (family Psephenidae) are small (1/4 inch/6 millimeters) grazers on streamside vegetation, but their larvae, from which comes the common name, fix their rounded flattened shapes to stones in the current, feeding on algae and detritus. Pried gently from the rock with a knife blade and held to the sunlight, the half-inch wafer of corroded copper becomes a glowing gem of lotic life. Whirligig beetles (family Gyrinidae) are among the stream's most conspicuous coleoptera. With eyes divided to scan above and below the waterline, these predators zigzag across the surface capturing small flying insects that have inadvertently become trapped in the surface film.

In the Piedmont's pasturelands we met the beetle family Hydrophylidae (see page 69), some members of which burrow into fresh manure. The family's common name is water scavenger beetles, reflecting the life mode of the adults and larvae of most species. Partially decomposed vegetable matter, such as current-borne detritus, is the dietary staple. Some of the larvae are predaceous on aquatic insects and may be found among bottom sediments in quiet parts of streams and on vegetation. Adults move freely above and below the surface, trapping air beneath their elytra (chitinous wing covers which are really the forewings modified as shields to protect the hindwings and body and are common to all coleoptera) with which to replenish their oxygen while submerged. The bubbles of air which beetles and other diving insects and arachnids take with them beneath the surface capture molecules of free oxygen dissolved in the water to replace what is used, so a given bubble supplies more oxygen than it initially contains.

The predaceous diving beetles (family Dytiscidae) and the burrowing water beetles (family Noteridae) occur in quiet backwaters of any Piedmont stream, but they are most strongly associated with lentic habitats and are discussed under pond life systems (see page 325). A final coleopteron which we must presume present in the Piedmont's streams is the beaver parasite *Platypsyllus castoris.* Although one may never expect to see the 1/8-inch (3-millimeters) beetle, it is known to be a common ectoparasite living in the fur of beavers, the Piedmont's largest rodent (see page 304).

Several families in the order Neuroptera are distinctly associated with flowing water. The best known and most fearsome is the larva of the dobsonfly (family Corydalidae) called the hellgrammite. It is three inches long and armed with jaws that can inflict a memorable pinch—the hellgrammite is probably the most powerful predator among our streams' aquatic larvae. The abdomen is spiked with eight pairs of lateral filaments and tipped with one pair of anal prolegs with which the benthic beast anchors against the current. Hellgrammites live beneath current-swept stones, preying on the naiads of mayflies, stoneflies, and even dragonflies. Adults are 2 to 3 inches long with large, clear wings held longitudinally at rest. Awkward fliers, they are generally found resting on streamside vegetation, but they may be attracted to lights at night. The similar but smaller alderflies (family Sialidae) are distinguished as larvae from the hellgrammites by their seven (instead of eight) pairs of lateral filaments and by their single caudal filament in place of anal prolegs. The alderfly larvae, too, are predaceous. A final neuropteran family peculiar to streams is the spongilla flies (Sisyridae) whose larvae feed on fresh-water sponges such as grow on submerged rocks in well-aerated rapids. A turquoise-bodied spongilla fly larva is occasionally seen burrowing open channels in the velvety jade of a green fresh-water sponge in the swift parts of a few clear Piedmont streams. We know about this association, but the sponge itself is infrequently found, for its water quality requirements are stringent. The telltale glimmer of green beneath the rapids is an occasional reward to the patient explorer of the hidden lotic world.

It is no surprise that mosquitoes (family Culicidae) breed in the quieter parts of streams and contribute to the flow of protoplasm in these near-lentic circumstances, principally by falling prey to small fish and to other aquatic larvae. Mosquito larvae are known for their habit of breathing through a tube at the end of the abdomen which they thrust through the surface film.

There are other dipterans in the stream. Midges are small (1/4 inch or less), mosquito-like insects representing several families of Diptera who spend their larval stages in the Piedmont's streams. Some attach to the rocks and vegetation, some live in tubes constructed of vegetable debris, others feed on the bottom side of the surface film. As adults, midges do not bite. The net-winged midges (family Blephariceridae) and the "midges" (Chironomidae) are firmly associated with flowing water. So too are the blackflies (Simuliidae) whose adults bite viciously and carry disease, but happily, they are not numerous in the Piedmont.

Reptiles

A rich variety of reptiles and amphibians attend the streams and their associated bottomlands. The dominant reptile in number, and sometimes in size, is a turtle, the river cooter, *Chrysemys concinna concinna*, often seen basking on the rocks and logs of watercourses. Younger individuals may have yellow markings on the head and may bear a large *C* on the second costal scute (lateral plate on the shell). Markings fade with age, and mature cooters appear uniformly dark. Several varieties are recognized, some of which may hybridize with the Florida cooter, *Chrysemys floridana floridana*, in the southern Piedmont, and possibly with the yellow-bellied turtle, *C. scripta scripta*, or with escaped exotic pets. The taxonomy of *Chrysemys* species is confusing to the point that any broad, flat (that is, low-profile), smooth-shelled turtle seen in the Piedmont's watercourses is best referred to simply as "cooter" or "slider." *Chrysemys* turtles are mainly vegetarians.

The stinkpot, *Sternothaerus odoratus*, so named for the musk it emits when captured, is common in Piedmont streams. Two yellowish stripes run the length of each side of the head, one above and one below the eye. The carapace (upper shell) is rounded. Patrolling the bottom for plant and animal matter, the stinkpot resembles a torpidly mobile stone. The plastron (lower shell) is small and does not protect the legs, nor does it restrict them, leaving the stinkpot able to climb onto trees to bask. Notes the great herpetologist Roger Conant, "If a turtle falls on your head or drops into your canoe it probably will be this one."

The Piedmont's most renowned reptilian predator is the snapping turtle, *Chelydra serpentina*, whom Conant calls "ugly in both appearance and disposition." The snapper can grow to 35 pounds (16 kilograms) and 1 1/2 feet (45 centimeters) in diameter in the Piedmont's rivers and ponds. It is the nemesis of ducks, fish, frogs, snakes, and mammals. Vegetable matter and carrion are also important in the diet.

The snapper rarely basks; it is most often seen resting or creeping on the bottom in shallow water or projecting its fearsome, hooked snout from the surface. The carapace has three longitudinal keels and a jagged rear edge. The tail, as long as the carapace, is saw-toothed at the base. The plastron is small and does not enclose the legs or tail.

88. Northern water snake, *Nerodia s. sipedon*

The streams of the Piedmont have robust populations of the northern water snake, *Nerodia sipedon sipedon,* replaced in the southern Piedmont by the Midland water snake, *N. s. pleuralis,* whose dark bandings and thickened body lead many to confuse this harmless serpent with the cottonmouth—which is confined to the extreme southeastern Piedmont if it penetrates the province at all. Older individuals may loose their markings. The water snake is often seen swimming, fish-in-mouth, holding its prey out of water to suffocate it. When molested, the water snake dives. My personal observations on the scope of water snake predation include bass, catfish, and eel; frogs and other fish are also known to be eaten. The eastern ribbon snake, *Thamnophis sauritus sauritus,* also occupies flowing water, taking small fish, frogs, and salamanders. The slender body is black with three thin yellow stripes running the entire length. The ribbon snake swims with alacrity but does not dive. The eastern kingsnake, *Lampropeltis gentulus gentulus,* black but embroidered with a bold chain of white, lurks on the banks of streams in wait for water snakes and in search of turtle eggs. The yellow rat snake, *Elaphe obsoleta quadrivittata,* principally a Coastal Plains reptile, is colored a deep gold with four wide longitudinal stripes of dark brown. It searches the stream banks and overhanging trees for rodents.

Two generalities can be advanced about any snake seen swimming with the Piedmont's flowing waters: (1) it is almost certainly harmless to humans, and (2) it is probably a northern water snake.

Amphibians

Among the Piedmont's frogs, the green frog, *Rana clamitans melanota*, the leopard frog, *R. utricularia*, and the pickerel frog, *R. palustris*, are associated with flowing water. All feed on flying insects which they snap from the air. Their relationship with water is based on their aquatic breeding requirements and on their use of the water for protection. When danger threatens they leap into the water and swim for cover on the bottom. The leopard and pickerel frogs wander some distance from water in summer (see page 179), but the green frog rarely quits its streamside post.

The sandy deposits on the inside edges of bends in the streams and the muddy seeps and moist embankments provide habitats favored by salamanders. The northern red and eastern mud salamanders, *Pseudotriton ruber* and *P. montanus montanus*, confuse even experienced observers; the older individuals of both species are dark and blotchy. The red salamander has a golden iris; the mud salamander's is brown. The northern two-lined salamander, *Eurycea bislineata*, coppery above a dark lateral line, hides under rocks and debris along stream banks, usually favoring small tributaries from which the larger predatory fish are absent. The salamanders found in streams feed mostly on worms and aquatic insect larvae.

In the alluvial bottomlands adjacent to flowing water is a terrestrial habitat of moist sandy soils, fallen deadwood, and temporary pools favored by other amphibians. The marbled and spotted salamanders, *Ambystoma opacum* and *A. maculatum*, gather briefly to breed in these pools, the former in autumn, the latter in early spring. At other times they remain underground or beneath vegetable debris, but they are out and about on warm rainy nights. The marbled salamander bears a distinctive pattern of transverse white (on the male) and gray (female) markings across the back. The background color is black. The spotted salamander wears yellow or orange spots in irregular dorsolateral rows against a black background. Both have obvious costal grooves (vertical creases along the sides).

Of the northern dusky salamander, *Desmognathus fuscus fuscus,* Conant notes that this amphibian "seldom wanders far from running or trickling water." It is in the family Plethodontidae, the lungless salamanders, distinguished by their immovable lower jaws, and, in the absence of lungs, their practice of breathing through the skin. It is abundant along the sides of Piedmont streams and brooks where rocks and flood debris harbor insects. The markings are brownish against a lighter background and the patterns are highly variable.

Equally variable is the red-spotted newt, *Notophthalmus viridescens viridescens,* found in mesic, but moist, forest litter in its larval, or eft, stage, during which the color is reddish to orange. Adults are aquatic, often occupying bottomland pools or the quieter parts of streams. Adults have vertically flattened tails to aid in swimming; the young have rounded tails. Reddish spots accent yellow to greenish background coloration. Newts are in a family (Salamandridae) separate from other salamanders. Their skin is rough rather than slimy.

The American and Fowler's toads are at home in the bottomlands which they utilize as nurseries for their young, placing eggs in temporary pools which provide homes for the tadpoles. The immatures after transformation and the adults may remain in the bottomlands, except during times of flood, or they may range into mesic habitats (see page 100). At the first warm rains of spring, the spring peepers, *Hyla crucifer,* give voice to their single notes which, heard in chorus, are said to suggest sleigh bells. As the Latin name suggests, there is a cross-shaped marking on the back that contrasts with the otherwise gray or olive skin. This is a small frog averaging 1 inch (25 millimeters). It is identified with woodland pools, most of which are in alluvial soils. The eastern gray tree frog, *H. versicolor versicolor,* is roughly twice as large. It forages in the trees for insects, protected by its bark-colored warty skin; it shows a preference for trees surrounded by standing water. Conant likens the call to that of the red-bellied woodpecker. The open bottomlands ring with the prolonged *crreeek* of the upland chorus frog, *Pseudacris triseriata feriarum,* singing from the standing pools and grassy swales as the peepers jingle in the forest. The skin coloration of this one-inch frog is tawny with highly variable dark markings or stripes on the back.

Birds

Hydric habitats attract the richest avifauna of any of the Piedmont's communities. Birds from diverse orders occupy fishing posts on the rivers' rocks and snags, more feed on the lotic invertebrates, and others eat the vegetation associated with flowing water and with the bottomlands. Wading birds (order Ciconiiformes) are large, obvious, and common on the watercourses. In summer the green heron, *Butorides striatus,* most numerous of the Piedmont's wading birds, skulks along the banks or perches on low overhanging limbs to skewer small fish. At rest the neck is retracted, but it can be extended in a quick thrust to almost the body's length. This crow-sized bird appears dark from a distance; field glasses reveal greenish wings, a russet neck, and orange legs. The single screeching protest note is distinctive.

The great blue heron, *Ardea herodias,* stands more than a yard high and spans six feet in flight. It spears sunfish, chubs, suckers, and catfish, beats them senseless against a rock, then swallows them whole, headfirst. Known fondly as *GBH* among birders, this awesome fisherman flies, like all herons and unlike the cranes (of which the Piedmont has none), with the neck retracted. The general blue–gray appearance is accented by black and white head markings and rusty thighs. In addition to taking fish, the GBH eats amphibians, snakes, crawfish, and large insects from the streams. It is a common bird on secluded waterways, but it is not numerous; pairs and individuals maintain large territories which, when focused on flowing water, tend to be linear. The bird lives in the Piedmont year-round and is most abundant on flowing water in the dead of winter when the ponds are frozen. The protest is a series of gutteral croaks uttered when the great bird is flushed.

There are two white herons seasonally present in the Piedmont's watercourses: the great egret, *Casmerodius albus,* formerly called the common egret, and the little blue heron, *Florida caerulea.* Both nest primarily in the Coastal Plain but venture inland and northward after the breeding season, generally following the rivers and creeks. The adult is slate gray of body with a warm burgundy head and neck. The hatch-year juveniles are pure white. The bill of birds in all plumages is light with a black tip. The great egret is plain white with black legs and a yellow beak. It is much larger than the little blue (55-inch/1.4-meter wingspan versus 40 inches/1 meter). The seasonal wanderings are comparable. Both dine on the fish, crustaceans, amphibians, and snakes of the Piedmont's streams. The terms "heron" and "egret" are synonymous.

Noisiest and flashiest of the rivers' piscivores is the belted kingfisher, *Megaceryle alcyon,* big-headed, crested, lustrous blue above and white with a blue "belt" below. The female sports a rusty band below the blue belt, making her one of the world's few avian females more colorful than her mate. The kingfisher scans the stream from an overhanging snag then dives headlong into the water to catch a dace or minnow. The plunge is occasionally made from hovering flight. To nest, pairs burrow deep tunnels into the steep embankments of alluvial soil which the stream long ago deposited and more recently cut. Kingfishers patrol the watercourses year-round, protesting intrusion with a sustained, raucous rattle. The flight is strong and swift and is characterized by irregular wingbeats.

89. Spotted sandpiper, *Actitis macularia*

90. Northern water thrush, *Seiurus motacilla*

The naturalist's curiosity is piqued by a behavioral quirk of two small birds who forage for insects, worms, and crustaceans in the shoals and streamsides. Similar in size, shape, and markings but from different orders, they are the spotted sandpiper, *Actitis macularia,* and the northern water thrush, *Seiurus motacilla.* Evolution has brought these two birds, no more closely related than the ostrich and the hummingbird, to the same feeding niche on the same streams in the Piedmont and over much of the rest of North America and has converged their appearance and behavior to nearly the same mold. Both are 5 to 6 inches (13 to 18 centimeters) long, brown above and have white breasts with dark spots or streaks. The water thrush is actually a wood warbler (family Dendroidae) and the sandpiper a charadriid who flies with wings held stiffly downward. While picking among the pebbles and on the banks, both carry their tails higher than their heads, and both perform a continuous, swaying dance, hulalike, with the head held stationary and the hindparts undulating. I've neither heard nor read an explanation, but my own speculation is that the motions approximate moving reflections from the surface of flowing water, a continuous kaleidescope of light and shadow playing on the stream banks. Such flickering illumination might render a stationary object more conspicuous than one in motion.

The bald eagle, *Haliaeetus leucocephalus,* lives mostly by fishing, though it takes some carrion. In spite of its rareness and imperiled status in the contiguous 48 states, it is occasionally sighted along the major rivers of the Piedmont.

Alert observers on the Susquehanna, the Potomac, and the James add sightings to seasonal bird counts. Four eagles, two adults and two juveniles, wintered on the Roanoke River in 1975 and could be seen scavenging for fish minced by the turbines of the John H. Kerr Dam. The osprey, *Pandion haliaetus*, also fishes the Piedmont's rivers. It is most readily observable on spring and fall migration but may be present in any season and probably nests in secluded fluvial habitats such as islands. During the nesting season in 1978 I saw a single osprey fly from an island in the Potomac a few miles upstream of Great Falls, about 15 miles (25 kilometers) above Washington, D.C. Two summers earlier I saw an osprey in the company of a bald eagle at the Kerr Dam. Conceivably, it is still possible in the Piedmont to witness the grand drama of an osprey yielding its hard-earned catch to aquiline piracy.

The red-shouldered hawk, *Buteo lineatus*, is the most common daytime raptor on Piedmont rivers. The bird's allegiance to the bottomlands is strong—I have seen them outside the alluvial forest mostly on migration. Their diet consists chiefly of reptiles and amphibians of the rivers, creeks, sloughs, oxbows, and temporary pools. Bottomland snakes, frogs, fish, turtles, crayfish, and some insects and rodents are also eaten. The unclosable shells of the cooters and stinkpots (see page 290) leave them particularly vulnerable to the red-shouldered hawk's hooked bill. On all but hatch-year birds the rusty shoulder patch may be glimpsed at rest or in flight, but the dark bands alternating with white in the wings and tail are the diagnostic markings. Backlit by the sun when the bird is in flight, the white bands become windows of translucence. *B. lineatus* is slightly smaller than its mesic counterpart, the red-tailed hawk, and the underparts are generally darker. The call is a series of short, high-pitched *kee-os*.

Field guides refer to the red-shouldered hawk as common, which it is in the southern coastal swamp forests, but in the Piedmont the bird's requirement for mature alluvial and swamp forests restricts its breeding territories. The recent and ongoing menace of dam-building is destroying many of the Piedmont's remaining bottomland forests and creating the vast, man-made lakes which proliferate south of the James. As a result, the red-shouldered hawk is disappearing from the province. It does not adapt to the lakeside mesic woods which remain after the flowing water is impounded.

South of the James the barred owl, *Strix varia*, shows a strong preference for the bottomlands, although in the northern Piedmont its attentions are divided between mesic and hydric forests. The exoskeletons of crayfish are common in the barred owl's regurgitated pellets, attesting to riparian tastes. The barred is a large owl—the wings span almost 4 feet (120 centimeters)—though not so massive as the great horned, into whose gullet the barred occasionally disappears. The head is rounded, almost spherical in appearance, and below the head is a zone of horizontal barrings for which the bird is named. Below the barrings, the breast is vertically streaked with brown. The principal visible distinction between the barred and the great horned owls is the ear tufts, absent on the barred, but which lend a catlike silhouette to the head of the great horned. Both these owls are heard much more often than they are seen and the regular cadence of two sets of four hoots with an extra syllable appended to the eighth hoot is the most common means of verifying the presence of the barred. The horned owl's call is a softer series of hoots, less regular in spacing and number, but often ending in three quick notes, *hoo-hoo-hoot*. The horned owl hunts anywhere and everywhere; the barred is a forest bird whose fear of the horned usually precludes open-country hunting.

Along the Piedmont's major rivers, we note that some of the crows give rather short nasal notes rather than the usual rich *caw*. Although all but identical in appearance to the common crow, the author of this adenoidal *ank* is the fish crow, *Corvus ossifragus*. The bill is slightly thinner than that of the common crow and the body slightly smaller, but these nuances would probably be difficult to detect even with the two birds in hand (heaven forbid). Voice is the best identifier. The fish crow may wander at large through the Coastal Plain, but in the Piedmont it forages along the watercourses almost exclusively, a habitat preference from which it departs only to visit urban areas. The diet is chiefly scavenged aquatic fauna, wharf pickings and the like. In the city it shifts its attention to garbage and to the eggs and young of starlings and English sparrows, tastes which make the fish crow a valuable bird indeed.

Several species of ducks and ducklike birds may be seen in the Piedmont's creeks and rivers, particularly in winter. The mallard, which is common, is a rather large duck with a 36-inch (90-millimeter) wingspan, which feeds on the surface and in the shallows by tipping forward until the body is vertical. The male is handsomely patterned with an iridescent green head, rusty breast, gray back, and black rump. The female is mottled with brown on white, appearing brown at a distance. Both sport a speculum of blue in the secondaries. The common merganser, *Mergus merganser*, is a common river bird in winter from the James northward. The hooded merganser, *Lophodytes cucullatus*, is a casual visitor from the Coastal Plain across the Piedmont. The male common merganser has a green, uncrested head; his cinnamon-headed lady is crested as are both sexes of the hooded merganser. The male hooded has a black head with a conspicuous wedge of white in the crest. Mergansers are fish-eating ducks, catching small fish with their thin, serrated bills. The Piedmont's most colorful duck, and to some the continent's most beautiful bird, is the wood duck, *Aix sponsa*. Wintering in Piedmont streams from the Potomac southward and nesting throughout the province, the wood duck is a fixture where water flows through woodlands. The head of the male is ornately plumed with dark iridescence and trimmed in white. The eye and bill are red. A rusty breast and a collage of greens and blues over light sides completes the plumage. The female is of subdued coloration, her dark eye shining from within a white mask. Aquatic plants, insects, and acorns from the bottomland oaks comprise the wood duck's diet. The elongated tail is an in-flight identifier. The voice is a plaintive squeal given when flushed.

The wood duck is arboreal no less than its name suggests; it may be seen sitting on limbs overhanging water. It is a happy conjunction that the wood duck and the pileated woodpecker are close enough in size for the duck to use this woodpecker's excavations in streamside trees for nesting. The pileated, a greater presence in the alluvial forests than in any other Piedmont habitat, hammers into living beech, oak, and sycamore to hollow out its own roosting and nesting cavities, sometimes taking advantage of decayed heartwood. These excavations provide the main source of natural nesting space for the wood duck and are important also to the screech owl, to honey bees and, after enlargement by squirrels, to raccoons and other mammals. The alluvial and swamp forests are, of course, home to all of the Piedmont's woodpeckers.

The woodcock, *Philohela minor,* probes into the moist alluvial soils of bottomland forests, taking earthworms with its prehensile 3-inch (8-centimeter) bill. Compact, plump, and exquisitely camouflaged, this nocturnal sandpiper permits close approach, then bursts from the forest floor on whistling wings. In late winter and early spring it rends the evening air in the southern Piedmont (James southward) with loud nasal *peents* followed by elaborate courtship flights which culminate in aerobatics and eerie warblings. A close relative, the common snipe, *Capella gallinago,* probes by day into the muddy alluvium of earlier, generally herbaceous or shrubby phases of the hydrosere. Flushing with a single *gzzrk* note of protest, the snipe quits its swale with a zigzagging burst of speed.

Canoeing the wooded creeks in winter, one glimpses the winter wren, *Troglodytes troglodytes,* a 3 inch (8 centimeter) brownish mite with only a hint of a tail held erect, bobbing and darting among the exposed tree roots along stream banks. After a few protest notes the bird simply vanishes, mouselike, into the overhanging lip of soil. This smallest North American wren leaves the Piedmont in March and April to nest in northern forests. Landward, but low in the alluvial forest, the wintering hermit thrush, *Chatarus guttatus,* gestures with its rusty tail. The hermit is the only spot-breasted thrush to spend the cold months in the Piedmont.

Southward of the Potomac, the eastern phoebe, *Sayornis phoebe,* plain gray above and light below, winters in the Piedmont's woodlands, particularly the bottomland forests. It is fond of nesting beneath old bridges and other shelters near water. The habit of flipping the tail at intervals easily identifies the phoebe. In spring, the phoebe is joined by another tyrannid, the Acadian flycatcher, *Empidonax vires-*

cens. Because it is in the genus *Empidonax*, one abandons all hope of visual identification but relies on its close association with flowing water and its call—a shrieked *peet-suh*—to spot this gray-green flycatcher.

Enumerating the birds associated exclusively with the bottomlands does scant justice to the avifauna of this habitat. A great many, probably most, of the birds from drier regimes are abundant in the low woods, making the habitat the province's best for year-round birding. But it is to witness the spectacle of the spring warbler migrations that seasoned birders make their way into the streamside forests. To see 15 different species in the bottomlands in a season is not exceptional. The redstart, Canada, Kentucky, palm, pine, chestnut-sided, black-throated blue, black-throated green, bay-breasted, blackburnian, black-and-white, blackpoll, yellow-throated, cerulean, Cape May, and myrtle warblers move through the Piedmont's bottomlands, to some extent following the watercourses where they have a significant north–south component. The Wilson's warbler passes through, spring and fall, taking insects from the foliage of alluvial shrubs such as swamp dogwood and willow. The yellow warbler nests in willows throughout the Piedmont and is one of the most frequent victims of the female brown-headed cowbird's egg-laying visits. Prominent nesting warblers in the Piedmont's bottomland forests include the prothonotary, parula, Kentucky, hooded, and redstart. The white-eyed vireo is common, pale above yellow–buff underparts and wearing conspicuous wing bars. To me, the voice suggests the syllables, if not the tones, of the chuck-will's-widow's call—*chip-whew-whew.* The chuck-will's-widow and its slightly more northerly congener, the whippoorwill, are both important summer birds in the bottomlands and other habitats. They are nocturnal catchers of flying insects who rest by day in the leaf litter of alluvial and mesic forests. Their calls, from which the names derive, are integral to our perceptions of spring in the Piedmont. The chuck-will's-widow, *Caprimulgus carolinensis*, is not often heard north of the Roanoke; *C. vociferus*, the whippoorwill, calls endlessly in spring throughout the province.

Mammals

A number of the mammals which occur casually (or even prominently) in drier habitats reach their greatest concentrations in the hydric forests. This may be because the rich alluvial soils and plentiful moisture support a greater biomass per acre than grow in the uplands, or it may relate to the fact that the wetlands are our wildest places and therefore are more attractive to the animals as a refuge from human encroachment. Raccoons and opossums, for example, forage in mesic habitats but are principally mammals of the streamsides in the Piedmont. The same may be said for the gray fox, the white-tailed deer, the eastern cottontail and, to a lesser extent, for the eastern gray squirrel, the flying squirrel, the masked, southeastern, and short-tailed shrews, the eastern mole, the meadow vole, the long-tailed weasel and several forms of myotis and pipestrel bats.

The bobcat, *Lynx rufus*, the only native wild cat now present in the Piedmont—the puma having long since been extirpated—rarely ventures outside the bottomland forests. The size of a small-to-medium dog, (20 to 35 pounds/45 to 80 kilograms), the bobcat is recognized by its 5-inch (13-centimeter) tail with a dark tip. It is perhaps three times the size of a normal house cat, but with a tail half as long. Prowling the bottomland forests at night, the bobcat takes small mammals and birds, sometimes carrion. Efforts to cover the shotgun-shell-size scats vary from thorough and compulsive to perfunctory. Any scat at which a burying gesture has been made is probably that of a bobcat.

The presence of the black bear, *Ursus americanus*, in the Piedmont is uncertain. I have seen scats in the Sumter National Forest in South Carolina which I took to be those of black bear, and there were uncomfirmed reports that the bulldozer drivers who cleared the Haw River bottomlands in North Carolina prior to the impoundment of the B. Everett Jordan Lake sighted bear. If the great beast is to be found anywhere in the province, it will surely be in some wide bottomland forest where it can paw unmolested into the alluvium for grubs and tubers, sample berries, beetles, and honey, and scavenge for carrion. It seems ironic that predation plays a relatively small part in the diet of this great carnivore. Birds and small mammals are the occasional prey.

Whether the bobcat and the black bear truly prefer the bottomland forests as their only habitat or are confined there by the press of civilization is conjecture. Both occupy higher, dryer habitats in the Appalachians and in the West, so it might seem that isolation from humans rather than wetness is the requirement. There are mammals whose adaptations *are* indisputably oriented to moist alluvial soils or to the flowing water itself. The eastern harvest mouse, *Reithrodontomys humulis*, is one. This small brown rodent with grooved upper incisors is firmly associated with marshy areas, wet meadows, and other alluvial habitats in the herbaceous phases. Another is the star-nosed mole, *Condylura cristata*, found mainly north of the James. An insectivore (that is, a member of the order Insectivora), the starnose has a poor sense of smell but senses worms and aquatic insects with fleshy tentacles on its snout. It is a good swimmer and spends some of its time rooting under stones for aquatic larvae.

Among the rather few truly riparian mammals are two mustelids, the mink and the river otter. Neither strays far from water other than to migrate between watersheds. They are rarely seen—both are nocturnal and extremely secretive—but they are more plentiful than is generally supposed. The naturalist who has had otters, *Lutra canadensis*, circle his canoe in the Okefenokee and the Everglades, snorting in curiosity and making spectacles of themselves in full daylight, will dispute my claim of the animal's wariness. Outside the great parks and refuges, however, these creatures with the sleek pelage have a history of experience with man which justifies every caution. Until the last few decades the otter was trapped out of much of its range. Returning now to the Piedmont's watercourses, it is glimpsed occasionally but is more often detected by its scats, glistening with fish scales and conspicuously placed on rocks or logs. The den entrances are below water level on stream banks; slides are worn into the snow or wet mud where exuberant energy, characteristic of mustelids but concentrated in the river otter, is expended at play. Otters also spend considerable time rolling in the snow or alluvial grasses in designated places. They have been observed floating through difficult passages of white water, head held erect and manueuvering with their webbed hind feet and long thick tail. Prey includes fish, frogs, crayfish, and salamanders, all of which, judging from the content of the scats, are eaten in their entirety. The otter is 2 to 2 1/2 feet (60 to 75 centimeters) long, weighs up to 25 pounds (55 kilograms) and bears a uniform brown pelage.

The mink, *Mustela vison,* with its lustrous dark brown fur, resembles the much larger otter in shape and in color. Although the feet are not webbed, the mink is an enthusiastic swimmer in pursuit of frogs, fish, and crayfish. It takes small mammals, including muskrats—of which it may well be the most important controller. Birds and the eggs of birds and reptiles augment the diet. The mink is nocturnal and secretive, much more often detected by its dark, thin, irregularly folded scats—which consist mostly of fur—than by direct sighting. Saturation density for the species may be one pair per linear mile of stream. The bottomlands must be wooded to be occupied by mink.

North America's largest rodent, the beaver, *Castor canadensis,* shares with man the behavioral trait of systematically altering its vegetative environment to produce a more desirable food crop. There being no natural lakes in the Piedmont, it is the historical function of the beaver to change, on a localized basis, lotic habitats to lentic. That is another way of saying that beavers build dams on small streams to create ponds. We explore the construction of ponds by beavers as a lentic phenomenon because flowing water is the raw material present when the beavers begin their task and because any stream to be dammed must have adequate year-round flow to sustain a constant water level in the impoundment.

In creating an impoundment, the beavers also change a forest habitat into a community of shrubs and hydrophytic herbs. The transformation of the habitat from unobstructed stream flowing through alluvial forest to a mature beaver pond may take thirty years or more and it involves a specialized instance of plant succession. The cycle begins when a family of beavers arrives at a site worth homesteading and builds a dam to raise the water level and flood some of the surrounding alluvial forest. They gnaw down trees up to 2 feet (60 centimeters) in diameter, opening the canopy and letting in sunlight. Other trees are killed by sustained flooding. At this stage in the development of the beaver pond, the beavers eat the bark and cambium of the upper branches and twigs of felled trees becoming, to a certain wildlife management mentality, pests; i.e., destroyers of board-feet of timber. Their dining habits seem wasteful in that huge trees may be felled and only a portion of the bark eaten. But with the exception of the hydrophytic oaks, the trees cut have little or no commercial value: in the Piedmont bottomlands such as sweet gum, tulip tree, dogwood, river birch, green ash, black haw, and woody plants.

When the time-consuming process of land clearing is completed—and Piedmont clearings are generally limited to 30 feet (10 meters) or so beyond the beaver pond's shoreline—sunlight is available to sustain the growth of herbs and shrubs, plants which reflect the beavers' true dietary preference. Buttonbush, swamp dogwood, the hydrophytic viburnums, golden club, cattails, and the tubers of water lilies are favored. Contrary to an early misconception that the beavers move on to deforest elsewhere after the trees are gone, the primary productivity of these beaver-arrested hydric shrub–herb communities is so high in the Piedmont that year-round food for a family of beavers is easily available. Beavers are highly territorial, a single pair of adults and their young of the year being the only occupants of the pond, their territorial unit. The sad fact is that the trapping of one or both adults can eliminate a beaver family, giving the impression that the feckless beavers, of which there must have been dozens to do such damage, have taken umbrage and decamped after the loss of one or two members.

Perhaps no other Piedmont mammal has been managed so extensively as the beaver. The nation awoke from the orgy of wildlife slaughter that characterized the nineteenth century to discover that there were no beaver left east of the Mississippi River. In the 1920s beavers live-trapped in the upper Midwest and released in Pennsylvania staged a suitable comeback under strict protection. Subsequently,

Virginia and North Carolina have received beaver transplants from the Pennsylvania populations, and reproductive performance of these imports has been splendid throughout the northern two-thirds of the Piedmont. Regrettably, the hand of management fidgets again to harvest some of the great rodents. Man is late to appreciate that the beaver is the sponsor of the Piedmont's native pond life, which includes some organisms that have survived and evolved there only under the beaver's aegis. Allowing the beavers to create such small ponds as they will in the bottomland forests would greatly enrich the diversity of the habitat.

The beaver also readily adapts to uncontrollably lotic habitats. The Delaware, the Potomac, the James, and the Haw are too powerful to be dammed with mud and sticks, but they are not devoid of beavers. Here, the beavers build no lodge heaped high with mud and sticks as they do in their ponds but burrow into the alluvial soils of the river banks, entering and leaving by submerged passages. Riverbank beavers are detected by the presence of felled trees, gnawed driftwood, and larders of edible branches floating in eddies near the den. The scats of beavers are rarely seen because they are deposited in water and, consisting mainly of sawdust, quickly disintegrate. Sightings are uncommon because the beaver is nocturnal and very shy. The flattened, scaly tail is dragged behind when walking, usually obscuring the tracks. Beavers are 35 to 40 inches (90 to 100 centimeters) long, tail included, weigh as much as 60 pounds (130 kilograms), have webbed hind feet, brown fur, and yellowed, chisel-size front teeth.

The muskrat is found in the Piedmont's stream communities but is more common in ponds so discussion is deferred. The muskrat is traditionally an opportunist arriving to occupy beaver ponds as soon as they are constructed and to eat the herbs the beaver labored to produce. See page 305.

Still Water: The Lentic Habitats

Of the Piedmont's hundreds of lakes and thousands of ponds, none are the result of glacial action, underground collapse, landslide across a stream or any other purely geological phenomenon. Of the lakes—a lake being larger than a pond—all are man-made, generally by constructing a dam across a previously free-flowing stream or river.

A few (less than ten) thousand years ago, there did exist large glacial lakes in the Piedmont of northern New Jersey, formed when huge blocks of ice melted within heaps of glacial debris, causing the surface to sink and collect standing water. Glacial Lakes Passaic and Hackensack once shimmered where the "meadows" of northern New Jersey now stand. Lake Hackensack, at the site now occupied by Hackensack Meadows, was connected with the sea by way of the Hackensack River and Newark Bay and was variably under the influence of tidal, saline waters. Consequently, the vegetative character and ecology of Hackensack Meadows is more akin to the coastal marshes than to a freshwater habitat. Farther south, the advance of the Wisconsin Glacier stopped in a line crossing the Watchung Mountains in the vicinity of Milburn. That left roughly the southern half of the Watchung Basin free of ice, and in that basin collected the waters of Glacial Lake Passaic. The lake shrank (the fate of all lakes) until it became the shallow standing waters of the Great Swamp, now a national wildlife refuge. The northern, or glaciated, portion of the Watchung Basin accumulated drainage and meltwater as the ice retreated and the lake or lakes which resulted are now occupied by Troy Meadows and Great Piece Meadows—low areas of moist soil and some standing water.

Of the province's ponds—year-round bodies of standing water of, to pick an arbitrary figure, not more than 100 acres or, using depth rather than surface area, shallow enough to support rooted plants throughout—all but a few are farm ponds and other man-made impoundments. A handful have been constructed in the last decade or two by beavers. Both types add diversity to the Piedmont's matrix of life, and their effect on wildlife has been largely beneficial. Those constructed by beavers offer the greater benefit because the builders are not obliged to observe the guidelines of the numerous federal and state agencies which require, for example, that pond sites be stripped of all standing trees, have waterlines cut with bulldozers, and otherwise be sterilized at the outset.

All bodies of still water are temporary. The cause of their inevitable demise is the process of being filled in by deposition of organic matter and by siltation. Water entering an impoundment from the watershed upstream bears inorganic silt and organic detritus which settle onto the bottom as the water slows. Vegetation in and around the pond or lake completes its life cycle and adds its substance to the organic accumulations on the bottom. The water's depth decreases as the buildup continues; the plants close in at the perimeter. Lakes become ponds, ponds resolve to swamps, and swamps to marshes and meadows. Ultimately, this reclamation process and its associated vegetative and edaphic succession produce alluvial soils which will sustain alluvial forests similar to those lining the rivers. Since very few of the Piedmont's man-made impoundments antedate the bulldozer, we have yet to see this reclamation process turn full cycle. Yet in most ponds and lakes it is well under way and readily observable soon after construction is completed. Indeed, it is the phases of lentic succession which define the life zones in a body of standing water at any moment. These phases, each constituting a definable habitat, are no less dynamic than their dry-land counterparts.

Energetics

The primary producers in lotic life systems live principally outside the water; they are the plants growing throughout the watershed which, upon dying, yield their partially decomposed detritus into the runoff. In lentic waters, the reverse is true; most of the plants that use solar energy to fix atmospheric carbon in their tissues live in the water itself. Further, the same plants, in performing their photosynthetic function, release oxygen into the water, making the pond or lake habitable by animals. This close co-location of the still-water autotrophs with their dependent heterotrophs (see Glossary) leads some to view the lentic habitats as "closed" systems capable of perpetuating themselves, given adequate sunlight and water. In Piedmont lakes and ponds, this concept is weakened by the intercourse of visiting birds, mammals, and insects who circulate energy and nutrients between the systems they visit and by the arrival of detritus and mineral nutrients in the runoff that feeds water into the impoundments. There is also a considerable flushing effect in bodies with some overflow, which serves to purge the acids produced by decomposition and to eliminate excessive populations of

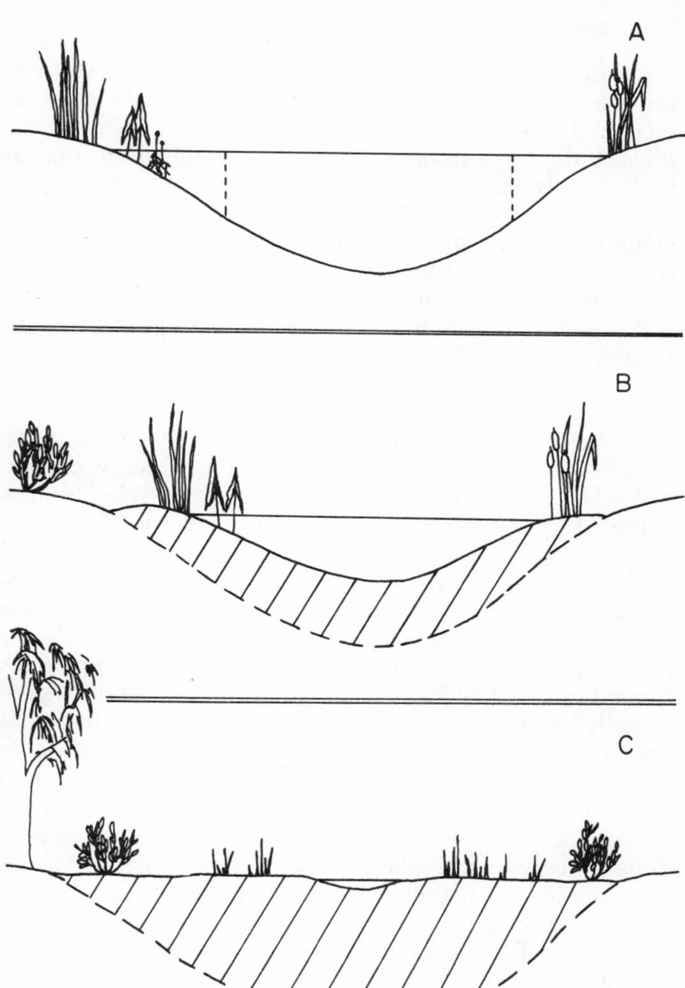

91. Edaphic filling of ponds

microorganisms such as algae. Still, a pond is undeniably more closed in its biology—more self-sufficient and self-contained—than a stream.

Zonation

Limnologists recognize a series of life zones in still waters which are related to depth and demarcated by the vegetation. The littoral zone extends from the shoreline to the greatest depth penetrated by sunlight adequate to support rooted vegetation. The littoral is the zone of greatest biological activity and it is the part of the impoundment undergoing the most rapid physical change. It is here that the aquatic plants are at work producing the life system's food and fiber, and it is in these shoreward shallows that the waters—due to presence of photosynthesizing plants—are richest in oxygen. With food and oxygen in abundance, the lentic fauna also are necessarily most concentrated here.

Viewed in greater detail, the littoral region can be further classified by the stature of its vegetation. Nearest the margin is a zone of emergent vegetation, which is characterized by such familiar forms as cattail, *Typha* spp., sedges (family Cyperaceae), and rushes, *Juncus* spp. The bulrushes, *Scirpus* spp., are actually in the sedge family, and probably merit their common name by the sessile reproductive structures which protrude, rushlike, from the side of the foliate spike. Another cyperaceous genus, *Eleocharis*, called spike rushes, is characterized by solitary spikelets topped by conelike fruiting structures. The spike rushes are actually sedges. The cattail is the tallest of the shoreline aquatics, sending its linear, erect leaves more than 6 feet (2 meters) high. The male flowers bloom above the female on a fertile spike, then disintegrate and leave the tightly packed seeds and pappuses resembling a skewered hot dog. Sedges are grasslike plants with triangular stems; rushes have round stems. Both grow in clumps in the zone of wet soils or standing water up to a few inches deep. Shoreward, these wet-soil aquatics yield to woody shrubs (buttonbush, swamp dogwood, and *Viburnum dentatum*) and trees (willows and alders).

In deepening waters, the emergent plants are not so tall as those at land's edge. Bur reed, *Sparganium americanum*

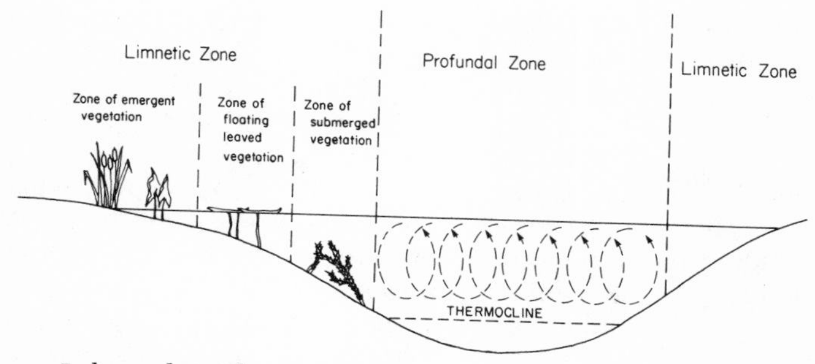

92. Lake and pond communities

(see page 243), is commonly tallest at about 3 feet (1 meter). Arrowhead, or duck potato, *Sagittaria* spp., mostly *S. latifolia*, may protrude up to a foot or two (30 to 60 centimeters), particularly if water levels have receded, holding a cluster of upward-pointing sagittate leaves clear of the water. Water plantain, *Alisma subcordatum*, carries its elliptic to ovate leaves similarly, though the leaf tips generally do not rise more than a few inches above the surface.

In depths generally between 1 and 4 feet (30 to 120 centimeters) there grows a community of plants whose leaves do not rise appreciably above the surface. This is the zone of floating-leaf vegetation. Water lily, spatterdock and water shield are the principals here, their rounded leaves buoyed by the surface tension. The leaves of water lily, *Nymphaea odorata*, are the largest, at a foot (30 centimeters) or more in diameter; the numerous white to pink petals show that this is not a true lily—lilies have three petals and three sepals. The leaves of spatterdock, *Nuphar luteum*, may float or may be submersed and they are less wide than long, the long axis being perhaps one foot (30 centimeters). The flowers are compact, golf ball-size and usually bright yellow. Floating, circular, almost paper-thin leaves 3 inches (75 millimeters) in diameter identify water shield, *Brasenia schreberi*, a very common aquatic in the Piedmont's ponds.

In the interstices between these plants with flat, floating leaves are aquatics bearing submerged leaves as well as some emergent foliage. Such plants are usually dimorphic; that is, the submerged leaves differ from the floating.

Pondweed, *Potamogeton pulcher*, is such a plant—the floating leaves are elliptic; the submerged foliage is linear. *P. diversifolius*, a pondweed widely distributed in the Piedmont, bears strongly dimorphic foliage, as the scientific name suggests. In many ponds the zone spanning 1 to 4 feet (30 to 120 centimeters) in depth is occupied by what appears to be the tightly spaced tips of small conifers protruding a few inches from the surface. This is the emergent foliage of parrot feather, *Myriophyllum brasiliense*, common in the littoral zones of ponds throughout the Piedmont. The submerged majority of the stem is frilled with featherlike foliage which waves gently with the water's motions but which is droopy and appressed when lifted from the water. Several species of *Ludwigia*, particularly *decurrens*, *uruguaensis*, and *alternifolia*, may also be seen in the same zone, with the upper 6 to 12 inches (15 to 30 centimeters) protruding to show widely spaced alternate leaves, generally entire and elliptic or lanceolate in shape. Occasionally, Piedmont ponds may be found to support the carnivorous aquatic, bladderwort, *Utricularia biflora* (normally a Coastal Plain plant but scattered widely inland), from whose submerged stolons dangle very small bladders with trap doors to capture minute aquatic animals. Bladderwort betrays its presence in summer by sending up thin stems, each bearing a single yellow flower.

The deepest band in the littoral zone is occupied by plants who do not reach the surface. They are characterized by long, delicate stems and short, usually frilly, foliage. Water nymph, *Najas* spp., is a common submerged plant, with sheathlike leaves occurring in widely spaced opposite pairs. Waterweed, *Elodea* spp., sold commonly as an aquarium plant and noted for its copious production of oxygen in the form of streams of visible bubbles, bears 1/4-inch (6 centimeter) leaves in dense whorls of four leaves each. Hornwort, *Ceratophyllum demersum*, has forked or branched leaves with serrated margins, 1/2 inch (12 millimeters) long in brushlike whorls. A pleasant characteristic of these plants of the inner littoral zone is their habit of releasing minute bubbles of oxygen so that the water sparkles with effervescence around the leisurely swimmer whose legs touch the vegetation.

In other physiographic provinces in the eastern United

States, particularly in the Coastal Plain, numerous species of plants have adapted to life as free-floating organisms on the surface. In the Piedmont, only the duckweeds, *Lemna* spp. (see page 242), fill this niche, though they may grow in profusion in selected waters, particularly the shaded headwaters where streams debouch into lakes.

Throughout the littoral zone, there may be simpler plant forms such as the stoneworts, *Chara* spp. and *Nitella* spp., whose tissues are not differentiated into roots, leaves, and stems as in the vascular plants. These large algae attach themselves to other plants and to objects on the bottom and may be incorporated in the botanical confusion pulled up with any handful of aquatic plants, particularly in the floating mats of vegetation which commonly accumulate offshore of the zone of floating-leaf plants. Like the vascular plants, the stoneworts are important primary producers in the Piedmont's still-water habitats.

At the edge of the littoral region we encounter a biological distinction between lakes and ponds. In lakes, where depths increase beyond the tolerance of rooted plants, there exists a botanically barren area called the profundal zone. A pond is botanically active throughout. In some small lakes there is enough diffusion of oxygen into the waters offshore of the zone occupied by rooted plants to sustain fish and invertebrates, or in very large lakes, wave action may circulate the waters oxygenated in the littoral zone so that they are habitable by fish far from shore. There is no question, however, that the biologically richest part of any body of still water is where the rooted plants grow, producing food and oxygen for other forms of life.

Animals of the Lentic Habitats

Microorganisms

In addition to the visible plants, there is another source of primary production and oxygenation in the lentic habitats: the microorganisms. There are numerous—nearly numberless—forms of life, too small to be seen individually, which contribute indispensibly to the lentic life processes. Protozoa, algae, diatoms, desmids, filamentous green algae, and blue–green algae are almost certainly present in any of the Piedmont's still-water habitats. Some of these organisms also fix nitrogen in their tissues, making it available for the construction of proteins in higher life forms. Their photosynthetic activity is great enough that they must be considered among the foremost primary producers of the habitat. There is uncertainty over the taxonomy of these groups, for some exhibit such animal characteristics as motility along with the decidedly plantlike property of photosynthesis. Their biological functions are viewed with alternating fear and favor, particularly in farm ponds. Excessive blooms in the populations can result in discoloration and opaqueness of the water, and when the rate of death and decomposition exceeds the rate of production, there occurs a condition known as eutrophication, wherein the oxygen demanded by the decomposition process leaves little available to sustain other life forms.

As in terrestrial habitats, there is a limited supply of certain critical elements (including phosphorus and potassium) available to the still-water communities, and these materials must be recycled. If they were allowed to remain in the tissues of the dead, there would soon be no means of sustaining life of any kind in the habitat. Also, if the carbon and nitrogen captured from the air by the primary producers were allowed to accumulate, it would quickly fill and choke the impoundment (although this is exactly what happens to the still-water habitats over the long term in spite of the best efforts of the decomposers). Some of the elements and nutrients are recycled by organisms feeding on the remains of the dead and fixing in their own tissues a portion of the materials eaten. Materials which go off as gases (NO_2 and CO_2) rather than being captured in the tissues of the scavengers must reenter the life cycle through the autotrophs (see Glossary) and work their way into the higher forms. Scavenging keeps the scarce life elements at a high trophic level.

Bacteria and fungi bear the responsibility for decomposition. Basically, the process is the reverse of primary production, in that oxygen is consumed and inorganic oxides of nitrogen and carbon are released into the atmosphere. Some bacteria, the anaerobics, can function, as the name suggests, without oxygen, a valuable adaptation in the oxygen poor sediments at the bottoms of ponds and lakes. The fungi and the aerobic bacteria require oxygen. No aquatic life system is free of bacteria, and the presence of these organisms need not signify impurity. Even the coliform bacteria produced in the digestive tracts of animals—fish and salamanders as well as people—are natural and unavoidable components. The fungi are said to be more effective than the bacteria at breaking down the more complex organic compounds like the lignin in wood and the chitin in the exoskeletons of insects. Certain fungi known as water molds (*Saprolegnia* spp.) infect, and often kill, fish and insects.

The zoologic composition of each lake or pond is unique, owing to the immense number of invertebrate species available for inclusion in the community. No listing could be complete. It suffices to be aware that there are countless animals below the visible threshhold of the field observer that feed on the minute primary producers previously discussed. Many are planktonic; that is, they simply drift where the water takes them. Others are necktonic—able to swim by means of flagellate hairs or cilia. The principal phyla of subvisible invertebrates in the Piedmont's still-water habitats are:

Protozoa (single-celled, or with multiple, undifferentiated cells)

Porifera (the sponges)

Coelenterata (of which the hydra are the most important freshwater representatives)

Platyhelminthes (the flatworms)

Nemathelminthes (the roundworms, perhaps the most abundant animals on earth)

Rotifera (the rotifers or "wheel animalcules")

Bryozoa (the moss animalcules; some form visible, gelatinous colonial masses)

Gastrotrichia (minute, poorly understood freshwater animals)

Visible to the unaided eye are other legions still. The Annelida include earthworms and leeches. The Arthropoda, a phylum embracing the crustaceans and the insects, are the most numerous. Another phylum, the Mollusca, is represented in the still waters of the Piedmont by snails and mussels. The ultimate phylum, Chordata, includes, but is not limited to, the vertebrates. We will devote no further attention here to the subvisual phyla beyond acknowledging their indispensibility as primary heterotrophs in still-water habitats.

Worms and Leeches

With the phylum Annelida, or segmented worms, we encounter the least sophisticated life forms subject to field observation. In addition to a few species of earthworms specializing in life on the bottom of still-water habitats, there are the delicate, generally transparent-bodied genera *Aeolosoma* (clear with red speckles), *Chaetogaster* (ice blue and bristly), *Deros* (clear and pinkish), and *Tubifex* (blood red), together comprising a taxonomic group called the oligochaetes, a reference to the relatively few bristles on each tuft on the body segments. All burrow into and eat the organic matter on lake and pond bottoms, performing the very important function of turning the litter and aiding in its decomposition. The tubifex worms are particularly important in oxygen-poor waters where they may be among the few animals capable of surviving. Building vertical tubes, from which they wave the crimson, hemoglobin-rich upper ends of their bodies to garner what oxygen is available, the tubifex worms "cultivate" the substrate, taking organic material out of the impacted realm of the anaerobic bacteria and bringing it to the surface (of the bottom) where it can be processed by the more effective aerobic types. This function contributes measurably to lengthening the life of any lake or pond, simply by increasing the time it takes to fill with organic debris.

The leeches are the other principal annelids present in still waters inland. They have flattened, segmented, elongate bodies with suckers at both ends and primitive eyes in their heads. Some swim, others move by "looping" with their suckers. There are many species, some who scavenge, some who eat whole prey, and others who suck the blood of vertebrates. They are frequent parasites of fishes, turtles, frogs, and bathing or wading humans, though, aside from being somewhat repugnant in form and habit they seem to be harmless. In attaching themselves to animals, they make a painless incision and inject an anticoagulant so that they may, at leisure, fill their elastic bodies with a meal which may last a year or more.

Some species of flatworms (Platyhelminthes) are as much as 1 inch (25 millimeters) long and are clearly visible. Flatworms occur in taxonomic listings prior to some of the subvisible phyla, with the implication that they have enjoyed less evolutionary advancement. This may be so, for their digestive tracts have only one opening. On the other hand, the creatures have eyespots in their head regions, an advancement lacking in groups of higher taxonomic rank. Some types of flatworms, such as the flukes and tapeworms, are parasites. One subdivision, the tubularians, are free-living creatures found in the littoral zones of ponds and lakes.

Mollusks

One of the distinctions between the chemistries of still and running water is that the presence in still waters of large masses of decaying vegetation lowers the pH. Acidic waters cause the shells of mollusks to deteriorate because the structures are composed largely of lime. Therefore, the abundance or lack of snails and bivalves in ponds and lakes is in part related to the vegetative strength of the littoral zone and to the rate of vegetative decomposition.

Generally, the mollusks are not so well represented in still waters as they are in streams. Snails (gastropods) seem to fair better than the mussels (pelecypods), and all mollusks appear to be limited to the littoral zone. Both pulmonate and gilled snails (see page 281) may be present, the two groups being easily distinguished by the bony operculum, or trap door, on the foot of the gilled snails. There is evidence that snails, particularly the pulmonate types, are more tolerant than mussels of the reduced amounts of dissolved oxygen characteristic of ponds. Moreover, snails feed on algae and on vascular aquatic plants, and are therefore better able to avoid the choking effects of heavy siltation on the pond bottom.

Mussels (see page 279) may be present in still waters,

though more often in lakes than in ponds. The most vigorous populations occur in the upper reaches where the inbound stream provides aeration and detritus, and more importantly, where there is considerable flow and flushing effect. The bivalves can live, even thrive, in sandy-bottomed stream mouths where the slowing waters release first their heaviest particles. Farther into the lake the finer silts settle onto the bottom and create a mucky substrate generally intolerable to mussels. Lest we conclude, however, that all impoundments are destructive to the bivalves, it is an inescapable observation of the Piedmont canoeist that the shorelines where streams flow into lakes are often littered with the freshly opened shells of the largest *Anodonta* (see page 280) to be found in the province. One before me on my writing desk, which I collected from atop a beaver lodge on the Rocky River where it flows into an old, and small, hydroelectric impoundment, is 6 1/4 inches (16 centimeters) in length. On the other hand, pill clams, *Pisidium* spp., half the size of an aspirin tablet, lace themselves into the tangles of floating vegetation of ponds and are so small and inconspicuous that they become visible only under careful scrutiny.

Crustaceans

The largest and most obvious crustaceans are those species of crayfish that specialize in still-water habitats. Some are totally aquatic, while others may be terrestrial or semiterrestrial and given to burrowing into the dams and banks of farm ponds. Some terrestrial forms build vertical chimneys of mud in moist, pondside soils. Due to the thickness of vegetation in the still-water habitats, pond crayfish are not so easily observed as their counterparts in running water, though their forms and functions as predators and scavengers are comparable (see page 281).

Many lakes and ponds, particularly in the southern Piedmont, contain representatives of the second largest type of still-water crustacean, the true freshwater shrimps. They are filter feeders who weave their thin bodies, 1/2 inch to 1 inch (12 to 25 millimeters) in length, through the vegetation, pursuing smaller crustaceans and other invertebrates.

While the larger crustaceans are the more familiar forms, it is the primitive orders which, by virtue of their diversity and sheer numbers, dominate the class in the Piedmont's lakes and ponds. Most are quite small, but with patience they are visible to the naked eye. The slight magnification of a watch glass or an inexpensive botanical magnifier reveals astonishing detail. These are best observed by scooping a small quantity of pond water into a shallow, preferably clear, dish and resting it on a stable surface to let the sediment settle while the eyes become accustomed to the scale. Some of the "particles" will remain in motion and the watch glass will reveal them to be members of several prominent crustacean orders. Fairy shrimp (order Anostraca), with about fifteen pairs of appendages, and tadpole shrimp (Notostraca), with forty to sixty pairs, swim with continuous, fluid motion. The seed shrimp (Ostracoda) and clam shrimp (Conchostraca) enclose their bodies in bivalved shells, revealing, generally, only a few appendages—antennae and leglike, tufted swimmerets—which seem to flail vigorously relative to the motion they produce. Most individuals could fit comfortably into a lowercase *o* in this typeface. The water fleas (Cladocera) are remarkable for their stunning diversity. Most are measured in fractions of a millimeter and are recognized by rounded, translucent bodies and prominent antennae, which are used as organs of propulsion. *Bosmina* and *Daphnia* are important genera, the latter being noteworthy for the seasonal changes in the shape of their heads. The movement of their appendages forces a continuous flow of water through the feeding filters, which are of such fine mesh that they can strain from the water any object larger than 1/25,000 of an inch. That includes all algae and protozoa as well as most bacteria. The copepods have radially symmetrical, torpedo-shaped bodies usually less than one millimeter in length. They swim with abrupt, discontinuous motions like players in an old silent film. Collectively, these primitive and diminutive crustaceans, most of which are filter feeders, are fundamentally important in cleansing the quiet waters of debris and in converting the algae, bacteria, protozoans, and lower invertebrates into food for the next level of predation. Isopods (order Isopoda) and side-swimmers (Amphipoda) are considerably larger than the foregoing orders of crustaceans, and they generally feed by scavenging on plant and animal remains. They are more often found among samples of plant debris and organic material from the bottom than from water near the surface. *Asellas* is a important isopod genus; the members are distinguished by flattened, spiderlike bodies (though with seven pairs of legs) nearly 1 inch (25 millimeters) long. The scuds, or side-swimmers, are laterally flattened, like fleas, and might be mistaken on casual observation for fairy shrimp. Most are 1/4 inch (6 millimeters) long or less.

Fishes

A dam obstructs the upstream travel of most fish, so those species that normally ply the Piedmont's rivers and streams on their way to or from the sea are normally absent above the dams. The herrings and shad of the Piedmont's open rivers, for example, are not commonly seen above the first dams. In addition to obstructing anadromous fish, impoundments offer less diversity of aquatic habitat than do the open watercourses, which, in the gradients of the Piedmont, feature alternating stretches of swift and quiet waters.

The diversity of fish populations in the Piedmont's still waters are further reduced by man, who employs programs that systematically exterminate some naturally occurring fish and replace them with a few species popular with anglers. When new impoundments are constructed, they are almost invariably stocked with bream, or sunfishes, and bass. The strategy is that the bream—usually the bluegill, *Lepomis macrochirus*—will eat the very small fish, including their own young, and will in turn provide food for the bass. The bass, as chief piscine predators, are to prevent overpopulation and stunting in the subdominant species. Managers point out that this arrangement results in very high productivity—perhaps 200 pounds (440 kilograms) of fish per acre—and larger fish more attractive to fishermen. It is claimed that individual fish attain greater size, on the average, in lakes and ponds than they do in free-flowing rivers, a generality most often applied to largemouth and smallmouth bass (see page 284). Such is the case, although still-water bass over 2 pounds (1 kilogram) are generally held to be inedible.

To supplement the paucity of species occasioned by the impassable dams, wildlife managers in the southern Piedmont have introduced a marine fish, the striped bass, *Morone saxatilis*, into some of the larger lakes. Ironically, this fish requires running water in order to reproduce, so fingerlings must be released each year if the populations are to be maintained. In the Piedmont, the striped bass, or striper, is said to be reproducing naturally only in Kerr Lake, on the Virginia–North Carolina border, an extremely large impoundment (the shoreline is more than 600 miles/960 kilometers long) with upstream access to flowing water. Lake Gaston, immediately downstream of Kerr lake on the Roanoke River, is unsuitable because Kerr Lake blocks the access to flowing water. Another consideration is that the striped bass requires as food fishes that school. The gizzard shad, *Dorosoma cepedianum*, fills this requirement and does not need access to the sea in its reproductive cycle. Present in Kerr Lake, the gizzard shad appears to complete the needs of the striped bass to maintain a presence in this one impoundment. In lakes devoid of gizzard shad, managers stock the threadfin shad, *D. petenense*, as forage for the striper. The hand of man also provides the province's still waters with bluegill and bass.

As a rule, however, the Piedmont's ponds and lakes are more diverse than the managers intend. (Their hope is to protect the introduced fish from predation and competition.) Eels are hard to stop. They apparently slither past the dams and make their way to the farthest headwaters, perhaps reaching adulthood in a farm pond astride a wet-weather ditch. Catfish show equal enterprise. Some of the native sunfishes such as the pumpkinseed, *Lepomis gibbosus* are commonly present in the older farm ponds and lakes. Certain topminnows and killifish, finger-length fishes that feed on insect larvae (including mosquito wrigglers) are nearly ubiquitous. Perches are present, particularly in impoundments fed by year-round streams.

Seeing the vast expanses of surface in some of the larger lakes, one wonders what creatures live in the depths. In summer, this is home for the larger fishes who occupy certain strata in search of an agreeable balance of temperature and dissolved oxygen content. Anglers troll in hopes of finding the right depth, the level where the piscine community has gathered to beat the heat and to prey on one another. Almost all fish are predaceous, the larger ones generally on other fish. There may be much activity just above the thermocline, that level at which an abrupt change in water temperature stops the circulation of convection currents. Below the thermocline there may not be enough oxygen to sustain even tubifex worms; just above it conditions may be optimal in the warmer months for the support of active fish populations at some distance, geographically and ecologically, from the sources of primary production.

Insects and Arachnids

Of the insect orders discussed under lotic habitats, all but the stone flies are present in lakes and ponds, though some of the orders are not well represented. The Megaloptera, for example, which embraces the dobsonflies and fish flies of streams, is represented in still waters only by the alderflies (see page 289). Broadly, those orders and families which are common to both habitats are represented by different species in each because the differences in the oxygen and mineral content generally exceed the adaptive span of any one form.

Mayflies are plentiful in the littoral zones, feeding on detritus and possibly on very small invertebrates, as do their lotic relatives (see page 285). Forms found in still waters, I observe, have prominent bristles on their abdominal segments, and these tend to collect algae and organic matter. Thus festooned, the mayflies may be less visible to fish and other aquatic predators. The caddis flies are well represented, though again, by species peculiar to still waters. The genera *Limnophilus*, *Phryganea*, and *Triaenodes* are the most frequently encountered. Each builds a characteristic chamber of whatever materials are available; the still-water types are commonly limited to twigs, seeds, and other vegetative matter, the polished sand grains of stream bottoms being unavailable. Further, the still-water species tend to greater mobility than those of lotic conditions, which are often obliged to cement their cases to the rocks to prevent being swept away.

The odonates (order Odonata) are abundant both as aquatic larvae in the Piedmont's lakes and ponds and as predaceous adults patrolling the airspace over the littoral zones, the open waters, and the adjacent fields. The adults are as prominent and conspicuous as any animal in the habitat; the naiads are rapacious predators in the Piedmont's still waters no less than in the streams. Mosquitoes, as larvae and as adults, are important in the diet of the odonates; "mosquito-hawks," in fact, is one of the folk names applied to the order. The dragonflies (suborder Anisoptera, see page 286) are dominated by the darners (Aeshnidae), with colorfully marked bodies up to 3 1/2 inches (9 centimeters) long, and the common skimmers (Libellulidae), with banded wings. The spread-winged damselflies (suborder Zygoptera, family Lestidae, see page 287) hold their wings rigidly outward at rest, as do the dragonflies, and care must be taken to observe the long, thin body and wider-than-body head to avoid misidentification. The narrow-winged damselflies are the most numerous pondside group (family Coenagrionidae). The family includes the bluets, *Enallagma* spp., with metallic blue abdomens, and the violet dancer, *Argia violacea.* Generally, male damselflies are more brilliantly colored than the females. All are raptorial as naiads and as adults.

Regarding the true flies (order Diptera), the noted limnologist R. E. Coker, who lived and worked in the Piedmont, wrote, "Among organisms of the bottom, midge larvae play a paramount role in converting organic detritus and microscopic organisms into the form of their bodies and they afford one of the chief foods of fishes." Some live in the mud and debris on the bottom, others in tubes suggestive of caddis fly tubes. The important midges in the Piedmont's ponds are in the family Chironomidae, which includes the red-bodied "blood-worms" favored by anglers. The latter build tubes into the bottom sediments and plow the substrate in a manner similar to the tubifex worms. It is estimated that midge larvae constitute one-tenth of the food supply for young fish in some lakes and ponds. One study in Iowa in 1919 by Dr. Emmeline Moore established that a certain midge, feeding exclusively on filamentous algae, was the most important food source for the young largemouth bass in the study area. Mosquitoes (family Culicidae) are the other dipterous family of importance in the still waters of the Piedmont. The larvae (see page 289) are aquatic and are the chief prey of certain small fishes including the topminnows, *Gambusia* spp., and the newly hatched fry of bass, bluegills, and other stocked pond fish.

It is unlikely that even a casual glance at the surface of a pond or lake in the Piedmont will fail to reveal a water strider (family Gerridae), one of the true bugs (order Hemiptera) with aquatic preferences. The broad-shouldered water striders (Veliidae) however, require running water and are absent from the lentic surface. Under the surface, the backswimmers (Notonectidae), half-inch bugs with keeled backs, swim upside-down with deliberate strokes of their hindmost legs, carrying a silvery package of air trapped on the abdomen. Backswimmers are highly predaceous on tadpoles, insects, and the fry of fish, piercing and sucking the juices from their prey with the same, typically hemipteran, proboscis, which can inflict a painful bite on the careless handler. A bug similar in appearance, but which swims upright, is the water boatman (Corixidae), a grazer on algae and a scavenger on plant and animal remains. Both are abundant in farm ponds and the littoral regions of lakes.

More than 2 inches (5 centimeters) long and slender as a walking stick is the water scorpion, *Ranatra* spp. This hemipteran hangs head down near the surface with its two, grooved caudal filaments clasped to form a breathing tube thrust through the surface film. Seizing a variety of prey with its raptorial front legs, the water scorpion is one of the major predatory insects of the still waters. The most fearsome of the lentic hemiptera are the giant water bugs (Belostomatidae), some exceeding 3 inches (8 centimeters) in length and capable of inflicting a painful wound on the human hand. This predator has a broad, flat body and a characteristic triangular shield (scutellum) on the back; it regularly devours tadpoles, frogs, small fish, and virtually any insect, holding the victim with powerful front legs and draining its fluids. A specific observation records a giant water bug capturing and killing a grass pickerel 3 5/8 inch (9.2 centimeters) long.

The beetles (Coleoptera, see pages 287–88) are the final order of still-water insects of interest to the field naturalist.

The whirligig beetles (see page 288) are common to the Piedmont's running and still waters. These predators enliven the surface of every pond, looking simultaneously above the surface film for danger and below it for food. A family of very small beetles (Halipidae) inhabits the bottom sediments foraging for algae and detritus. The family which comes more readily to the limnologist's mind when thinking of the still-water beetles is Dytiscidae, commonly called the "predaceous diving beetles." *Dytiscus* and *Cybister* are two genera important for the size of the individuals involved, if for no other reason; at 1 1/2 inches (almost 4 centimeters), they are among the larger predaceous insects. Other dytiscids are quite small, though all are enthusiastically predaceous. As adults, the dytiscids trap air beneath their elytra (wing covers) to fuel their underwater forays. By casual observation they appear similar to the giant water bugs, and their niche in the flow of still-water energetics seems comparable. The larvae, too, are highly predaceous, earning the name "water tigers." The complete range of the larger invertebrates as well as fish and frogs is believed to be included in the dytiscid diet. Other significant beetle families include the water scavenger beetles (Hydrophilidae, see page 288), scavengers as adults but predaceous as larvae, and the leaf-eating beetles (Chrysomelidae), whose larvae burrow into the roots and stems of waterlilies and other aquatic vegetation.

Although the arachnids are adapted primarily to life on land, some spiders can take advantage of submerged feeding opportunities. Pausing to watch the activity on the emergent and floating vegetation, one of the first movements to catch our eye might be the predatory dash of the fishing spider, *Dolomedes* spp. They make short work of the leaf hoppers and other insects that populate the plants, and they are capable of diving to capture tadpoles and the fry of fish. Jumping spiders are also common on the water shield's floating platforms, leaping in pursuit of the same insects. Web-spinning spiders apparently do not find the habitat suitable, although the fishing spiders have been known to weave submerged, horizontal webs in which they store transported bubbles of air for their own use during hibernation or for their newly hatched young.

Mites are differentiated from spiders in that they do not

have a cephalothorax segmented from the body. Several families of mites are entirely aquatic and are loosely grouped under the heading of water mites. These colorful little arachnids swim frenetically in any sample of water taken near the surface. Many are predaceous on other arthropods; some are parasites on mussels and insects.

Reptiles

The turtles occupying the Piedmont's still waters include those that live in the streams and rivers (see pages 290–91). The cooters and sliders, *Chrysemys* spp., with their confusing taxonomies and hybridizations, are the most commonly seen turtles basking on logs and other objects surrounded by the haven of water to which these flat-bodied reptiles take recourse at the slightest provocation. The stinkpot (see page 290) is present in ponds and lakes, creeping at its deliberate pace through the bottom vegetation. The eastern mud turtle, *Kinosternon subrubrum subrubrum*, occupies the shallow waters of small ponds and marshes south of the Roanoke River; it often wanders far from water. It is a small turtle, 3 to 4 inches (8 to 10 centimeters) long with a smooth, rounded, unmarked carapace, double-hinged plastron, and spotted head and neck. In similar habitats north of the Savannah River, the spotted turtle, *Clemmys guttata*, may be found, basking individuals being readily identified by the polka-dot pattern of yellow spots on the shell. The attractive red and black margins of the shell identify the eastern painted turtle, *Chrysemys picta picta;* confirmation is provided by the placement of scutes in straight rows across the reptile's back. With yellow streakings on the head and neck, the eastern painted is the most colorful of our aquatic turtles. The length is generally under 6 inches (15 centimeters). Most of the diets of the foregoing turtles involve only minor predations and concentrate instead on scavenging and on vegetable matter. The reverse is true regarding the snapping turtle (see page 290), responsible for the disappearance of ducklings and many other creatures from still waters as well as streams.

The northern water snake (see page 291) is the most abundant aquatic snake in the Piedmont, and it favors ponds no less than streams. It is a harmless reptile, as (in my experience) are all other snakes seen voluntarily swimming in the Piedmont's waters (a copperhead who takes to the water when molested in his pondside shady spot in late summer is not to be considered a voluntary swimmer). The water snake's prey is aquatic vertebrates. The eastern ribbon snake and the eastern kingsnake (see page 291) are seen occasionally on the vegetated banks of ponds. Conceivably, any terrestrial snake may be seen at a pond's edge, for the rushes and hydric shrub communities are rich in birds, insects, and small mammals. The queen snake, *Natrix septemvittata*, is less a still-water inhabitant than a lover of rivers and streams, but this slender, coffee-hued serpent with yellow underparts may be found anywhere it can satisfy its somewhat refined taste for freshly molted crayfish.

Amphibians

Among the Piedmont's salamanders, there are few that depart from the rills and seeps to venture into larger waters inhabited by predatory fish. Some, such as the dusky salamander (see page 293) finds shelter in the sandy and rocky bottoms of streams. A few make do with a special still-water habitat which these and other amphibians make worthy of the naturalist's attention: small, mostly woodland pools that are free of predatory fish. Most such pools are found in the alluvial soils adjacent to rivers, but they may occur in the uplands as well. It is here that we encounter the spotted salamander, *Ambystoma maculatum*, to 8 inches (20 centimeters) in length and black with large yellow to orange spots; the adults of the red-spotted newt, *Notophthalmus viridescens viridescens*, the efts of which are terrestrial; and, occasionally near the fall line, the eastern tiger salamander, *Ambystoma tigrinum tigrinum*. The tiger salamander also inhabits larger bodies of water such as beaver ponds and farm ponds which have not been stocked with fish. Dappled yellow on a dark background, the tiger salamander (8 to 12 inches/20 to 30 centimeters in length) is one of the Piedmont's largest salamanders. The still-water salamanders prey on earthworms, leeches, very small mollusks, and other aquatic invertebrates. Being rather slow moving, they fall prey to just about any carnivorous organism, including the larger predatory insects; they survive by carefully selecting habitats that provide the protection they require. Still-water impoundments, lacking the sheltered ledges and caves common on the burnished bottoms of streams, are suitable to only a few salamanders.

Frogs, on the other hand, are considerably more agile and can risk greater exposure with confidence that an arcing leap will probably carry them to safety. The Piedmont's still-water habitats contain the province's most varied array of frogs and toads. The American and Fowler's toads (see page 100) wander the uplands as adults, but they are strongly aquatic as larvae, the tadpoles occupying every category of still water, including ponds and lakes. The diminutive northern cricket frog, *Acris crepitans crepitans*, is peculiar to the shoreline of well-vegetated littoral zones. Throughout the warmer months they gather in clusters to feed in the sun-warmed shallows and, at intervals, to engage in a group vocalization which sounds like a sack of marbles poured slowly onto a concrete floor. Cricket frogs are not easily seen, even when singing intensely at the observer's feet. Spring peepers (see page 293) are present in wooded, shallow reaches of lakes and ponds; the upland chorus frog (see page 294) sings simultaneously in spring from the open shallows of ponds and vernal pools. The gray tree frog (see page 293) sings from the trees above the wooded pools favored by spring peepers.

The amphibian most closely associated with lakes and ponds is also the largest—the 7-inch (18-centimeter) bullfrog, *Rana catesbeiana*. The bullfrog has a greenish head, gold–green body and coppery tympanum. It preys on a great variety of flying insects and probably on small mice, other frogs, and small snakes. Dr. Alexander Wetmore, the great ornithologist and Secretary of the Smithsonian Institution, records a bullfrog leaping to catch a sparrow in full flight, then swallow the bird whole. The bullfrog's deep, resonant *mmwonnng* echoes from the banks of nearly every lake and farm pond on summer nights. The green frog (see page 292) averages about half the bullfrog's bulk and is differentiated at sight from the larger amphibian by the dark lateral stripe connecting its tympanum and haunches. The green frog is more closely associated with flowing water than still, but it is absent from few impoundments in the Piedmont. The leopard frog (see page 292) may also be present, particularly in its subadult stages, for it breeds in pools and ponds, though it wanders into the mesic realm in summer. These true frogs—members of the genus *Rana*—are the largest of the frogs and are the providers of frogs' legs for some of the Piedmont's more discriminating predators.

Birds

The extent to which the conversion of flowing water to still and the loss of the alluvial forest affects birdlife is the object of inquiry by birding groups and other naturalists in the Piedmont. Obviously, the forest birds are forced to abandon the site. Among the more regrettable losses are the lowland wood warblers, including the water thrush (see page 296), the majority of the pileated woodpeckers, and the red-shouldered hawk.

Many, however, remain and others are added. The spotted sandpiper (see page 296) is present in the nesting season. The greater and lesser yellowlegs, *Totanus melanoleucus* and *T. flavipes*, and the solitary sandpiper pass through, and occasionally linger, on migration between their arctic breeding grounds and their wintering quarters thousands of miles to the south. The yellowlegs, pigeon-size birds with light brown upper parts and white rumps, are not easily distinguished visually, but the greater yellowlegs' ringing four or five-note call and the lessers' subdued couplets are a help. With the forest gone, the woodcock is absent but is replaced by the common snipe, *Capella gallinago* (see page 300), who probes in the marshy soils at lake's edge, showing particular interest in the mats of floating vegetation and recently created muck associated with the successional process by which the impoundment is filled in. The kingfisher (see page 295) is present year-round, sitting in snags and on nearby fence posts and diving into the still waters for fingerlings. A burrowing nester, the kingfisher probably finds suitable facilities upstream from the lake, rather than adjacent to the impoundment. The osprey (on migration), the willow-loving yellow warbler, together with its nesting parasite, the cowbird, and the eastern phoebe, all discussed earlier as part of the lotic habitats, are also associated with impoundments.

Among the wading birds, the still and moving-water populations are comparable except that the cattle egret, *Bubulcus ibis*, is substituted south of the James for the immature little blue heron, *Florida caerulea*. Many farm ponds in the southern Piedmont offer attractive conditions to this African immigrant—marshy shorelines and nearby livestock. The cattle egret is a crow-sized, short-legged wader with pinkish legs and bill. Males in breeding plumage are tinged with cinnamon on the breast, back, and crest. The great blue heron (see page 294) may be seen on Piedmont lakes and ponds so long as they are free of ice. The great egret (see page 294) visits casually in summer. The green heron (see page 294) is the most common wading bird of the habitat. The American bittern, *Botaurus lentiginosus*, and the least bittern, *Ixobrychus exilis*, are not uncommon but are notably elusive in the reedy pondsides. Both are protectively colored. The American is brown-on-white with vertical striping on the breast (to simulate vegetation) and the least bittern is generally cinnamon colored with a greenish back. The least is confusible with, but smaller than the green heron. The American bittern is larger than a crow. Similar in size and shape is the black-crowned night heron, *Nycticorax nycticorax*, a nocturnal forager for fish and frogs in the Piedmont's ponds and lakes whose single croaked note, given in flight, adds mystery to the night. Generally, the wading birds are more numerous in the eastern and southern parts of the Piedmont than toward the mountains.

The red-winged blackbird, *Agelaius phoeniceus*, is one of the main beneficiaries of dam-building. A relative of the orioles, this handsome singer favors the reedy shorelines, foraging in adjacent pastures for grasshoppers and other insects. Females wear protective brownish coloration. First-year males are similar but with red shoulder patches faintly visible. Breeding males, of two years and older, are glossy black with crimson epaulets edged in yellow. Because first-year males are not attractive to females, there is always an excess of breeding females, and many adult males are polygamous. Breeding males put all their effort into singing and defending territory—they do not contribute to the rearing of the young beyond escorting the ladies on their foraging errands about the territory and pestering them, seemingly unnoticed, with tireless displays of color and song.

As a group, the waterfowl probably benefit most from the building of lakes and ponds. Almost any loon, grebe, duck, or coot using the Piedmont as a migratory flyway may be seen, at least temporarily, in an impoundment. Most are quite temporary. The mallard and the wood duck (see page 299) are the most frequently sighted nesting waterfowl, though the wood duck, of course, requires its waters, still or flowing, to be sheltered in woodlands. Hence, it is typically limited to the wooded upper reaches of larger lakes. In winter, Canada geese, whistling swans, and other large anatids rest on still waters out of the sound of gunfire, and some may be induced to stay if grain is made available. Man-made Gaddy's Pond, in Anson County, North Carolina, provides a well-known example of how large numbers of geese and ducks can be habituated to wintering in a Piedmont impoundment. A dangerous dependency on humans sometimes results of this practice; we induce waterfowl to alter their ancestral migratory habits at their peril.

The ring-necked duck, *Aythya collaris,* includes the southern Piedmont (James River southward) in its natural wintering range and may thrive on lakes and larger ponds with well-developed littoral vegetation. The bird is identified by its purple head, green breast, dark back, and contrasting whitish flanks. The common merganser (see page 299) winters as far south as the James and in that vicinity is replaced as a wintering species by the hooded merganser (see page 299). The ruddy duck, *Oxyura jamaicensis,* a squat, compact diving duck with tail held erect, visits on migration and may linger in winter on ponds toward the outer Piedmont. The pied-billed grebe, *Podilymbus podiceps,* small and compact of body like the ruddy duck and also given to diving, nests on impoundments throughout the Piedmont if littoral vegetation is adequate. The food is small fish and other aquatic animals.

The marsh hawk, *Circus cyanus,* is the diurnal bird of prey peculiar to the open wetlands associated with the bottomlands and the successionally reclaimed edges of lakes. The male is silvery with dark wingtips; the female is brown with contoured markings beneath her wings. Both have a conspicuous white rump patch. The wings are held high to clear the vegetation as the marsh hawk tirelessly quarters the lowlands in search of rodents. It rests on the ground or on lookouts no higher than a muskrat house or fence post.

Mammals

As the vegetation, combined with the deposition of silt, gradually converts an impoundment to a terrace of alluvial soil, one finds, not surprisingly, that the animal life of the pond's edge is similar to that of bottomland soils in comparable seral stages. The meadowlike zone outside the ring of shrubs surrounding farm ponds, rich in rushes and sedges, is haven for many of the same mice, voles, moles, and shrews found in the bottomlands (see pages 302–3). The rice rat, *Oryzomys palustris*, present south of the Potomac in moist swales common along the edges of older lakes and ponds, is smaller than the hispid cotton rat (see page 107), and has a scaly tail and short soft fur. These small mammals are the prey of the long-tailed weasel (see page 108), the mink, the barred owl, and the marsh hawk.

The opossum and the raccoon are not particularly discriminating in their selection of habitat, but find the shoreline of lakes and ponds very favorable for their nocturnal patrols. The bobcat is a forest creature and may visit the wooded upper waters of certain lakes if the smell of humanity is sufficiently faint.

We have noted the three-way relationship between the beaver, the muskrat, and the mink. The beaver builds the ponds and clears land to grow the herbs favored by the muskrat who, in turn, is prey for the mink. In many cases, the construction of a lake by humans bypasses the beaver's role in this troika, but not in all. Near the headwaters of some lakes which have stabilized water levels and are fed by streams of sizable flow, the beavers find a zone of flooded alluvial soil suitable for the growth of wetland herbs and shrubs. Vast acreages of alluvial forest suitable as beaver habitat may have been destroyed downstream to achieve this effect, but in this ecotone where flowing waters meet the lake the beavers find suitable conditions without having to go to the trouble of building a dam. It is in this zone, too, that the *Anodonta* mussels (see page 280) reach such great size, offering lavish meals to the muskrat, raccoon, and otter. The beavers need only build their lodge in the shallows or swim upstream and tunnel into the bank. The muskrats accommodate themselves similarly, building houses of heaped vegetation in the shallows.

The muskrat, *Ondatra zibethica*, is present in aquatic habitats throughout the Piedmont. It is present in rivers and streams, though in modest numbers because of the relative paucity of aquatic herbs. Herbaceous plants from the riverbanks and adjacent floodplain are taken, but at the risk of considerable exposure to predators. In the richly vegetated littoral zones of impoundments, the muskrat can swim and forage beneath the surface and may use the mats of floating vegetation for quick refuge. Of its principal predators, perhaps only the mink can pursue it through the water.

Feeding voraciously on aquatic plants, the muskrat materially retards the filling-in process. It occupies the Piedmont's impoundments at all depths, including the deepest sections at the dam into which it tunnels to a depth of 6 or 8 feet (180 to 240 centimeters). These tunnels can cause leaks and failures in an earthen dam and can be a hazard to livestock. Yet whether the rodent is viewed as a blessing or a curse, it is on the lakes and ponds to stay. In man the muskrat has found a bigger, better beaver, a provider of still waters beyond its wildest evolutionary dreams.

The muskrat is about 12 inches (30 centimeters) long (tail excluded). The coat is of rich chestnut, uniformly dark and lustrous above, silvery below. The tail is nearly as long as the body, naked, scaly, and flattened vertically. When swimming at the surface, the top of the body is out of the water for its entire length; the otter, in contrast, swims at the surface with only the head visible. A serpentine motion of the muskrat's tail helps propel it through the water. The muskrat is the only small mammal likely to be seen swimming for some distance in a straight line at the surface as if traveling from point to point. The otter continually bobs at the surface and dives, following an erratic course. The beaver rarely permits itself to be seen in daylight.

PART IV

Special Places

CHAPTER EIGHT

Rare Botanical Communities

THE PRECEDING CHAPTERS describe the successional phases typical of the Piedmont's habitats. Few examples of specific locations are cited because the habitat types are common in the province. This chapter presents habitats in the Piedmont which are atypical, even unique: the granite outcrops, the serpentine barrens, the highly developed north-facing slopes with their associated rich woods, and, in the extreme northern Piedmont, the postglacial swamps and meadows. We'll also look at some of the rare xeric, mesic, and hydric habitats, particularly those at maturity.

Some of these sites are in public ownership and access to the public is unrestricted or at least available on a controlled basis. At the other end of the spectrum are locations whose owners, for reasons of privacy or to protect their property, prohibit all access. In the great majority of cases where access to private lands is restricted, there is legitimate reason, usually to prevent an uninformed public from damaging the habitat. The author urges that in seeking access to the Piedmont's natural habitats, rare or common, the right of landowners be scrupulously observed. If you or I owned a north-facing slope that shelters an isolated population of dutchman's breeches, we too would be protective. One has only to look at the beer cans, broken glass, and motorbike tracks through the endemic plants in the vernal pools of some of the Piedmont's granite outcrops to wish that some landowners were even more protective. Remember, if a site is privately owned, it is the visitor's responsibility to secure permission to enter. And that may not be easy.

Many of the sites are in obvious need of some kind of

protection. Most—though not all—are priority targets for acquisition by state or local agencies for park lands. Some are so fragile that unrestricted public use would destroy them, so some means of private preservation is necessary. The Nature Conservancy has chapters in every state to focus the efforts of residents and interested visitors. The special places of the Piedmont and the organizations which work to preserve them need our help. The near term will be decisive in the future of our natural heritage, for what we fail to find, recognize, and preserve in the next few years will almost certainly be destroyed.

Granitic Outcrops

As the crustal plates of Africa and North America approached one another half a billion years ago, the friction of their leading edges grinding together caused volcanic violence in the protopiedmont. Veins of molten rock coursed near the surface, some to rupture and blow their white hot minerals skyward. Others found their way into deep vaults of rock from which there was no escape, regardless of the intense temperatures and pressures. Great bubbles of the hot, fluid magma forced their way into older rock masses deep in the earth's crust and cooled, very gradually, in place. The slow-cooling magma formed a particularly obdurate granite, in many cases more resistant to weathering than the surrounding rock.

Later, perhaps in the upward thrusts resulting from the full impact of the continents roughly 450 million years ago, the entire lithic structure of what is now the Piedmont was elevated to great heights. A new erosion cycle began; others followed. Ultimately, the country rock containing the granite was exposed; then it weathered away, leaving the Piedmont's granitic domes.

"Exfoliation domes" is preferred by some geologists. The supposition, not beyond debate, is that the granite bubbles cooled in such a way as to produce concentric layers of rock, onionlike in structure. The rhythms of expansion and contraction with seasonal temperature change caused these layers, ranging in thickness from a few inches to several feet, to peel off, or exfoliate, littering the edges of the higher domes with skirts of scree. The exfoliated rock varies in size from granular sand to house-size boulders. The rate of exfoliation is so slow and the vigor of the Piedmont's plants is so great that the scree is much more likely to be buried under soil and vegetation than to be exposed like the piles of talus beneath cliffs in the western United States.

All the major exposures of granite in the Piedmont are south of the Roanoke River. Those in the western part of the province are generally tall, like the other topographic features in that region, and merit the term "dome." Stone Mountain, near Atlanta, and the helmet bearing the same name in Wilkes County, North Carolina, are among the best developed of the Piedmont's domes, being approximately hemispheric and hundreds of feet high. Eastward, the granitic outcrops are reduced in size, in many cases not rising significantly above the other crests in the rolling terrain. In the eastern Piedmont, where the peneplaining is most advanced, the outcrops may be worn quite flat and are referred to locally and scientifically as flatrocks, or pavements.

The geologists have their slant on the outcrops; the life scientists and naturalists have another. It is particularly true in the east that, in the absence of conspicuous domes, the outcrops are differentiated from the landscape biologically more than geologically. A casual glance at the passing countryside may not reveal the terrain features which signal a granitic outcrop, but the endemic outcrop flora, if caught in flower, will stop all traffic. The plant communities associated with the outcrops are peculiar to these few isolated locations; they are distinct from all other habitats, radically different from the surrounding mesic slopes. Some of the rarest plants in the Piedmont are found on the outcrops; some are known from a single location. The botanical rarities and the far-flung, disjunctive, and sparse occurrence of these granitic landforms make it a near certainty that each outcrop community you visit is a unique habitat.

The initial glance at the surface of a granitic outcrop gives the impression that the relentless forces of plant succession are suspended on these seemingly barren surfaces; this is an illusion. In fact, there may be no habitat in which

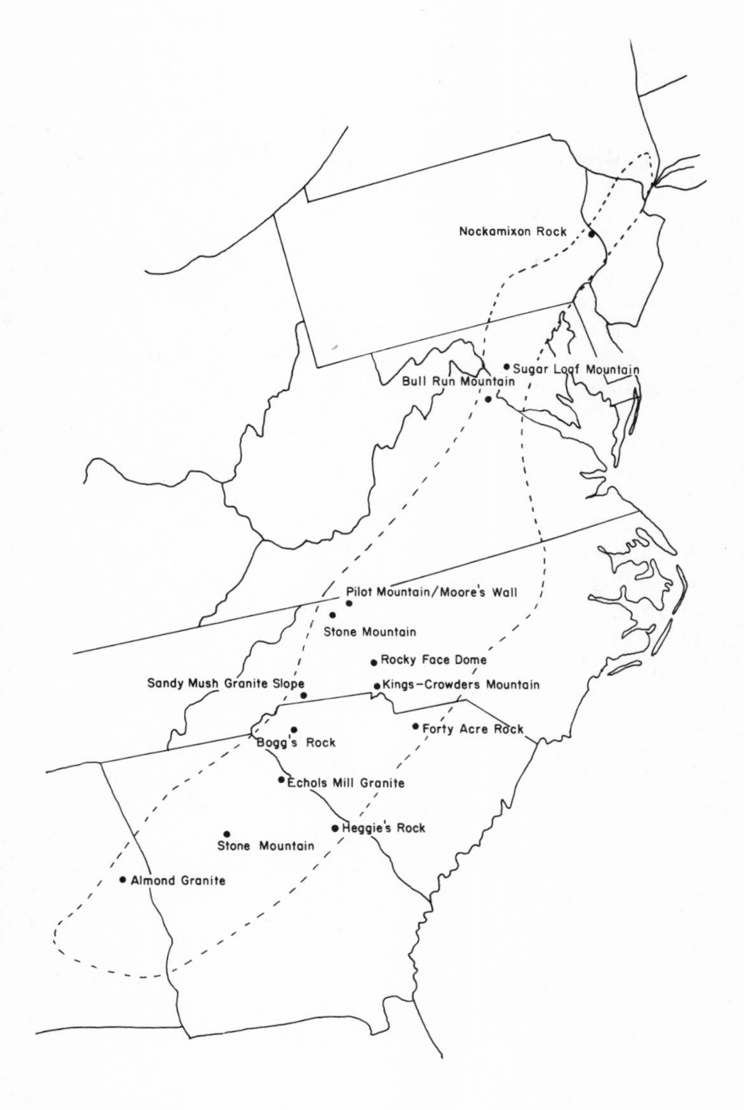

93. Major rock outcrops

succession works with such grinding determination, though the process is slow. On the outcrops, the soils, to the extent that they exist, are deeply xeric, yet the term used to describe the successional process here is not xerosere, but petrosere, implying the additional dryness, acidity, and thinness of soil associated with life on a bald rock face. Outcrops do not retain moisture well but are highly efficient conductors of heat; unmitigated exposure to sun and wind make the outcrops in the Piedmont almost like deserts. The difference is that the rainfall is plentiful enough; it's simply a matter of letting enough time pass for erosion to flatten the rock so that soil can accumulate.

Since we are talking about the accumulation of soil on previously barren surfaces, the successional process is primary, not secondary. Endemic to the petrosere on the Piedmont's outcrops of granite is a community of pioneer annuals well-suited to advance the province's harshest sere. But first, a word about their predecessors, the lichens and mosses.

The lichen is a very ancient form of life, simple in the sense that its structure is nonvascular, but complex because it is really two plants—an alga and a fungus—living symbiotically together. In fact, algae and fungi that live together as lichens are not known to exist separately. Their life cycles are interdependent to the point that they are, in appearance and function, a single plant. The alga contains chlorophyll enabling it to photosynthesize carbohydrates, some of which are used by the fungus. The fungus excretes acids which dissolve rock-bound minerals used by the alga. The little splashes of lichen that seem at first to be simply part of the color and texture of the granite are powerful agents of erosion. Water seeps into the acid-etched crevices beneath the lichens, then freezes, expands, and flakes off bits and chunks of rock. Beneath some of the Piedmont's helmets of granite lie heaps of shards, some still wearing the lichens that chiseled them from the parent rock.

Life has the best chance of gaining a purchase on the bare rock when the combined actions of lichens and weather scoop out a small depression into which a fine sand of granite particles may combine with organic debris. Here the lichen may be joined by certain mosses and ultimately by the vascular plants. It is the herbaceous phase of the petrosere on the Piedmont's granitic outcrops which attracts so much botanical attention, for the species individually are uncommon or rare, and the seral communities of zonated bands marking the changes in soil depth are among the province's great spectacles. Beyond the herbaceous phase, a stunted forest of cedar and dry-soil broadleaf trees may develop after the soils are established to a few inches depth. Ultimately, the site may be smoothed and buried under a mesic forest undistinguishable from those cloaking the surrounding slopes.

We come to the outcrops mainly to see the herbs. We recall that the spring ephemeral wildflowers of the alluvial forests and north-facing mesic slopes compensate for the stress of being deprived of the sun during the growing season by carefully timing their appearance and by practicing a form of ephemeralism. On the outcrops the stresses are considerably more severe. The herbs of the granitic outcrops—like plants in a desert—span the periods of stress between the times favorable for growth and reproduction as seeds rather than as underground storage organs. Occupying the shallow depressions and vernal pools on the bare rock face, the stonecrop, *Sedum* spp., and sandwort, *Arenaria* spp., deposit their seeds in the thin—sometimes only a few millimeters—collections of fine gravel and organic matter in the depressions. When (and if) moisture and mild warmth arrive again in the vernal weeks, the seeds germinate and the depressions and organic collections explode into life. The rare *Sedum smallii* erects columnar, succulent, bright red stems and leaves topped by four-part white flowers. The entire plant may be 3 inches (8 centimeters) tall, but in the view of the ant, the saltacid spider, and the botanist the little zones of stonecrop are a thriving crimson forest. Where the "soils" deepen a millimeter or two, the sandworts—on selected outcrops the very rare *Arenaria uniflora*—bristles an inch or more above the stonecrop. The white flowers, a 1/4 inch (6 centimeters) across, are shaped by five recurved petals.

Amphianthus pusillus is possibly the Piedmont's rarest plant. A. E. Radford writes in *Manual of the Vascular Flora of the Carolinas:* "[It occupies] shallow pools on granite outcrops, extremely rare; Lancaster Co., S.C., and Ga.

Apparently seed germination occurs only during certain, probably wet, years. The plants are thus intermittent in their specialized environment. A monotypic genus endemic to the two states listed." Resembling a miniature water lily, *Amphianthus pusillus* sends its floating leaves, two per stem, on lax stems spanning the inch or two between the bottom of its vernal pool and the surface. A minute white flower opens at the surface between the two floating leaves. Another set of flowers, said to be cleistogamous (meaning 'hidden marriage,' hence, closed and self-pollinating, in case external pollination of the surface flowers fails), clusters in a basal rosette at the bottom of the pool. One's visit must be timed precisely to see the *A. pusillus*, as well as the sandwort and the stonecrop. Try early April and be prepared to visit again in a few days if these treasures of the petrosere are a bit late. A week after conditions are optimal is generally too late to see anything but a few dried stems.

Other outcrop endemics occupy soils of an inch or so in depth. *Portulaca smallii* and *P. coronata* are succulents having leaves radiating from the stems like the bristles of a brush. The 12-inch (30-centimeters) tall, yellow-flowered *Senecio smallii* (and possibly *S. tomentosus*) often grows in a zone bordering that of the sandwort. Daisylike and clearly asteraceous, the outcrop senecios are sometimes colloquially called Confederate daisies, a common name sanctioned by taxonomists only in the genus *Viquiera*. The blue-flowered *Phacelia maculata* blooms in soils almost deep enough to sustain shrubs and trees amid the grass *Panicum lithophilum* and the sedge *Cyperus granitophilus*. Where soils deepen still further, the vegetation endemic to the outcrops melds with the xerophytic plants common in the region.

The following are some of the more prominent granitic outcrops. There are, of course, others, mostly of lesser stature botanically and geologically, but interesting nonetheless. Locating the smaller, locally known outcrops can be rewarding, particularly if a visit can be timed to catch one or more of the major sites at its seasonal peak.

Almond Granite Outcrop

LOCATION: Randolph County, Alabama; approximately 1/2 mile south of Almond; lat. 33°08′32″ N, long. 85°37′45″ W.

DIRECTIONS: State Route 77 passes through the outcrop area, a 400-acre (160-hectare), roughly circular tract bordered by the town of Almond on the north and by Cedar Creek on the south. Park on Route 77 and walk south or take the county road southwest from Almond and park 1/4 mile south of Route 77.

OWNERSHIP: Southern Union State College, Wadley, Alabama.

SIGNIFICANCE: This is a large outcrop with various slope and drainage conditions. Substantial accumulations of soil result in storage and subsequent seepage of water, accounting for the presence of sweet bay, *Magnolia virginiana*. This is one of two known sites for *Arenaria alabamensis; A. uniflora* is also present. *Sedum smallii*, a stonecrop, is an ephemeral of the shallow depressions. For comparison of moisture requirements, prickly pear, a cactus, occupies somewhat deeper, wetter depressions. The carnivorous *Utircularia cornuta*, a yellow-flowering terrestrial bladderwort, grows profusely in the boggy seeps, especially where mosses are present. *Senecio smallii*, *S. tomentosa*, and *S. Pauperculus* grow in the deepening soils at the edges of the rock.

Blakes Ferry Flatrock

LOCATION: Randolph County, Alabama; lat. 33°16′30″ N, long. 85°38′30″ W.

DIRECTIONS: From Wedowee go 8 miles west on Route 48 and take the second left after the bridge over the Tallapoosa River. Go left at the next intersection (about 1 1/2 miles). The flatrock is on the left.

OWNERSHIP: Private

SIGNIFICANCE: This is where *Arenaria* (formerly

Minuartia) alamabensis was discovered in 1971. *Sedum smallii, Arenaria uniflora,* and the *Senecio* spp. are among the outcrop endemics present. The main granite exposure is a pavement about 3000 feet (900 meters) long and 900 feet (275 meters) wide. Dirt bikes and four-wheel drive vehicles are damaging the site, and a 1975 report indicates bulldozing and tree removal. Several smaller pavements in the area are more secluded than the main surface and are largely undisturbed.

Echols Mill Granite Outcrop

LOCATION: Oglethorpe County, Georgia; lat. 33° 58′00″ N, long. 83°00′10″ W.

OWNERSHIP: Unknown

SIGNIFICANCE: Radford and Martin, in their report to the National Park Service on *Potential Ecological Natural Landmarks, Piedmont Region,* write, "Even though this area is being actively disturbed, it still has the greatest plant diversity of *all* the granitic outcrops." (Italics mine). The disturbance is, of course, people amusing themselves driving motorcycles and jeeps through the vernal pools of rare plants. The representative outcrop-endemic plants are here in force. One of the attractions is American pillwort, *Pilularia americana,* described in the natural landmarks report as "small and very elusive," ranging "irregularly from central Oregon . . . to southern California, and disjunct to Kansas, Arkansas, Georgia and Chile."

Heggies Rock

LOCATION: Columbia County, Georgia; lat. 33°32′30″ N, long. 82°15′05″ W.

DIRECTIONS: From Augusta go northwest on Route 104 to Friendship church and turn left (southwest) onto Tubman Road. At 3.5 miles take a left at a delapidated barn on the left covered with cow itch vine. Go 2 miles southeast to a road marker which indicates a bend in the road to the left. Turn left onto the dirt road just beyond the road marker. The road almost immediately divides into three forks. Take the middle fork, drive as far as you can and park. A sign indicates that an "Ecological Reserve" is ahead, approximately 1/2 mile.

OWNERSHIP: Private

SIGNIFICANCE: According to Radford and Martin, Heggie's Rock is the most important of the Piedmont's granitic flat rocks for (1) the number of granitic outcrop endemics, (2) the number of rare species, (3) the number of disjunct species, (4) the best community zonation on granite, and (5) being the least disturbed of all the granite outcrops. Eleven of the nineteen outcrop endemics occur here, including the Piedmont's crowning botanical gem, *Amphianthus pusillus,* found in a dozen or so pools in 1975. Forty Acre Rock in South Carolina is the plant's only other known location. Other rarities at Heggies Rock include:

Isoetes melanospora
Panicum lithophilum
Cyperus granitophilus
Rhynchospora saxicola
Juncus georgiana
Sedum pusillum
Draba aprica
Oenothera linifolia var. *glandulosa*
Oenothera fruticosa var. *subglobosa*
Phacelia dubia var. *georgiana*
Viguiera porteri
Cladonia caroliniana
Lindernia monticola
Houstonia pusilla
Senecio tomentosus
Botrychium lunaroides
Ophiglossum crotalophoroides
Lepuropetalon spathulatum
Belamcanda chinensis
Riccia dictyospora
Agrostis elliottiana
Scirpus kiololepis
Fimbristylis dichotoma
Tradescantia hirsuticaulis
Commelina erecta
Nothoscordum bivalve
Schoenolirion croceum
Talinum teretifolium
Portulaca coronata
Arenaria groenlandica var. *glabra*
Diamorpha cymosa
Trifolium carolinianum
Polygala curtisii
Crotonopsis elliptica
Forestiera lingustrina
Selaginella tortipila
Opuntia drummondii
Ipomopsis rubra

Stone Mountain Dome

LOCATION: Dekalb County, Georgia 20 miles (32 kilometers) due east of the center of Atlanta; lat. 33° 44′05″ N, long. 84°08′15″ W.

DIRECTIONS: Proceed east from Atlanta on U.S. 78. The dome is obvious.

OWNERSHIP: State of Georgia

SIGNIFICANCE: Stone Mountain is the largest granitic outcrop in the eastern United States. It is the "type location" (place of discovery) of the Georgia oak, *Quercus georgiana*, which is endemic to the outcrops of Georgia, and of the herbs *Viguiera porteri, Isoetes melanospora*, and *Juncus georgianus. Coreopsis saxicola* may be limited to Stone Mountain. The site is one of very great botanical and geological significance; a century ago, in its undisturbed state, it was acrawl with naturalists and scientists viewing and recording its biotic treasures. Today heavy recreational use and commerical exploitation have blurred Stone Mountain's natural history values. The historic figures whose likenesses are carved into the rock would no doubt be aghast at the defacement, though they might be consoled that time may one day erase the vandalism. Stone Mountain is an excellent and readily accessible place to get a taste of the granitic outcrop concept.

Bogg's Rock

LOCATION: Pickens County, South Carolina; lat. 34° 47′00″ N, long. 82°44′00″ W.

DIRECTIONS: Park 1/2 mile north of the limits of Liberty, South Carolina, on U.S. 178. The site is on the west side of the road. There are about 10 acres (4 hectares) of exposed flatrock in four separate areas.

OWNERSHIP: Private

SIGNIFICANCE: This is the type locality and the only known location for the granitic endemic *Aster avitus*.

Growing in thick clumps under two or three cedars at the edge of the rock, in places which at other sites might be occupied by *Sedum,* this extremely rare aster rises to 12 inches (30 centimeters) on a linear stem which branches at the top to accommodate pale blue inflorescences. The many short, stiff, linear leaves, 1/2 inch (6 centimeters) long, bristle from the vertical stem, giving it a brushlike appearance. *Sedum smallii, Arenaria uniflora,* and *A. glabra* are present. The well-developed xeric chestnut oak—heath woods landward from the scrub pines give an excellent accounting of the petrosere in its entirety. The site has been abused by trash-dumping.

Forty Acre Rock

LOCATION: Lancaster County, South Carolina; lat. 34°39′20″ N, long. 80°31′10″ W.

OWNERSHIP: Private

SIGNIFICANCE: This is a very large outcrop—the name accounts for only about one-quarter of the area of intermittently exposed rock and thin soils. The most accessible part near the road head is the vicinity where the outcrop endemic *Amphianthus pusillus*—an aquatic suggestive of a miniature water lily (see page 311—occupies a few pools. Motorcycle tracks and wine bottles frame some of the continent's rarest botanical treasures. A ravine 100 feet (30 meters) deep cleaves the exposure, bearing water from the seeps near the tops of the rock's many small domes. A convex waterfall cascading over one of the domes into the ravine adds splendor.

The pioneering herbs of the petrosere, blooming in April, make Forty Acre Rock a place of stunning beauty. Endemics include *Sedum smallii, Arenaria brevifolia, Cyperus granitophilus* (a sedge), *Portulaca coronata, Senecio tomentosus, Allium bivalve* (an odorless onion called false garlic) and the rare *Sedum pusillum.* In and around the edges of the xeric forests of chestnut oak over blueberries grow the herbaceous *Phacelia maculata* and the scrophulariaceous toad flax, *Linaria canadensis.*

Stunted, twisted cedars pry into the cracks in the rock.

The drainage off the granite flows into Flat Creek south of Forty Acre Rock. On the south side of Flat Creek slightly downstream of the outcrop is a steep, north-facing bluff underlain by a diabase dike, the soils over which are characteristically rich and basic. According to Radford and Martin, "This was, until it was cut over, the most impressive forest in the Carolinas. Today, its understory is still intact." It includes ginseng, *Panax quinquefolium*—rare (and protected by law)—as well as nodding trillium, *Trillium cernuum* (which is present at only one other South Carolina locality, Stevens Creek), and creeping phlox, *Phlox stolonifera*, rare in the Piedmont of the Carolinas. Rare woody plants may also be found on this slope: the yellow chestnut oak, *Quercus muehlenbergii*, and the shrub *Euonymous atropurpureus*.

Rocky Face Dome

LOCATION: Alexander County, North Carolina, five miles NNE of Hiddenite; lat. 35°58′00″ N, long. 81°10′00″ W.

DIRECTIONS: From Route 90 east of Taylorsville, go north on County Road 1422. Go right on County Road 1419, then left on 1425, 1½ miles to a jeep track leading up into the woods on the right. Park and walk one mile into the woods and up the slope, passing exposed areas of flatrock. The track ends at the summit, an obvious outcrop and viewing point.

OWNERSHIP: Private

SIGNIFICANCE: Rocky Face Dome is part of the Brushy Mountains, a low range in the Piedmont of western North Carolina. The splendid flowering herbs of the more southerly outcrops are absent, but Rocky Face is an excellent place at which to observe the earliest phases of primary succession on bare granite and gneiss. Numerous lichens are present. The rock spikemoss, *Selaginella rupestris*, forms mats over the lichens, enabling grasses and forbs to colonize. Deepening soils support a rare gymnosperm community of Virginia pine and table mountain pine, *Pinus pungens*, a diminutive, scrubby pine whose occurrence in the Piedmont is limited to the domes and craggy outcrops on the higher monadnocks. A montane heath community of mountain laurel, *Kalmia latifolia*, and purple rosebay, *Rhododendron catawbiense*, on the north face is uncommon in the Piedmont. The canopy is chestnut oak. Contrasting canopies of Virginia pine over yucca and *Talium tererifolium* (a rhyzomatous herb with a short tuft of pinelike foliage at the base of a single floral spike 6 to 10 inches/15 to 25 centimeters tall) grow on the southern slopes.

Sandy Mush Granite Slope

LOCATION: Rutherford County, North Carolina; lat. 35°16′00″ N, long. 81°51′20″ W.

DIRECTIONS: From Sandy Mush proceed south on Route 221A 1.7 miles to County Road 2149 and turn west. The granite slope is ½ mile on the right (the north side of Route 2149).

OWNERSHIP: Martin Marietta Aggregates, a stone company

SIGNIFICANCE: This outcrop is intermediate in topography between a dome and a flattop—it is similar to Forty Acre Rock and is considered to be an alternative candidate for landmark status if Forty Acre Rock is destroyed by recreational abuse. Pioneer herbs include sandwort and stonecrop plus the outcrop endemic *Crotonopsis elliptica*, a bare-stemmed annual with elliptical, axial leaves 1 inch (25 millimeters) long. The stems branch and divide repeatedly. Bladderwort (the terrestrial variety, *Utricularia cornuta*) grows in the seeps. One authority considers this the best-preserved granitic outcrop in the Carolinas.

Stone Mountain Dome

LOCATION: Allegheny and Wilkes Counties, North Carolina; lat. 36°23′00″ N, long. 82°02′35″ W.

DIRECTIONS: Go north on U.S. 21 from Elkin. Follow signs to Stone Mountain State Park.

OWNERSHIP: State of North Carolina (State Parks Division)

SIGNIFICANCE: Cresting at 2305 feet (703 meters) above sea level in the extreme western Piedmont of North Carolina, Stone Mountain affords a view of what the Piedmont's other outcrops looked like before they were reduced by weathering and exfoliation. The dome stands as a complete hemisphere 600 feet (180 meters) from top to bottom. Owing perhaps to the altitude and more westerly location, the outcrop endemic herbs are not well expressed here, but the woody petrosere is unexcelled. Virginia, pitch, and table mountain pines pry into the crevices and cling to the organic accumulations on the slope. *Talium teretifolium* and rock spikemoss, *Selaginella rupestris*, are representative herbs. Heath thickets of mountain laurel, purple rosebay, and black huckleberry, *Gaylussacia baccata*, grow beneath pine, chestnut oak, and sweet birch, *Betula lenta*, in accumulations at or near the summit.

The sheer face of the main dome at Stone Mountain is one of the most important recreational resources in the eastern United States for a sport gaining rapidly in popularity among outdoor athletes—rock climbing. Climbers come from all over the world to test themselves on the friction faces and exfoliation dihedrals on this dome. Climbing routes are established and controlled by the Park authorities.

Stone Mountain is one of the most meticulously kept and least-disturbed public recreation areas in the Piedmont. In addition to the awesome dome which dominates the Park, there is a smaller exposure ½ mile to the southwest near which vultures roost periodically. (No connection between the vultures and the climbing activities has been established.) A small herd of wild goats scampers about the dome, scrupulously avoiding humans. A splendid water fall is accessible by hiking trail east of the dome. Birds of prey seem especially visible from the dome, particularly during the migratory times. My own sightings include red-tailed hawks, sharp-shinned hawks, ospreys, vultures (both species) and broad-winged hawks without number. Ravens are present year-round—I have seen them at only three other sites in the Piedmont.

Non-Granitic Outcrops

There are major rock exposures in the Piedmont (a few of which I'll mention here for their exceptional scenic, geologic, and natural history values) which decompose to craggy, angular faces rather than to the rounded smoothness characteristic of granite. They are mainly of metamorphosed volcanic origin. The vegetation tends toward the extreme xeric and shows considerable montane influence. The vegetation mentioned here is probably all unique, each reflecting the chemistry of the underlying rock and its climatic history. In the Piedmont naturalist's mind they are united by their scenic splendors and by such rarities as the bear oak, the table mountain pine, and the raven.

Kings–Crowders Mountain Area

LOCATION: Gaston County, North Carolina; lat. 35°12′ 15″ N, long. 81°18′05″ W.

DIRECTIONS: From Interstate 85 turn south on 161 which passes between Kings and Crowders Mountains.

OWNERSHIP: Private

SIGNIFICANCE: This is the southernmost location for bear oak, *Quercus ilicifolia*, a shrub-size tree found on a few, widely scattered, rocky pinnacles in the Piedmont. The leaf is about 4 to 6 inches (10 to 15 centimeters) long with asymmetrical, pointed lobes and shallow sinuses. (The leaf margin looks as if it might be the creation of a preschooler connecting dots on a piece of paper—no curves, all angles.) The pinnacle is dominated by an open xeric forest of chestnut oak over sourwood, sassafras, persimmon, bear oak, and chestnut, *Castenea dentata*. Herbs include *Silphium compositum*, with its 12-inch (30-centimeter), deeply cleft leaves, and montane ferns such as *Vittaria* spp., *Asplenium montanum*, and *A. bradlei* (spleenwort ferns) and the sparsely distributed *Liatris regimontis*, a blazing star. Prostrate shrubs of the rare ground juniper, *Juniperus communis*, are recorded on the lower slopes.

Thrusting upward almost 1000 feet (300 meters) from the surrounding peneplain, the twin pinnacles of Kings Mountain and Crowders Mountain present a splendid vantage of the regional landforms. An important battle in the revolutionary war was fought here. The rock is friable and dangerous for climbing, though tempting in its vertical grandeur.

The Sauratown Mountains—Pilot Mountain and Moore's Wall

LOCATION: Stokes County, North Carolina; lat. 36°22′ 00″ N, long. 80°25′30″ W.

DIRECTIONS: Take Route 66 north from Kernersville approximately 20 miles (32 kilometers). Follow signs westward off Route 66 to Pilot Mountain, clearly visible. For Moore's Wall, take Moore Spring Road right off Route 66 to Charlie Young Road. Drive upward and toward the face of Moore's Wall as far as possible. Park and hike to the base, then left and scramble up the very steep trail to the summit. Or enter by trails from the Hanging Rock State Park side.

OWNERSHIP: Private, except for the summit of Moore's Wall, which is the western boundary of Hanging Rock State Park.

SIGNIFICANCE: The rock masses of Pilot Mountain and Moore's Wall are remnants of an earlier erosion cycle; the whole peneplain of the Piedmont was once level with their summits. Twenty-five miles (40 kilometers) east of the Blue Ridge scarp and rising 1000 feet (300 meters) or more above the surrounding terrain, they afford views unexcelled in eastern North America. The northwestern face of Moore's Wall presents some of the continent's most severe challenges in face climbing; some of its ascents are made only in the troubled dreams of the most addicted climbers. Ravens yodel encouragement, broad-winged hawks whistle in disbelief, and pileated woodpeckers cackle in mockery at the climber's struggle. It is the wildest, most spectacular place in the Piedmont!

The vegetation ranges from cove hardwoods on the lower slopes to xeric with montane touches at the summits. Bear oak and table mountain pine grow atop Moore's Wall, amid chestnut oak and sweet birch. Carolina hemlock is important in the canopy from the base to the summit. Cucumber trees, *Magnolia acuminata*, grow to 18 inches (45 centimeters) dbh beneath the sheer walls, towering over purple rosebay and mountain laurel. In the shrub layer of the xeric forest of the summits are *Clethra acuminata*, *Vaccinium constablaei*, *Ilex ambigua*, and *Rhododendron catawbiense*, all montane shrubs disjunct from the Blue Ridge Mountains visible to the north and west. Herbs of the crevices include columbine, sandwort (*Arenaria* spp.), stonecrop (*Sedum* spp.), and a saxifrage, *Saxifraga michauxii*. Whole meals can be made of the blueberries ripening in June and July.

Bull Run Mountain

LOCATION: Fauquier County, Virginia, between Thoroughfare Gap and Hopewell Gap; lat. 38°51′15″ N, long. 77°43′00″ W.

DIRECTIONS: Proceed westward on Route 55 through Thoroughfare Gap. Turn right on County Road 628, cross a railroad track and proceed north approximately one mile to the highest point in the road, where a jeep trail leads off to the right. Park and walk eastward.

OWNERSHIP: National Heritage Foundation.

SIGNIFICANCE: The lower western slope (the part you first enter following the directions above) is cloaked in a unique forest of immense tulip trees over spicebush.

Basic soils owing to underlying diabase rock account for this community. Splendid assemblages of ferns grow in the seepages, including royal and cinnamon ferns. Continuing upslope through locust groves, which probably represent nineteenth-century clearings, one encounters stone fences surrounding dry-mesic stands of shagbark hickory, black oak, and tulip trees to 24 inches (60 centimeters) dbh. These may be the successional inheritors of fields abandoned at about the time of the American Revolution. The grade steepens into an obviously xeric forest of bent chestnut oaks over mountain laurel. The summit is dominated by severely twisted and stunted chestnut oaks over sourwood, mountain laurel and *Vaccinium* spp. Some of the oaks 20 inches (50 centimeters) in dbh are not even 40 feet (12 meters) tall.

The quartzite cap of the ridge outcrops at the summit present a vista of the Fauquier and Loudoun valleys—horse country, quartered by white fences and hedgerows of osage orange, *Maclura pomifera*, and, as this goes to press, by an extension of Interstate 66 whose vanguard of bulldozers is laying waste the grand estates in a manner Sherman would have envied. Ravens play in the updrafts deflected by the quartzite crags.

The area is redolent of historic ferment. Lee and Longstreet brushed aside resistance at Thoroughfare Gap to relieve Jackson and maul the Federal Army of the Potomac at Second Manassas. The following summer Lee, now blind to the temptations of Thoroughfare Gap in favor of more ambitious prizes to the north, passed to the west of Bull Run Mountain, en route to his engagement at Gettysburg.

Nockamixon Rocks

LOCATION: On the south bank of the Delaware River in Bucks County, Pennsylvania, along State Route 32 between Narrowsville and Kintnersville; lat. 40°34′00″ N, long. 75°09′30″ W.

DIRECTIONS: Park along Route 32 and climb. Cautiously.

OWNERSHIP: Private

SIGNIFICANCE: Rising 100 feet (30 meters) or more above the Delaware, these cliffs of Triassic shale–sandstone are checkered with shelves and crevices which harbor an outstanding display of cliff vegetation. On shelves near the top of the cliff grows the arctic–alpine stonecrop, *Sedum rosea*, at what is thought to be its southernmost station in the United States. The rock is cleft by several small cascades, each of which supports a splendid flora likely to include columbine, *Huechera americana* (alumroot), *Saxifraga virginiensis*, and *S. pensylvanica*, walking fern, and rock cap fern. In some sections a canopy of trees is established on ledges and in crevices. Eastern hop hornbeam, hackberry, American elm, sugar maple, butternut *(Juglans cinorea)*, black cherry, witch hazel, Choke cherry *(Prunus virginiana)*, and ninebark *(Physocarpus opulifolius)* are important trees and shrubs on the cliff. This is an outcrop of contrasting acidic and basic rock which supports an uncommonly diverse flora. It is viewed as the best example of cliff vegetation in the Piedmont of Pennsylvania and New Jersey.

Serpentine Barrens

In the formidable alembics of volcanic activity, possibilities exist for the formation of numerous minerals. Magnesium is not a particularly plentiful element in the planet's makeup, but where it does occur it is probable that any volcanic activity in its immediate vicinity will help it to combine chemically with the ubiquitous iron and silicon in the earth's crust. A compound known as peridotite may result, and it, in turn, may subsequently be subjected to the requisite heat and pressure to metamorphose into the mineral called serpentine.

Deposits of serpentine, rich in magnesium and sometimes containing chromium and nickel, occur the world over, including the regions along the Appalachians and the mountains near the Pacific Coast. In the Piedmont, several major deposits are clustered in southeastern Pennsylvania and adjacent Maryland. A disjunct site exists in Columbia

County, Georgia. The soils that result from the residual decomposition of serpentine, like their parent rock, are enriched with magnesium and other metals and are poor in nutrients. The presence of magnesium, nickel, and chromium retards or eliminates most plants. Consequently, the edaphic climax on these soils produces a notably undiverse floral system whose relatively few members grow sparsely and generally fail to attain their genetic potential in size. Open parklands, or savannahs, of scattered stunted conifers amid the cover of a few hearty herbaceous species give the character of what botanists call a serpentine barren.

To survive in serpentine soils, a plant must be tolerant to the metals present, and its demands for nutrients must be modest. Some species may benefit from these conditions, however, if the soil excludes their competitors but falls within their limits of toleration. This is apparently the case with the eastern red cedar and *Andropogon scoparius*, which combine to form an arrested climax vegetation, their normal seral successors being excluded by the chemistry of the soil. This red cedar–broom sedge cover is typical of the harsher soils. Pitch pine may substitute for cedar. Grassy savannahs feature dropseed, *Sporobolus* spp., turkey foot, *Andropogon gerardii*, and tufted hair grass, *Aira* spp. Where moisture is adequate, scrubby deciduous canopies may develop, composed of dwarf chinquapin, blackjack, and post oaks. Turkey oak is a serpentine occasional.

At first glance, a serpentine barren may resemble an old field—a poor one—in the early stages of woody succession. But what you are seeing is a climax community. The barrens are too poor to have been farmed beyond light pasturage.

Burks Mountain Serpentine Slopes

LOCATION: Columbia County, Georgia; lat. 33°37′10″ N, long. 82°13′00″ W.

DIRECTIONS: Go south-southeast from Pollard's Corner on Route 104; turn left on the Old Middleton Road. Proceed northeast for three miles to the first topographic

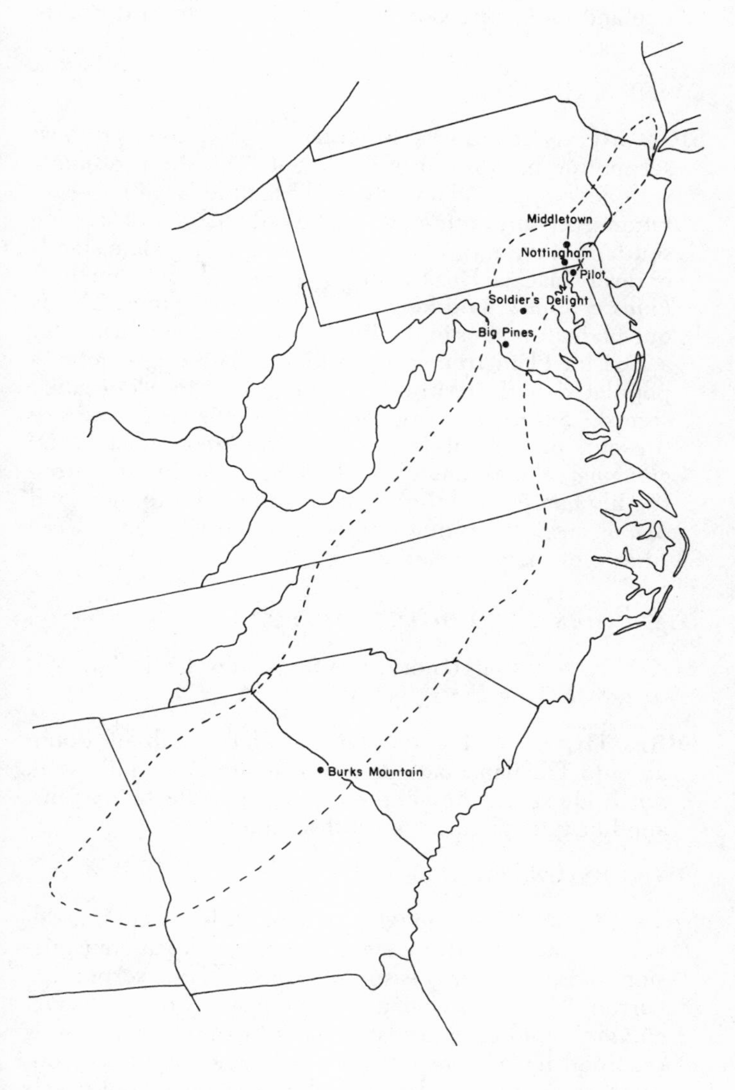

94. Serpentine barrens

rise and park. The slopes are on the left (north) side of the road.

OWNERSHIP: Unknown

SIGNIFICANCE: Burks Mountain is the southernmost serpentine barren in the Piedmont. The site presents a well-developed and undisturbed example of climax vegetation over serpentine and is the only habitat of its type south of Maryland. The aspect is of open parklike stands of dwarf blackjack oak with some post oak and shortleaf, *Pinus echinata*, and longleaf, *P. palustris*, pines. This is one of the few sites in the Piedmont where longleaf pine, a Coastal Plain tree, can be found. The open slope is populated with a shrubby mint (check the four-sided stems), *Satureja georgiana*. Numerous legumes are present. Rare plants present are the erect clematis, *C. albicoma*, the Solomon's seal, *Polygonatum tenue*, from the lily family, and the ornately and fragrantly flowered *Agave virginica* rising 6 feet (2 meters) over a basal rosette of succulent leaves.

Big Pines Serpentine Barrens

LOCATION: Montgomery County, Maryland; lat. 39° 04′30″ N, long. 77°13′30″ W.

DIRECTIONS: At Hunting Hill turn left (south) off Route 28 onto Travilan Road. At approximately one mile walk north along the pipeline clearing 1/2 mile to barrens; another tract is 1/2 mile south of the road.

OWNERSHIP: Private

SIGNIFICANCE: Although surrounded by suburban development and torn by electric power and gas transmission lines, this site is easily recognizable as a serpentine barren. There is an open canopy of scrub pines, *P. virginiana*, amid characteristic herbs. A similar barren was recorded at Gaithersburg in 1910; it is now a subdivision. Another serpentine barren may exist on a southwest-facing slope southwest of Glen Road and east of Greenbrier Creek. The societal value of the barrens is all the greater because of intense residential development.

Pilot Serpentine Barren

LOCATION: Cecil County, Maryland, about ½ mile east of the crossroads community of Pilot; lat. 39°42′30″ N, long. 78°11′30″ W.

DIRECTIONS: Turn west off Route 222 onto the county road which passes through Oakwood on an east-west heading. One mile west of Oakwood the road takes a dogleg north (a 90 degree right followed by a 90 degree left turn). The barren is north of the road at the dogleg and westward to the settlement of Pilot.

OWNERSHIP: Private.

SIGNIFICANCE: Experts suspect that careful investigation may show Pilot Serpentine Barren to be the most diverse of its habitat type in the Piedmont. In the broomsedge community between the sparse scrub pine and red cedar on the xeric slopes grows the serpentine endemic *Cerastium arvense* var. *villosissimum*, a type of chickweed. Herbs characteristic of the barrens include *Deschampsia caespitosa*, *Phlox subulata*, and *Arabis lyrata*.

Soldiers Delight Serpentine Barren

LOCATION: Baltimore County, Maryland; lat. 39°25′00″ N, long. 76°51′00″ W.

DIRECTIONS: From the Baltimore Circumferential Highway I-695, go west on Route 26 about 5 miles to the Deer Park Road. Turn right (north) and go through the village of Deer Park and one mile beyond to the intersection of the Deer Park Road and Wards Chapel Road, the approximate center of the barren.

OWNERSHIP: 1000 acres (400 hectares) or more by State of Maryland, with acquisition continuing.

SIGNIFICANCE: This is a very large serpentine barren, parklike in aspect with blackjack and post oak growing in open stands amid the broomsedges, *A. virginicus*, *A. scoparius*, and *A. gerardii*. *Cerastium arvense* var. *vil-*

losissimum, a chickweed endemic of serpentine barrens, *Polygonum tenue,* and *Arabis lyrata* are present. The fringed gentian, a rare plant of the high Appalachians, grows along the drainages at Soldiers Delight, its only known location in Maryland.

Middletown Serpentine Barrens

LOCATION: Delaware County, Pennsylvania; lat. 39°55′45″ N, long. 75°25′45″ W.

DIRECTIONS: From U.S. Route 1 take Pennsylvania Route 452 north to Lima. Proceed northeast on Barren Road to Penn Crest High School and an equal distance past. Barrens are on the immediate left of the road.

OWNERSHIP: Tyler Arboretum.

SIGNIFICANCE: An early report lists 11 serpentine barrens in Delaware County; this is the only one remaining, and it is fortunately under permanent protection. Urban sprawl has claimed the rest. Encompassing only 20 acres (5 hectares), this barren is neither large nor floristically diverse, but it is unmistakeably a serpentine barren, with parklike stands of blackjack oak and pitch pine growing in meadows of broomsedge. *Phlox subulata* is abundant.

Nottingham Serpentine Barrens

LOCATION: Chester County, Pennsylvania, 1½ miles southwest of Nottingham; lat. 39°44′00″ N, long. 76° 02′30″ W.

DIRECTIONS: From Nottingham, follow U.S. 1 south to signs for Nottingham Park.

OWNERSHIP: Chester County.

SIGNIFICANCE: For its size, characteristic flora and all-around natural history value, this is considered the most outstanding of the Piedmont's serpentine barrens. It is one of the most highly valued natural treasures in the province. Post oak and pitch pine grow in open stands over dwarf chinquapin oak and catbrier. Black locust is established in some areas. Large-toothed aspens line the drainages. Sassafras is common. Radford records that, "the serpentine endemics, *Aster depauperatus* and *Cerastium arvense* var. *villosissimum*, both occur in the prairie-like grass openings with *Arenaria stricta*, *Phlox subulata* and *Arabis lyrata*." My own notes record Joe Pye weed, thistles (*Carduus* spp,), winged sumac, dropseed (a grass), goldenrod, asters and, along the principal stream, tag alder. A kingfisher, in fact, was using one for a hunting perch. Deer, foxes, and rabbits were in evidence at my visit. It is not difficult to gain a sense of being within a vibrant and very special habitat at the Nottingham Barrens.

There are numerous other serpentine barrens in a band of chrome series serpentinized soils which straddle the Maryland–Pennsylvania border east of the Susquehanna. All are apparently in private ownership and access is difficult. Barrens west of Rock Springs (Lancaster County, Pennsylvania), northeast of Unionville (Chester County, Pennsylvania), and in the vicinity of Goat Hill (Cecil County, Maryland) are among the more prominent. The barrens at Unionville may be the ones where the vegetation responds most markedly to the effects of serpentinization. All the barrens have traditionally been waste places, useless for agriculture or even sylvaculture. Land prices have been correspondingly low, and in this region of intense urbanization, these seemingly useless areas have been ready victims of development. They are unique and priceless habitats which merit our interest and our protective efforts.

Unique Gymnosperm Communities

Special communities of conifers are found at several sites in the Piedmont. It is difficult to describe an edaphic theme which accommodates these communities for their underlying rock is quite varied. Serpentine, slate, and volcanic flow are recorded at the sites discussed in this section. The principal characteristic the sites bear in common is that they occupy steep, north-facing slopes along rivers (or creeks). Because conifers are associated with boreal climates, we

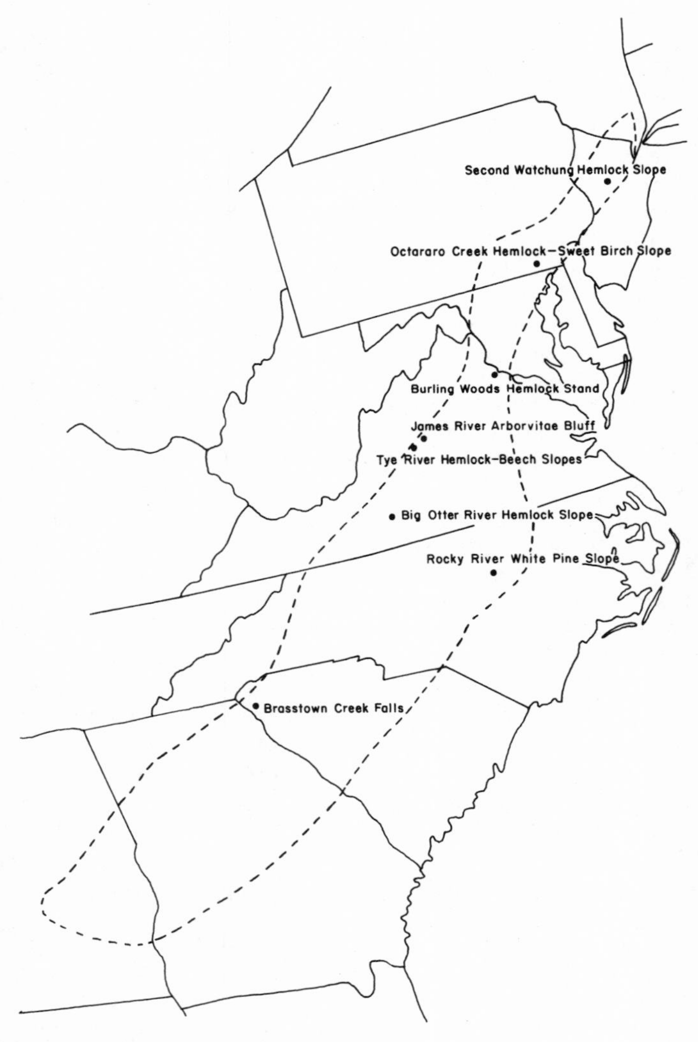

95. Unique gymnosperm communities

may infer that the gymnosperm communities sequestered from the sun on the north-facing bluffs today may be remnants of forests that lined both sides of the rivers during the Wisconsin glaciation. The steepness of the terrain has protected them from logging.

For the most part, the special gymnosperm communities are dominated by either hemlock *(Tsuga canadensis* or *T. caroliniana)* or white pine, both of which are sub-boreal or montane in their preferences and are generally uncommon in the Piedmont. In the unique case of the James River site, dominance is by arbor vitae, *Thuja occidentalis*. The boreal gymnosperms may share dominance or may at least share the site with broadleaf trees, also generally of northerly affinities. The same applies to the woody understory and to the herbs.

Brasstown Creek Falls

LOCATION: Oconee County, South Carolina; lat. 34° 42′00″ N, long. 83°18′20″ W.

DIRECTIONS: From Westminster go northwest on Route 76 into the Sumter National Forest to the first road on the left past the Brasstown Church (about ¾ mile). Turn left (south) on the Forest Service road paralleling Brasstown Creek and drive as far as you can without losing sight of the creek (4 miles). Park and hike downstream to the falls roughly one mile.

OWNERSHIP: U.S. Department of Agriculture. Sumter National Forest.

SIGNIFICANCE: Brasstown Creek tumbles through a deep gorge from the Blue Ridge Province into the Piedmont at this site, bringing with it some of the montane vegetation. In the mists from the falls and shoals grows a pristine community of white pines and hemlocks, some to 30 inches (75 centimeters) dbh, over mountain laurel, rosebay *(Rhododendron maximum)*, and galax. This community is unique in the Piedmont. An impressive display of riverweed, *Podostemum* spp., defies the rapids, and filmy fern, *Trichomanes petersii*, clings to some of the misted boulders. The oakleaf hydrangea,

Hydrangea quercifolia, grows in the shrub layer upstream of the falls, finding the northern limit of its range at this site. This is its only locality in South Carolina. Toward the top of the slopes are communities of chestnut oak and pines over ericaceous shrubs. There are cove hardwoods in the area.

Rocky River White Pine Slopes

LOCATION: Chatham County, North Carolina, slightly upstream of the confluence of the Rocky and the Deep rivers; lat. 35°37′20″ N, long. 79°09′30″ W.

DIRECTIONS: It is possible to reach the site through a maze of county and private roads but the simplest approach by far is to put a canoe in the Rocky River at Route 15–501 upstream and approach by water.

OWNERSHIP: Private.

SIGNIFICANCE: This stand of white pines, some of which were recently logged, is extremely disjunct. The nearest stand of its kind is 150 miles (240 kilometers) west in the Blue Ridge. In the Piedmont the nearest stand is 350 miles (560 kilometers) to the north in southern Pennsylvania and northern Maryland. Even there, white pines are more likely to be adjuncts than to dominate. Ericads form the shrub layer along with black haw, *Viburnum prunifolium*, and silverberry, *Elaeagnus umbellata*. American hop hornbeam is in the understory. Toward the bottom of the slope is *Hydrangea arborescens* over an impressive array of spring ephemerals including dutchman's breeches and the largest growths of spring beauty I have seen in the Piedmont.

Immediately downstream of the white pine stands is a disjunct community of chestnut oak over rosebay, *Rhododendron catawbiense*, also a montane disjunct, as is the dwarf dandelion, *Krigia biflora*, found at this site. At the river's edge is a shrub community of ninebark, *Physocarpus opulifolius*, a shrub to 10 feet (3 meters) distinguished by its shreddy bark, leatherwood, *Dirca palustris*, and swamp dogwood.

Tye River Hemlock–Beech Slopes

LOCATION: Nelson County, Virginia, on the south bank of the Tye River along State Road 739. The site is 7 miles (11 kilometers) upstream of the confluence of the Tye with the James.

DIRECTIONS: Park at the south end of the bridge and walk downstream along the slopes.

OWNERSHIP: Unknown.

SIGNIFICANCE: The steep, northwest-facing slopes are dominated by Canadian hemlock and beech. Chestnut oak and tulip tree are lesser presences. Northern red oak is present, and botanists hold white oak notable by its absence. Cherry birch, *Betula lenta*, grows near the crest. This appears to be the only stand of hemlock of consequence in the Piedmont of Virginia.

James River Arborvitae Bluff

LOCATION: Buckingham County, Virginia, on the south bank of the James opposite the confluence with the Tye River.

DIRECTIONS: Enter from State Road 606.

OWNERSHIP: Private.

SIGNIFICANCE: This is the only native stand of arborvitae in the Piedmont. The species requires basic soils and its presence here betrays the calcareous rock, in this case marble, beneath the surface. A band of this marble extends from the river's edge 90 feet (27 meters) up the precipitous bluff. Other northern or montane plants requiring limestone soils present are the bulbet bladder fern, *Cystopteris bulbifera*, and, from the turnip family, *Draba ramosissima*. Also present is the purple cliff-brake fern, *Pellaea atropurpurea*, and another rare calciphile, the stonecrop, *Sedum nevii*. Growing amid this collection of plants of strong basic affinities are some acid-loving ericads—mountain laurel and purple rosebay.

Big Otter River Hemlock Slope

LOCATION: Campbell County, Virginia, on southeast bank of Big Otter River 0.2 miles downstream of the U.S. 29 bridge.

DIRECTIONS: Park as close to the bridge as practical on the east side of the river and walk downstream.

OWNERSHIP: Private.

SIGNIFICANCE: Growing over a steep slope underlain by slate, the Canadian hemlocks at this site share dominance with mesic hardwoods—red oak, white oak, tulip tree, and sweet birch. The shrub layer is composed of a surprising mixture of acid-soil ericads (mountain laurel) and calciphiles (pawpaw, spicebush and hydrangea).

Octoraro Creek Hemlock—Sweet Birch Slope

LOCATION: Chester County, Pennsylvania, 1.1 miles west of Fremont; lat. 39°44′45″ N, long. 76°05′20″ W.

DIRECTIONS: Take Red Pump Road north out of Rising Sun. It is a dirt road maintained by the township. It crosses Octoraro Creek on a wooden bridge.

OWNERSHIP: Private.

SIGNIFICANCE: On this northwest-facing slope of about 30 to 45 degrees, a stand of hemlock is established over serpentine, a rock suggesting poor and acidic soils. The site, however, is anything but poor—it is one of the most stunningly beautiful spots in the Piedmont. On the south bank of the creek hemlocks to 36 inches (90 centimeters) dbh are mixed with scarlet oak, white oak and sweet birch. The understory is open, containing occasional witch hazel, downy arrowwood, and sparse spicebush. The herb layer includes hay-scented fern, *Dennstedtia punctilobula*, checkerberry, *Gaultheria procumbens* (actually a diminutive ericaceous shrub), pipsissewa, and running cedar, *Lycopodium* spp. The largest red maple I have ever seen grows at the water's edge 100 yards (40 meters) downstream of the bridge—it is 48 inches (120 centimeters) across at breast height and its hollows inside speak of the presence of raccoon.

Second Watchung Hemlock Slope

LOCATION: Somerset County, New Jersey, 2 miles (3 kilometers) north of Martinsville.

DIRECTIONS: Take Dock Watch Hollow Road north off Route 525 paralleling the creek through the water gap. Hemlock stand is visible on the right at about 2 miles (3 kilometers).

OWNERSHIP: Private.

SIGNIFICANCE: A few north-facing slopes over minor streams in the Watchungs harbor stands of hemlock, occasionally mixed with white pine. This is the best-developed stand I found in the region and it may be the only one of consequence in the Piedmont of New Jersey. The hemlock dominates in groves over sweet birch, beech, and red maple. Winter herbs are partridge berry and christmas fern.

Continuing up the slope toward the crest of the Second Watchung the canopy yields to a magnificent assemblage of red oak, beech, and sugar maple, surprisingly tall on the thin traprock soils. The understory is occupied by occasional hemlock, sweet birch, and chestnut oak over mapleleaf viburnum, flowering dogwood, witch hazel, and Christmas and hay-scented ferns. The whole effect is quite boreal, more suggestive of New England than of the Piedmont.

Unique Deciduous Habitats

There are numerous deciduous tracts in the Piedmont which are at or near seral maturity and could be considered examples of the climax forest. A few also host outstanding floral communities. It is these deciduous tracts to which we now turn our attention.

96. Unique deciduous zones

Alcovy Swamp Forest

LOCATION: Newton County, Georgia, one mile east of Covington and immediately south of U.S. 278; lat. 33° 35′30″ N, long. 83°48′30″ W.

DIRECTIONS: Turn right off 278 onto Elks Club Road, park, and walk west into the low woods.

OWNERSHIP: Private.

SIGNIFICANCE: Coastal Plain influence is strong in this bottomland forest, as evidenced by the dominance of tupelo, *Nyssa aquatica,* in many sections. Other parts are dominated by mature overcup, swamp chestnut, and willow oaks. This is one of the best-developed bottomland hardwood tracts in the Piedmont. Wildlife of interest includes Coastal Plain disjuncts, the bird-voiced treefrog, and the mole salamander.

Cooler Branch Slopes

LOCATION: Meriweather County, Georgia, one mile northeast of Nebula; lat. 32°52′00″ N, long. 84°41′ 00″ W.

DIRECTIONS: Go 2 miles due west from Manchester on the Nebula Road. Bear right at the Nebula intersection and parallel Cooler Branch on the right (north) side. The slopes are on both sides of the road as far north as the Franklin D. Roosevelt Highway.

OWNERSHIP: Private.

SIGNIFICANCE: The forest growing over quartzite at Cooler Branch mixes elements from the Coastal Plain and Blue Ridge with Piedmont endemics. Alongside Cooler Branch grow black oak and white oak to 24 inches (60 centimeters) dbh beside sweet bay, *Magnolia virginiana,* reaching to the canopy. Sweet bay is shrub-size in its normal range in the Coastal Plain. Tulip tree and red maple are present in the canopy. Redoubling the site's uniqueness is an unparalleled assemblage of shrubs including inkberry, *Ilex glabra* (rare in the Piedmont), horse sugar, *Symplocos tinctoria*, ti ti, *Cyrilla racemosa*, mountain laurel, flame azalea, *Rhododendron calendulaceum*, cane, and oakleaf hydrangea *Hydrangea quercifolia*. Downstream is a separate and quite montane community of chestnut oak over mountain laurel, witch hazel, and flame azalea but with a bottomland tincture added by sweet shrub, *Calycanthus floridus*. The upper slopes feature quercine canopies of Georgia, blackjack, Spanish, and white oaks over the shrubby Coastal Plain (specifically, Sandhill) endemic turkey oak, *Q. laevis*. Botanically, this is one of the most important forests in the Piedmont.

Fernbank Forest

LOCATION: DeKalb County, Georgia, in the northeast suburbs of Atlanta; lat. 33°46′40″ N, long. 84°19′20″ W.

DIRECTIONS: Turn north off of U.S. 29 immediately west of Seabord Railroad tracks.

OWNERSHIP: Fernbank, Inc.; leased to DeKalb County Board of Education for use as an outdoor educational facility.

SIGNIFICANCE: This is a mature deciduous forest dominated by very large tulip trees and, to a lesser extent, by oaks, beech, and hickory species. White basswood, *Tilia heterophylla*, is present. The proximity of this exceptionally mature tract to a large city lends social significance.

Monticello Bottomland Woods

LOCATION: Jasper County, Georgia, 2.5 miles (4 kilometers) south of Monticello; lat. 33°15′30″ N, long. 83° 41′05″ W.

DIRECTIONS: Take Route 11 south from Monticello 6 miles (10 kilometers) to the bridge crossing Cedar Creek. Park and walk west and south into the lowland woods.

OWNERSHIP: Private.

SIGNIFICANCE: This swamp forest, growing not in bottomland alluvium but in depressed residual soils, is the southernmost locality for Oglethorpe oak. Dominance is shared with pin, overcup, and swamp chestnut oaks and with American elm and shagbark hickory. *Viburnum* spp. and parsley hawthorn are strong in the understory, and dwarf palmetto, *Sabal minor*, adds a Coastal Plain touch. Ferns include *Isoetes englemannii* and *Ophioglossum vulgatum*.

Panther Creek Cove

LOCATION: Stevens County, Georgia, 1.1 miles (1.8 kilometers) southwest of Yonah Lake Dam; lat. 54°40′ 05″ N, long. 83°21′10″ W.

DIRECTIONS: Park 1 mile west of the dam. Hike across the creek (wadable at the jeep ford), and across the open bottomland into the coves.

OWNERSHIP: Private.

SIGNIFICANCE: The Panther Creek Cove hardwoods might have been translocated intact from any of the richer coves in the southern Appalachians. It is the most highly developed montane mixed mesophytic community in the Piedmont. The coves are dominated by buckeye, *Aesculus octandra*, beech, basswood, red oak, and black walnut. A rare vine in the magnolia family, *Schisandra glabra*, is abundant in one cove. Walking fern loops over the boulders and glade fern, *Athyrium pycnocarpon*, is an abundant ground cover. The rare and protected ginseng is present. *Collinsonia verticillata* is a rare endemic at the base of the Blue Ridge scarp. It is a perennial herb to 24 inches (60 centimeters) tall on a straight, four-sided stem, with leaves apparently whorled, but actually paired.

Pumpkinvine Creek

LOCATION: Bartow County, Georgia, between State

Route 293 and U.S. 41 on the north-facing slopes south of Pumpkinvine Creek; lat. 34°01′30″ N, long. 84° 44′30″ W.

DIRECTIONS: Park on the south side of Pumpkinvine Creek and walk into the woods, on the jeep trail.

OWNERSHIP: Unknown.

SIGNIFICANCE: This is a diverse cove hardwood community with a canopy consisting of beech, basswood, tulip tree, and (southern) sugar maple. Bottomland elements include shagbark hickory, hackberry, and ash. Chalk maple (a subspecies of sugar maple) and cucumber tree, *Magnolia acuminata*, are present. The exceptionally rich shrub layer features spicebush, pawpaw, leatherwood *(Dirca palustris)*, Allegheny spurge *(Pachysandra procumbens)*, and the rare prickly ash *(Zanthoxylum americanum)*. The seldom-seen liana, *Schisandra glabra*, is present here, as at Panther Creek Cove. The spring wildflowers here are said to create one of the significant botanical spectales of the Piedmont. Toward the top of the slope is a dry mesic or xeric community of post oak over at least four species of hawthorn.

John De La Howe Tract

LOCATION: McCormick County, South Carolina, 1.4 miles (2.2 kilometers) south of the De La Howe State School on the Clark Hill Reservoir; lat. 33°50′15″ N, long. 82°24′35″ W.

DIRECTIONS: Drive past the school compound to the end of the road. Park and walk in.

OWNERSHIP: State of South Carolina.

SIGNIFICANCE: This is an exceptionally diverse, maturing forest. The pines (shortleaf and loblolly) are old and very large; many are decadent and the replacement process is evident. White oak and Spanish oak with dbh greater than 36 inches (90 centimeters) are common. Mockernut, pignut, and shagbark hickory may be 24 inches (60 centimeters) dbh. Beech, sugar maple, and tulip tree give a montane cove flavor to the draws. There is a rich and diverse shrub layer of hawthorn, fringe tree *(Chionanthus virginicus)*, French mulberry *(Callicarpa americana)*, leatherwood, deerberry *(Vaccinium stamineum)*, and New Jersey tea *(Ceanothus americanus)*. The wildflower pennywort *(Obolaria virginica)* is exceptionally abundant.

Stevens Creek Rich Woods

LOCATION: McCormick County, South Carolina, 1.5 miles (2.4 kilometers) northeast of Clarks Hill; lat. 33° 41′00″ N, long. 82°09′00″ W.

DIRECTIONS: Take the secondary road northeast from Clarks Hill. Park at the south (near) end of the bridge over Stevens Creek and walk upstream. You are in the rich woods immediately upon leaving the road. The fabled north-facing slopes are on your left.

OWNERSHIP: State of South Carolina, by recent gift of the Continental Can Corporation, with the help of The Nature Conservancy, A. E. Radford, and others.

SIGNIFICANCE: This is one of the most diverse and botanically important sites in the Piedmont. The alluvium is canopied in mature bitternut hickory and sugar maple with occasional cypress, *Taxodium distichum*, at the edge of the creek. The slopes are dominated by shagbark, red oak, white oak, tulip tree, and slippery elm over the rare *Ribes echinellum*, not known from any other site outside Florida. The herb community is extraordinary. It features the largest population, by far, of *Isopyrum biternatum* (see page 257) together with Dutchman's breeches, shooting star *(Dodecatheon meadia)*, cromwell *(Lithospermum tuberosum)*, and three species of trillium. This unique juxtaposition of subtropical and montane plants causes Radford to consider "this climax community to be a relic of a once-widespread forest that grew in this area during the last glacial period in the Pleistocene."

York County Lowland Woods (also called Camassia Flat)

LOCATION: York County, South Carolina; lat. 34°50′ 00″ N, long. 81°07′30″ W.

DIRECTIONS: From U.S. 21, go west at Smith's Turnout to the third dirt road on the right, or approximately 2.0 miles (3.2 kilometers). Park 0.4 miles (0.6 kilometers) from that intersection and walk across the open field on your right toward the low woods behind it.

OWNERSHIP: Private.

SIGNIFICANCE: This is a poorly drained tract over a gabbro depression. It is dominated by a canopy typical of alluvial forests in the region—American elm, willow oak, green ash, and Shumard oak. A number of bottomland forest herbs are abundant here, including, *Allium* spp., *Cardemine bulbosa,* spring beauty, atamasco lily, quillwort, *Isoetes engelmannii,* and others. In late April the herb community is dominated by wild hyacinth, *Camassia scilloides,* a member of the lily family not found elsewhere east of the Appalachians. This lowland woods is also the only locality in South Carolina where pin oak, *Quereus palustris,* and swamp white oak, *Q. bicolor,* are found.

Fourth Creek Magnolia Slope

LOCATION: Iredell County, North Carolina, across a tributary of Fourth Creek east of the Statesville Country Club; lat. 37°47′15″ N, long. 80°50′15″ W.

DIRECTIONS: From U.S. 64 go south on County Road 2320, cross Fourth Creek and park on the south side of the bridge. Walk west along the creek. The magnolia slopes are on your left.

OWNERSHIP: Private.

SIGNIFICANCE: Growing in circumneutral soils over mica gneiss on a gentle northwest-facing slope is a mature canopy touched with a boreal or montane flavor—

dominance is shared by beech, tulip tree, and red oak. The subcanopy is occupied almost entirely by *Magnolia macrophylla* (one of several magnolias bearing the folk name "umbrella tree"), ranging in size from seedlings to 10 inches (25 centimeters) dbh. One botanical evaluation says, "This is the most outstanding beech–deciduous magnolia–mixed herb community seen in the entire Piedmont."

Eno River at Catsburg

LOCATION: Durham County, North Carolina, on the south bank of the Eno River 1 mile (1.6 kilometers) downstream of the Old Roxboro Road bridge.

DIRECTIONS: From Route 85 take the Braggtown Road exit northwest. At Braggtown, 2.5 miles (4 kilometers) from Interstate 85, turn right on the Old Roxboro Road. At Catsburg take the left fork and proceed to the bridge over the Eno. Park on the near side and walk downstream one mile.

OWNERSHIP: Private.

SIGNIFICANCE: This is the best developed rich woods and north-facing deciduous community I know of in the Piedmont of North Carolina. After walking a mile or so along a littered dirt road which was to have been the entrance to an industrial park, crossing a 20-acre eroded barren left from the aborted construction, you come to a cliff over the quiet little Eno. Beneath you is an amphitheater carved into the sandstone by the river. The canopy is white oak and shagbark hickory (one measures 48 inches/120 centimeters dbh) on the steep slope, shifting to Shumard and swamp chestnut oak, hackberry, and sugar maple (to 36 inches/90 centimeters dbh) on the alluvial soils. The understory is dominated by the buckeye, *Aesculus sylvatica*, "a shrub 1–3 meters tall, rarely a small tree." (Radford) Here the buckeye grows into the subcanopy, 40 feet (12 meters) or more, and some attain a dbh of nearly 12 inches (30 centimeters).

The most stunning aspect of the Catsburg site is the display of spring wildflowers. Dutchman's breeches grow in profusion along the lower slope with toothwort, *Dentaria laciniata*. Trout lily blankets the upper slopes. Spring beauty, golden corydalis, *Corydalis flavula*, and bloodroot are present in force. On the slight elevation left by the uprooting of a large tree now long decayed is a five square yard colony of *Isopyrum*. Ferns include the walking fern and maidenhair.

Pilot Mountain

LOCATION: Surry County, North Carolina, 2 miles (3.2 kilometers) northwest of Pinnacle; lat. 36°20′30″ N, long. 80°28′30″ W.

DIRECTIONS: From Route 52, take County Road 2053 to base of Pilot Mountain. Follow signs.

OWNERSHIP: State of North Carolina.

SIGNIFICANCE: Pilot Mountain and nearby Moore's Knob in Hanging Rock State Park contain splendid xeric montane communities. The summits are cloaked in pitch pine, and dwarfed chestnut oak with table mountain pine and bear oak as occasionals. Shrubs include *Pieris floribunda*, *Leucothoe recurva* and purple rosebay. Xerix herbs of interest on the rocky crevices are *Arenaria groenlandica* var. *groenlandica* and *Saxifraga michauxii*.

Raven Rock State Park

LOCATION: Harnett County, North Carolina, along south bank of the Cape Fear River, 5 miles (8 kilometers) upstream of Lillington.

DIRECTIONS: Turn north off U.S. 421 onto County Road 1250 and enter park.

OWNERSHIP: State of North Carolina.

SIGNIFICANCE: With its 150-foot (45-meter) high walls of granite facing north along the Cape Fear River, Raven Rock establishes perhaps the greatest distance between mountain laurel and its central range in the Blue Ridge. A rich ericaceous shrub layer beneath chestnut oak at the

Piedmont's outer margin is uncommon, since the terrain there is generally too gentle to simulate montane settings. The displays of rosebay, *Rhododendron catawbiense*, and mountain laurel are all the more noteworthy for their proximity to the Coastal Plain.

Rocky River Chalk Maple–Hop Hornbeam Slope

LOCATION: Stanley County, North Carolina, 1 1/2 miles west of State Route 200.

DIRECTIONS: From Route 200 take County Road 1122 west 0.8 miles (1.3 kilometers) to a bridge over a tributary of the Rocky River. Park and walk riverward. The tract is on the slopes along the tributary and upstream from it along the river.

OWNERSHIP: Private.

SIGNIFICANCE: Chalk maple, *Acer saccharum leucoderme*, is a subspecies of the sugar maple whose leaves are generally less than 4 inches (10 centimeters) long and have rounded tips and yellow-greenish undersides. The 8-inch (20-centimeter) dbh individuals along the Rocky River are large for the species. Crevices in the exposed rock support two hairlip ferns, *Cheilanthes lanosa* and *C. tomentosa*, and cliffbrake fern, *Pellaea wrightiana*, known at only one other locality east of Oklahoma.

Observatory Hill Slopes

LOCATION: Immediately west of the University of Virginia campus in Charlottesville, Virginia.

DIRECTIONS: Park on Stadium Road and walk westward uphill into the woods.

OWNERSHIP: University of Virginia.

SIGNIFICANCE: These slopes, facing generally eastward, present a gradient of drainage from the mesic accumulations at the foot of the slope to the dry mesic and xeric portions at the summit and on minor crests. Tall crowns of tulip tree and red, black, and white oak over black gum and dogwood cloak the lower slopes, grading into post oak, Spanish oak, and chestnut oak toward the summit. Growing on the drier, thinner soils are chestnut oaks whose pileated woodpecker excavations are occupied by gray squirrels. Observatory Hill is also a capital post at which to witness the movement of wood warblers and other migratory birds, spring and fall.

Big Otter River North Slope

LOCATION: Campbell County, Virginia, near confluence of Johnson Creek with Big Otter River, 2 miles (3.2 kilometers) southwest of Abingdon.

DIRECTIONS: Park on Route 682 on the south side of the Big Otter River and walk downstream. The tract is immediately downstream of the road.

SIGNIFICANCE: This is a well-developed, second-growth stand on a north-facing slope over basic rock. The canopy is basswood, a wet–mesic or cove hardwood species, over hop hornbeam, a shreddy-barked understory tree characteristic of dry upper slopes. Botanists hold the community to be unique in the eastern United States. Shrubs present are hydrangea and spicebush; herbs are walking fern, squirrel corn, Dutchman's breeches, bloodroot and wild ginger.

Randolph Macon Nature Preserve

LOCATION: Campbell County, Virginia, 1/2 mile southwest of Flat Creek Church; lat. 37°18′15″ N, long. 79°10′ 15″ W.

DIRECTIONS: Park on Route 622 where it is crossed by the utility line. The tract is to the north and east along 622.

OWNERSHIP: Randolph Macon Womens' College.

SIGNIFICANCE: There is nothing extraordinary about

this tract, but it offers a representative example of a chestnut oak–heath community such as is common on well-drained slopes, particularly in the western Piedmont.

Sweet Briar College White Oak Woods

LOCATION: Amherst County, Virginia, immediately southeast of Sweetbriar College.

DIRECTION: Proceed on foot from the Sweetbriar campus.

SIGNIFICANCE: This is an accessible example of a mature dry mesic forest dominated by white oak with the participation of tulip tree, mockernut hickory, beech, and others. It is said to be the best-developed tract of its type in the Piedmont. The understories and herb strata are also representative.

Accotink Creek Bottomland Woods

LOCATION: Fairfax County, Virginia, where Prosperity Road (State Route 699) crosses Accotink Creek.

DIRECTION: From Arlington Blvd (U.S. 50) turn south onto Prosperity Road. Park at either end of the bridge over Accotink Creek and walk into the surrounding woods.

OWNERSHIP: Fairfax County.

SIGNIFICANCE: This may be the most northerly outpost in the Piedmont for the bottomland endemic Shumard oak. The canopy is well developed, with most trees being between 12 and 36 inches (30 and 90 centimeters) dbh. Pin, swamp white, swamp chestnut, willow, and swamp red oaks are also present, along with three maples (box elder, red, and silver) and two elms (American and winged). The forest is conspicuous for its diversity. It effects a touch of primal wilderness in a densely populated area.

Burling Woods

LOCATION: Fairfax County, Virginia, bordered on the north by the Potomac River, on the east by Interstate 495, on the south by the Old Georgetown Pike, and on the west by Scott Run.

DIRECTIONS: Enter on foot from the Old Georgetown Pike.

OWNERSHIP: Fairfax County.

SIGNIFICANCE: Radford describes the tract as "the most diverse forest from a plant community and species standpoint seen in the entire Piedmont." Along sheltered ravines are stands of hemlock and tulip trees. Birch–maple stands grow on some slopes, chestnut oak on the better-drained areas. River birch, cottonwood, sycamore, and silver maple line the banks of the Potomac. This fall-line community is comparable to some recorded at the 2000 to 3000 foot (600 to 1200 meter) level in the Blue Ridge Province.

Hawlings River Regional Park

LOCATION: Montgomery County, Maryland, 0.7 miles south-southwest of Unity; lat. 39°13′00″ N, long. 77°04′ 30″ W.

DIRECTIONS: From State Route 91 take State Route 420 west to Unity and turn left. Track is 0.2 miles (300 meters) on left.

OWNERSHIP: Montgomery County.

SIGNIFICANCE: Much of this site is a dry–mesic to xeric woods with an exceptionally well-advanced chestnut oak canopy (to 48 inches/120 centimeters dbh) over ericaceous shrubs and flowering dogwood. Tulip tree, beech, and black, white, and scarlet oaks are prominent on the more sheltered parts. New York and cinnamon ferns are abundant on the lower slopes, and adder's tongue, *Ophioglossum engelmannii*, occurs around an old spring near a former homesite.

Long Green Creek and Sweathouse Branch Natural Area

LOCATION: Baltimore County, Maryland, 8 miles (13 kilometers) east-northeast of Towson; lat. 39°26′20″ N, long. 76°27′50″ W.

DIRECTIONS: Where Hartford Road (State Route 147) crosses Long Green Creek, park and walk 0.3 miles (500 meters) downstream. Tract begins there and extends to the Gunpowder River.

OWNERSHIP: State of Maryland.

SIGNIFICANCE: This is a richly diverse, mature upland forest on moderately steep slopes facing virtually all directions. Beech, tulip tree, black, white, red, scarlet, and chestnut oaks, mockernut hickory, and red and sugar maples are prominent on the mesic drainages. At streamside, under typical canopies of sycamore, sweet gum and box elder are rich flowering herb communities, 14 species of ferns, and horsetail, *Equisetum arvense*. The confluence of three watercourses flowing between well-protected slopes which have apparently been under protection for years are factors which help account for the maturity and diversity.

Brandywine Creek State Park

LOCATION: New Castle County, Delaware, 1 mile (1.6 kilometers) north of Rockland; lat. 39°49′30″ N, long. 75°33′15″ W.

DIRECTION: From Rockland walk upstream on Brandywine Creek, or walk downstream from Thompson Bridge Road.

OWNERSHIP: State of Delaware.

SIGNIFICANCE: Radford says, "The Brandywine Creek–Woodlawn forest is the most beautiful high-canopied woody community in the entire Piedmont. The trees are magnificent specimens mostly in the 36–42 foot range with some of record size." White oak, black oak, beech, and tulip tree are seen here in sizes perhaps unmatched. Mapleleaf viburnum is abundant under the tulip trees in the mesic areas and mountain laurel occupies the better-drained understory. Three hundred species of wildflowers, including skunk cabbage in a picturesque glade, are reported here.

Iron Hill Hardwoods

LOCATION: New Castle County, Delaware, at the intersection of Interstate 95 and State Route 896.

DIRECTIONS: Park along Route 896 as close to Interstate 95 as possible and walk west into the woods.

OWNERSHIP: New Castle County.

SIGNIFICANCE: This is an easily accessible site where the later stages of the mesosere may be observed. The stand is primarily red oak and beech over redbud, dogwood, spicebush, and mapleleaf viburnum. Black, white, and Spanish oaks are present, as are shagbark hickory and tulip tree.

Octoraro Creek Red Oak Slope

LOCATION: Chester County, Pennsylvania, on south side of Octoraro Creek, 1.5 miles (2.4 kilometers) upstream from the hemlock slope (see page 366).

DIRECTIONS: Walk upstream from the hemlock slope.

OWNERSHIP: State of Pennsylvania.

SIGNIFICANCE: This stand of red oak, growing over serpentine, may be a separate ecotype with distinctive light bark coloration. Red maple and sweet birch are occasionals in the canopy. There is a rich fern community which includes four species of woodfern, genus *Dryopteris*, in addition to hay-scented, Christmas, maidenhair, and interrupted ferns.

Otter Creek Natural Area

LOCATION: York County, Pennsylvania, 2 miles (3.2 kilometers) south of Safe Harbor on the Holtwood Reservoir.

DIRECTIONS: Enter from State Route 425, walking southwest to Otter Creek.

OWNERSHIP: Pennsylvania Power and Light Company.

SIGNIFICANCE: Otter Creek is one of several well-developed forests known collectively as the Holtwood Reservoir Natural Areas. Otter Creek and other nearby courses are tributaries of the Susquehanna, some of which have not been logged in this century. There are a few virgin tracts in the ravines. Shenks Ferry, Counselman Run, Kelly Run, and Oakland Run have been designated Natural Areas by the utility company and are afforded protection under company policy. Downstream are the Conowingo Islands, which are important for their geological and natural history values. Collectively, the utility-owned woodlands along this impoundment on the Susquehanna constitute the most important woodlands in the Piedmont of Pennsylvania.

French Creek State Park

LOCATION: Berks and Montgomery Counties, Pennsylvania, south of State Route 422 connecting Reading and Pottstown; lat. 40°12′00″ N, long. 75°45′00″ W.

DIRECTIONS: Enter the park on Route 345.

OWNERSHIP: State of Pennsylvania.

SIGNIFICANCE: French Creek is a large park with diverse habitats. Probably all have been logged, but some not for many years. Drainages range from the near xeric with canopies of chestnut oak over dwarf chinquapin and shoots of chestnut. On more mesic sites, canopies are of red oak and black oak. Near the lake is a lowland woodlot dominated by red maple and swamp Spanish oak to 24 inches (60 centimeters) dbh. Pawpaw is the dominant shrub. Spicebush, hazel nut, *Viburnum rafinesquianum*, and *Rhododendron nudiflorum* are also present over a rich herb community of jack-in-the-pulpit, fancy fern, wind flower, and the mint *Lycopus virginicus*.

Herrontown Woods

LOCATION: Mercer County, New Jersey, on the northern outskirts of Princeton.

DIRECTIONS: The main entrance is on Herrontown Road. The tract also fronts on Mount Lucas Road and Poor Farm Road.

OWNERSHIP: Stony Brook–Millstone Watersheds Association.

SIGNIFICANCE: This is an ideal location to observe the mesic successional process in the northern Piedmont. Some of the tract was abandoned seventy to eighty years ago; other parts are mature upland woods dominated by red, white, and black oak to 30 inches (75 millimeters) dbh. Old stone fences enclose younger stands of tulip, ash, and sweet gum, with occasional decadent cedar still visible. *Viburnum acerifolium* and spicebush are important shrubs. (See page 132.)

Hutcheson Memorial Forest

LOCATION: Somerset County, New Jersey, 1 mile (1.6 kilometers) east of East Millstone on the south side of State Route 514.

DIRECTIONS: Enter from the Amwell Road (Route 514) 0.3 miles (500 meters) east of the Penn Central Railroad crossing.

OWNERSHIP: Rutgers University.

SIGNIFICANCE: The eastern third of this mature mesic forest has reputedly never been cut. White, red, and black oaks to 36 inches (90 centimeters) dbh are in dominance. Beech and sweet pignut hickory are abundant to 2 1/2 inches (6 centimeters) dbh. The understory is predominantly dogwood over spicebush, as well as *Viburnum acerifolium*, and *V. prunifolium*. Poke grows beneath gaps in the canopy around standing deadwood riddled with woodpecker holes. Deer, pheasants, and foxes are copiously in evidence. (See page 132.)

The Glaciated Piedmont

Ten to fifteen thousand years ago the southward progress of the Wisconsin Glacier halted in the Piedmont of New Jersey on an east–west front approximating a line between Morristown and Millburn, thence southward to Plainfield, nearly to the Raritan, then east to Perth Amboy and the sea. To the north of this line the Triassic deposits are covered to a depth of about 20 feet (6 meters) with glacial till—unstratified, unsorted rocks and gravel deposited by the glacier. Along the line itself, the glacial debris is much thicker—up to 200 feet (60 meters) thick—lying in rolling mounds which constitute a terminal morraine. The morraine crosses the Watchungs along the Millburn–Morristown line.

As the ice began to melt, the Watchung basin south of the ice front filled with water and became what is now called glacial Lake Passaic. Later, as the ice front retreated northward, the northern part of the Watchung Basin also filled with meltwater and the Basin was apparently occupied entirely by Glacial Lake Passaic. There is a notch in the Watchungs near Millburn through which, geologists presume, there flowed a preglacial river which drained the Basin. Glacial deposits plugged the gap and it is not presently in use. The Watchung Basin therefore probably did not drain until the ice retreated northward far enough to uncover the gap at Patterson, through which the Passaic River now carries the Basin's runoff.

The draining of the glacial Lake Passaic and of the depressions in the Watchung Basin to the north of the morraine may not have been abrupt. Some very large chunks of ice apparently became buried in the glacial debris and

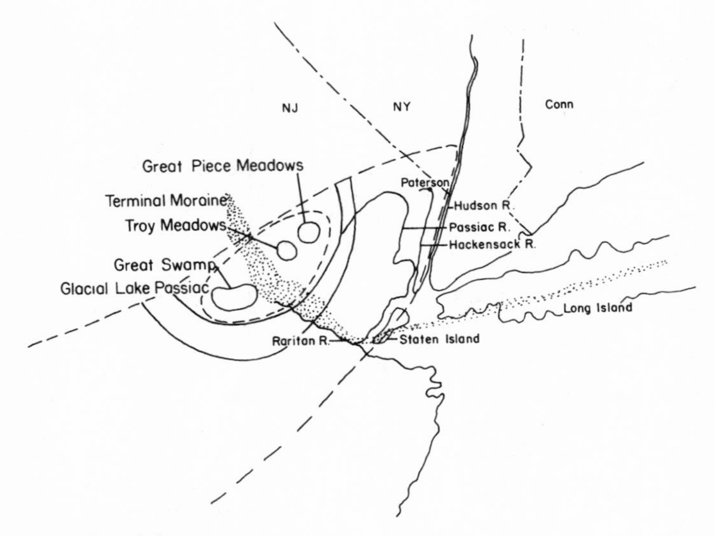

97. Glacial features

slowly melted, resulting in depressions that retained their meltwater. They probably never did actually drain but, like lakes everywhere, slowly filled in with accumulated silt and organic matter (see pages 307–8). The present reconstruction of postglacial events identifies glacial Lake Passaic as the progenitor of what is now the Great Swamp. Troy Meadows and Great Piece Meadows, north of the morraine, are the result of slow-draining depressions in the glacial till, probably involving the burial of masses of ice. Hackensack Meadows was formerly an arm of the sea, now filled with glacial debris to a depth of 100 feet (30 meters) or more. Because Hackensack Meadows is occupied largely by brackish marsh under tidal influence, its life systems are those of a coastal marsh rather than a Piedmont impoundment, and we omit their description.

The Great Swamp

LOCATION: Morris County, New Jersey, between the Reading Prong of the New England Province on the west and the Second Watchung Mountain on the south and east. The swamp is 7 miles (11 kilometers) south of Morristown, 30 miles (48 kilometers) west of Times Square.

DIRECTIONS: The Great Swamp National Wildlife Refuge Headquarters is on Pleasant Plains Road 2 miles (3.2 kilometers) east of Basking Ridge. The Morris County Outdoor Education Center (and boardwalk through the swamp) is reached from Southern Boulevard, on the eastern edge of the Wilderness Area.

OWNERSHIP: U.S. Government.

SIGNIFICANCE: The Great Swamp is a priceless pocket of wilderness in the nation's most heavily urbanized region. A variety of vegetative habitats differentiates the slight changes in elevation. Aquatics include duckweed, smartweed, pickerelweed, cattail, and buttonbush. Open bogs are occupied by buttonbush and hardhack, *Spiraea tomentosa*. Canopy dominance in the forested parts of the swamp is by special varieties of black gum, *Nyssa sylvatica* var. *biflora*, and red maple, *Acer rubrum tridens*, over *Vaccinium atrococcum* and white alder, *Clethra alnifolia* (which is identified by its simple alternate, oblanceolate leaves and terminal racemes of white florets). Pin oak and swamp white oak are present, as are beech and sweet birch over spicebush on the better drainages.

Great Piece Meadows

LOCATION: Essex and Morris Counties, New Jersey, in the lowlands on both sides of the Passaic River, 0.5 mile (0.8 kilometers) southeast of Towaco.

DIRECTIONS: The ideal visit would be by canoe on the Passaic River, putting in perhaps at Horse Neck Bridge where the river flows into the Meadows. Numerous roads circumscribe the Meadows, including Interstate 80 on the south.

OWNERSHIP: Unknown.

Troy Meadows

LOCATION: Morris County, New Jersey, south of Interstate 280, north of Route 10 (Mt. Pleasant Ave), east of Troy Hills and west of Hanover Neck.

DIRECTIONS: From Troy Hills, take Troy Hills Road south into the meadows.

OWNERSHIP: Wildlife Preserves, Inc.

SIGNIFICANCE: Both these meadows are glacial relics which were probably deeper lakes shortly after the retreat of the Wisconsin ice. They are divisible botanically into open aquatic communities of emergent and floating plants such as duckweed, wild rice, and bur reed; cattails; sedge swales populated by sedges, *Carex* spp., rushes, *Juncus* spp., ferns, and loosestrife, *Lythrum salicaria;* and sedge–shrub communities where tussocks and soils accumulated by the sedges support buttonbush, willows, swamp dogwood *(Cornus amomum),* marsh rose *(Rosa palustris), Spiraea* spp., and ultimately alder, *Viburnum dentatum,* American maple, and green ash. On the better-drained elevations the sere begins with milkweed and asters, then progresses through elder *(Sambucus candensis)* and staghorn sumac to climbing buckwheat *(Polygonum cuspidatum)* and into an early forest of locust, American elm, and box elder.

Appendices

Table A–1

Name	Food and Habitat of Larvae
Black swallowtail *Papilio polyxenes asterius*	Parsnip, Queen Anne's Lace; open weedy sites, mostly mesic
Spicebush swallowtail *Papilio troilus*	Spicebush, sassafras; early and mid-successional woods, moist and mesic
Tiger swallowtail *Papilio glaucus*	Wild cherry, tulip trees; fencerows, early woody sites
Zebra swallowtail *Papilio marcellus*	Pawpaw; bottomland hardwood forests; larvae are cannibalistic
Clouded sulfur *Colias philodice*	Clover; pastures, hayfields, roadsides
Falcate orange tip *Anthocaris genutia*	Plants of the turnip family (Brassicaceae); open fields, pastures, gardens
Monarch *Danaus plexippus*	Milkweed; abandoned fields in herbaceous phase, roadsides
Satyrs family Satyridae	Grasses; feeds at night in fields, grassy places
Fritillaries genera *Boloria* and *Speyeria*	Mostly violets; damp and mesic open places
Question mark *Polygonia interrogationis*	Elm; mature bottomlands, early woody sites in N Piedmont

Selected Lepidoptera of the Piedmont

Food and Habitat of Adults
Apiaceae, clover and other low-blooming flowers; flies and feeds near ground in meadows, weedy fields
Shady woody, open fields; flies low, feeds on flowers of Asteraceae and other families
Many asteraceous and fabaceous flowers; woodlands and open spaces, abandoned fields
Many flowers of open and wooded sites
Flowers of clover, alfalfa; fields, roadsides
Flowers near the ground in damp, open woods
Flowers at medium and upper levels in abandoned fields; large migrating flocks traverse Piedmont in fall
Visits flowers avidly in many habitats
Wide variety of flowers in weedy fields, roadsides, open spaces
Tree sap from sapsucker wells, forest flowers; deciduous woodlands

Selected Lepidoptera of the Piedmont

Name	Food and Habitat of Larvae
Comma *Polygonia comma*	Nettle (*Urtica* spp.), elm, hops (*Humulus* spp.); alluvial woods
Buckeye *Precis lavinia coenia*	Plantain (*Plantago* spp.), gerardia; open weedy places, fields soon after abandonment
Painted lady *Vanessa cardui*	Thistles and other Compositae; open weedy places
Viceroy *Limenitis archippus*	Willow, poplar; early woody bottomlands; hibernates in rolled-up leaf
Abbot's sphinx *Shecondia abbotti*	Grape leaves; alluvial and mesic forests, usually at considerable height
Twin-spotted sphinx *Smerinthus geminatus*	Cherry, birch, willow; alluvial and mesic early woody succession sites
Walnut sphinx *Cressonia juglandis*	Walnut, pecan, hickory; mature mesic woods, fencerows, and early succession sites in north
Pandora sphinx *Pholus satellitia*	Leaves of grape and Virginia creeper; mature and early woody mesic sites
Nessus sphinx *Amphion nessus*	Leaves of grape and Virginia creeper; mature and early woody mesic sites

continued

Food and Habitat of Adults
Sap, flowers, decomposing material in wooded bottomlands
Numerous flowers of the open spaces; a highly variable species with numerous forms
Flowers of thistle and other Compositae; open weedy places, pastures, roadsides
Flowers, decomposing material in open fields
Nectar
Nectar
Nectar; nocturnal
Nectar; nocturnal
Nectar; flies at dusk

Selected Lepidoptera of the Piedmont

Name	Food and Habitat of Larvae
Cynthia moth *Samia cynthia*	Ailanthus; found near cities; naturalized from China
Cecropia moth *Hyalophora cecropia*	Cherry, elder (*Sambucus* spp.), willow, maple, birch; wooded bottomlands
Polyphemus moth *Antheraea polyphemus*	Oak, hickory, elm, maple, birch; mature alluvial and mesic forests
Tulip tree silk moth *Callosamia angulifera*	Tulip tree *(Liriodendron tulipifera)*; moist and mesic woods
Promethea moth *Callosamia promethea*	Spicebush (*Lindera* spp.), sassafras, tulip tree, cherry; alluvial woodlands
Luna moth *Actias luns*	Sweet gum, persimmon, hickory, walnut; mesic early and mature woodlands
Io moth *Automaris io*	Plants in Rosaceae and other families; weedy fields; larvae move in long trains
Buck moth *Hemileuca maia*	Blackjack oak; dry wooded hilltops, poor soils bearing blackjack oak; like Io, larvae have stinging spines

continued

Food and Habitat of Adults

Mouth parts poorly developed; adult does not feed; nocturnal

Same as Cynthia moth

Same as Cynthia moth

Same as Cynthia moth

Same as Cynthia moth

Same as Cynthia moth; two brooded: first brood has yellow wing margin, second, has pink–purple

Same as Cynthia moth, single-brooded

Same as Cynthia moth

Selected Lepidoptera of the Piedmont

Name	Food and Habitat of Larvae
Royal walnut moth *Citheronia regalis*	Hickory, sumac, sweet gum; mesic early and mature sites
Imperial moth *Eacles imperialis*	Oak, maple, pine, sycamore, sweet gum, sassafras; varied wooded habitats
Great leopard moth *Ecpantheria scibonia*	Plantain; early herbaceous mesic sites
Isabella moth *Isia isabella*	Plantain; weedy grasslands; called "wooly bear"; legend says amount of black predicts severity of winter
Dogbane tiger moth *Cycnia teneria*	Dogbane *(Erigeron annuus)*; aster fields
Bella moth *Utethesa bella*	Rattlebox (*Crotalaria* spp.); abandoned fields, wood margins, waste places
Underwing moths *Catocala* spp.	Walnut, hickory, oak; mature woodlands
Locust underwing *Euparthenos nubilis*	Locust; mesic sites of early woody succession
White-veined dagger *Simyra henrici*	Cattail; pond edges, marshes
Tentacled prominent *Cerura multiscripta*	Willow; hydric early woody sites

continued

Food and Habitat of Adults
Same as Cynthia moth
Same as Cynthia moth
Nectar
Nectar
Probably nothing
Probably nothing
Nectar
Nectar
Nectar
Nectar

Selected Lepidoptera of the Piedmont

Name	Food and Habitat of Larvae
Sumac caterpillar *Datana perspicua*	Sumac; mesic overgrown fields; roadsides
Variable oak leaf caterpillar *Heterocampa manteo*	Oaks in white oak group; mature mesic forests
Elm leaf caterpillar *Nerice bidentata*	Elm; early woody mesic sites in north; mature alluvial forests in south
Eastern tent caterpillar *Malacosma americanum*	Apple, cherry; orchards, overgrown fields in early woody stage, fencerows; makes tent webs in tree crotches, hides by day, feeds at night
Wild cherry moth *Apatelodes torrefacta*	Cherry *(Prunus serotina* and *P. avium)*; overgrown fields; does not attack domestic cherry
Cherry scallop shell moth *Hydria undulata*	Cherry (*Prunus* spp.); folds leaves together and hides in them during the day
Other Geometridae	Variety of deciduous trees and shrubs
Bagworm moth *Thyridopterix euphemeraeformis*	Principally conifers (pine, cedar); also locust, sweet gum; early woody succession
Red-spotted purple *Limenitis arthemis astyanax*	A variety of trees and shrubs, incl. cherry, willows, plum, apple, hawthorn, *Carpinus*

continued

Food and Habitat of Adults
Nectar
Nectar
Nectar
Nectar
Nectar
Nectar
Nectar
Unclear
Nectar, dung, carrion; open, scrubby woody, forest edges, early woody succession sites

Selected Lepidoptera of the Piedmont

Name	Food and Habitat of Larvae
Tawny emperor *Asterocampa clyton*	Hackberry (*Celtis* spp.); river bottoms and swamp forests, wherever hackberry grows
Red admiral *Vanessa atalanta*	Nettles, hops and other Urticaceae; alluvial and mesic woods
Gray hairstreak *Strymon melinus*	Hops, hawthorn, mallow, cultivated beans; alluvial woods, dry hillsides
Banded hairstreak *Strymon falacer*	Oak, hickory; mesic woodlands
Pine elfin *Incisalia niphon*	The "hard" pines, *Pinus virginiana*, *P. rigida*
American copper *Lycaena phlaeas americana*	Sheep sorrel and other *Rumex* spp.; open weedy places, mesic abandoned fields in herbaceous phase
Harvester *Feniseca tarquinius*	Wooly aphids on alder and other trees and shrubs; early woody bottomlands, streamsides
Silver-spotted skipper *Epargyreus clarus*	Woody Fabaceae, incl. locust, honey locust, wisteria, and some herbaceous legumes
Indian, leonard's, and other skippers *Hesperia* spp.	Grasses; open grassy places

continued

Food and Habitat of Adults

Plant secretions on twigs and bark, nectar; visits muddy spots for moisture and minerals; fast, erratic

Nectar; often seen in aster fields, open woodlands; favors butterfly bush (genus *Buddleia*)

Nectar; usually seen in open, but can be found in a wide range of habitats

Nectar; in or near maturing woodlands

Nectar from flowers including rabbit tobacco, locust; overgrown fields, mesic early woods; N of James

Nectar; flowers of open fields; pugnacious—buzzes all intruders, other butterflies, and humans; prominent in N Piedmont

Honeydew from aphids

Nectar; flies in swift, powerful sweeps close to ground in open spaces; pugnacious

Nectar; open grassy and weedy places

Selected Lepidoptera of the Piedmont

Name	Food and Habitat of Larvae
Catalpa sphinx (moth) *Ceratomia catalpae*	Catalpa; hedgerows, early woods
Elm sphinx *Ceratomia amyntor*	Elm, birch; alluvial woods, mesic wooded hillsides
Pawpaw sphinx *Dolba hylaeus*	Pawpaw; alluvial woods
Great ash sphinx *Sphinx chersis*	Ash, privet; early woods, overgrown fields
Wild cherry sphinx *Sphinx drupiferarum*	Wild cherry, *Prunus serotina*; hides during the day, unlike most sphinx larvae
Saddleback caterpillar *Sinine stimulea*	Variety of woodland shrubs, incl. *Euonymus* and pawpaw
Solitary oak leaf miner *Cameraria hamadryadella*	Leaves of oak; mesic woodlands
Goldenrod spindle gall moth *Gnorimoschema gallaesolidagnis*	Stems of goldenrod; mesic abandoned fields
Mimosa webworm *Homadaula albizziae*	Mimosa, honey locust; early woody mesic sites
Yucca moth *Tegeticula yuccasella*	Seeds of yucca; mesic and xeric sites prior to woody takeover

continued

Food and Habitat of Adults
Nectar; nocturnal
Nectar; nocturnal
Nectar
Nectar; nocturnal
Nectar
Nectar
Nectar
Nectar
Nectar
Yucca, nectar; this moth is essential to the pollination of yucca

Table A–2

Name	Food
Snapping turtle *Chelydra serpentina*	Fish, waterfowl, invertebrates, vegetation, carrion
Stinkpot *Sternotherus odoratus*	Aquatic vertebrates and invertebrates
Eastern mud turtle *Kinosternon s. subrubrum*	Aquatic insects, small animals
Spotted turtle *Clemmys guttata*	Mixture of aquatic plant and animal matter
Eastern box turtle *Terrapene c. carolina*	Worms, insects, snails, fruits
Eastern painted turtle *Chrysemys p. picta*	Aquatic vegetation, insects, crayfish, carrion
River cooter *Chrysemys c. concinna*	Aquatic vegetation and animals, dead fish
Green anole *Anolis c. carolinensis*	Insects and spiders
Northern fence swift *Sceloporus undulatus*	Insects
Six-lined racerunner *Cnemidophorus sexlineatus*	Insects, spiders
Ground skink *Sincella laterale*	Insects, insect eggs

Reptiles of the Piedmont

Habitat and Habits	Range in Piedmont
Streams, lakes, ponds, rivers; not aggressive in water but vicious when handled	Throughout
Ponds, lakes, rivers, streams	Throughout
Ponds, lakes, ditches; likes still, shallow water	S of Roanoke R.
Marshy meadows, ditches, ponds	N of Rappahannock R; largely coastal plain
Moist open woodlands, near water; male has concave plastron, red eye	Throughout
Muddy bottomed ponds, ditches, streams	Throughout
Streams; endemic to the Piedmont	S of James R. (central Va.)
Fences, trees, vines, often found sunning; arboreal and terrestrial; changes colors through a range of greens and browns	S of Yadkin R.
Deciduous and coniferous woods, sunny spots; arboreal; only spiny lizard in Piedmont	Throughout
Very fast, prefers open dry areas	S of Potomac R.
Forest floor litter; does not climb; easily captured; delicate	S of Potomac R.

Reptiles of the Piedmont

Name	Food
Five-lined skink *Eumeces fasciatus*	Insects
Broad-headed skink *Eumeces laticeps*	Insects
Southeastern five-lined skink *Eumeces inexpectatus*	Insects
Slender glass lizard *Ophisaurus attenuatus*	Insects
Northern water snake *Nerodia s. sipedon*	Fish, frogs
Queen snake *Nerodia septemvittata*	Molted crayfish
Northern brown snake *Storeria d. dekayi*	Slugs, worms, soft-bodied insects
Red-bellied snake *Storeria occipitomaculata*	Slugs, earthworms, soft-bodied insects
Eastern garter snake *Thamnophis s. sirtalis*	Frogs, toads, tadpoles

continued

Habitat and Habits	Range in Piedmont
Cutover woods, farm buildings, rotting stumps; inveterate basker; sheds tail at slightest tug; young have blue tails	Throughout
Principally arboreal but may be found on rocks, stumps, in buildings in open	Throughout
Varied habitats, tolerates dry conditions such as xeric hilltops	S of James R.
Grasslands, wet meadows, pine woods	Piedmont of N.C. and S.C.
Most aquatic habitats in the Piedmont; harmless, but ill-tempered when handled	Throughout (replaced by midland water snake in Piedmont of S.C.)
Aquatic in flowing water; doesn't bask much; seen swimming or under debris at shore	Throughout
Any terrestrial habitat, moist or dry, which provides water, sun, and refuge; a secretive species	Throughout
No identifiable habitat preference; secretive	Throughout
Damp soils and mesic sites, wooded or open; needs access to moisture; harmless	Throughout

Reptiles of the Piedmont

Name	Food
Eastern ribbon snake *Thamnophis s. sauritus*	Salamanders, frogs, small fish
Rough earth snake *Virginia striatula*	Earthworms, soft-bodied insects
Smooth earth snake *Virginia valeriae*	Earthworms, insects
Eastern hognose snake *Heterodon platyrhinos*	Toads, occasionally frogs
Ringneck snakes *Diadophis punctatus* (subspp.)	Salamanders, earth-worms, lizards, frogs
Eastern worm snake *Carphophis a. amoenus*	Earthworms, soft-bodied insects
Northern black racer *Coluber c. constrictor*	Rodents, birds, lizards, frogs, snakes
Eastern coachwhip *Masticophis f. flagellum*	Rodents, lizards, small snakes
Rough green snake *Opheodrys aestivus*	Spiders, caterpillars, orthoptera
Corn snake *Elaphe g. guttata*	Rodents, birds

continued

Habitat and Habits	Range in Piedmont
Streams, ponds, boggy places	Throughout
Mesic open places, under stones, boards	S of James R.
Abandoned fields, deciduous woods; harmless, secretive, stays underground	Throughout
Loose mesic soils, principally open spaces; gives frightful display but is harmless	Throughout
Cutover woods, rocky wooded slopes	Throughout
Rarely seen above ground; burrows after its prey, earthworms, which it resembles	Throughout
Open and wooded sites; partially arboreal; harmless but disagreeable to handle	Throughout
Woodlands and brushy areas; fast-moving and ill-tempered; not venomous, but teeth can lacerate	S of Yadkin R.
Found in shrubs and vines at woods' edge, thickets, fencerows; gentle, tame, harmless	Throughout
Mesic sites; spends much time underground in rodent burrows; kills by constricting; harmless to man	Throughout

Reptiles of the Piedmont

Name	Food
Black rat snake *Elaphe o. obsoleta*	Rodents, birds
Eastern kingsnake *Lampropeltis g. getulus*	Other snakes, turtle eggs
Scarlet kingsnake *Lampropeltis elapsoides*	Small reptiles, mice, soft insects, earthworms
Scarlet snake *Cemophora coccinea*	Young mice, small snakes, reptile eggs
Mole snake (Brown kingsnake) *Lampropeltis rhombomaculata*	Rodents, other snakes
Southeastern crowned snake *Tantilla coronata*	Centipedes, underground larvae of insects
Southern copperhead *Agkistrodon c. contortrix*	Mice, frogs, birds, insects
Northern copperhead *Agkistrodon c. mokeson*	Mice, frogs, birds, insects

continued

Habitat and Habits	Range in Piedmont
Virtually any terrestrial habitat across the spectra of drainage and vegetation; unpredictable disposition	Throughout
Stream banks, marshes; follows the water courses across the Piedmont	Throughout
Pine woods under logs, stumps, loose bark; coloration mimics coral snake; harmless, secretive, nocturnal	S of Roanoke R.
Any terrestrial habitat suitable for burrowing	S of Potomac R.
Mesic woods, fields, thickets; burrows after its prey	S of Potomac R.
Any terrestrial habitat; rarely seen in open, usually found under something; rear-fanged but harmless	S of Roanoke R.
Prefers lowlands but does visit mesic sites; notably slow to anger; lethal	S of Yadkin R.
Many mesic habitats, including wooded hillsides, open fields; lethargic, slow to anger but lethal	N of Yadkin R.

Reptiles of the Piedmont

Name	Food
Eastern cottonmouth *Agkistrodon p. piscivorus*	Frogs, fish, salamanders, baby alligators
Timber rattlesnake *Crotalus h. horridus*	Rodents, birds, small mammals
Canebrake rattlesnake *Crotalus h. atricaudatus*	Rodents, small mammals, birds

continued

Habitat and Habits	Range in Piedmont
Rare in Piedmont but occupies some water courses and swamps in Ga. and Ala.; vicious, pugnacious, lethal	Ga. and Ala.
Wooded slopes, particularly where rodents are plentiful; rare except in remote highlands	Higher elevations throughout
Bottomlands and moist places of eastern and southern Piedmont; uncommon	S of Roanoke R.

Table A–3

Name	Food
Neuse River waterdog *Necturus lewisi*	Worms, other aquatic invertebrates
Marbled salamander *Ambystoma opacum*	Earthworms, other invertebrates
Jefferson salamander *Ambystoma jeffersonianum*	Earthworms, other invertebrates
Spotted salamander *Ambystoma maculatum*	Worms, larvae of forest insects
Red-spotted newt *Notophthalmus v. viridescens*	Insects, leeches, worms, crustaceans
Northern dusky salamander *Desmognathus f. fuscus*	Worms, larvae of forest floor insects
Red-backed salamander *Plethodon c. cinereus*	Ants, beetles, worms, other invertebrates
Slimy salamander *Plethodon g. glutinosus*	Same as red-backed
Four-toed salamander *Hemidactylium scutatum*	Same as red-backed

Amphibians of the Piedmont

Habitat and Habits	Range in Piedmont
Entirely aquatic; lurks under rocks, darts out for food; neotenic, 6 to 9 in (15 to 22 cm) long; rare and endangered by habitat destruction	Neuse and Tar R. systems of eastern Piedmont
Wooded moist and mesic sites; leaf litter, rotting logs, stumps	Throughout
Moist wooded lowlands, hill-sides, streams; burrows into soil and litter; rarely seen	N of James R.
Moist woods; breeds in forest ponds after heavy rains	Throughout
Quiet waters along streams, ponds, ditches; skin is rough, not slimy; eft is terrestrial	Throughout
Wooded mesic and bottomland sites, near streams, seeps, springs; lungs absent, breathes through skin; eft aquatic	Throughout
Moist and mesic woodland habitats; ubiquitous	N of Yadkin R.
Moist wooded coves, hillsides; needs more moisture than red-backed but often coexists where ranges overlap; secretes sticky substance from skin when frightened	Throughout
Sphagnum bogs; larvae are aquatic; distribution spotty	Throughout

Amphibians of the Piedmont

Name	Food
Eastern mud salamander *Pseudotriton m. montanus*	Larvae of aquatic insects, invertebrates
Northern red salamander *Pseudotriton r. ruber*	Worms, insects, insect larvae
Northern two-lined salamander *Eurycea b. bislineata*	Same as northern red
Three-lined salamander *Eurycea longicauda guttolineata*	Aquatic invertebrates
American toad *Bufo americanus*	Insects and other invertebrates
Fowler's toad *Bufo woodhousei fowleri*	Insects and other invertebrates
Northern cricket frog *Acris c. crepitans*	Insects
Eastern spadefoot toad *Scaphiopus holbrooki*	Insects
Spring peeper *Hyla crucifer*	Insects

continued

Habitat and Habits	Range in Piedmont
Muddy bottoms of springs, streams, quiet pools, usually in wooded tracts	S of Roanoke R.
Woodland springs, seeps, streambanks under moss, rocks	Throughout
Very small brooks, tributaries where fish are absent or at minimum; larvae aquatic	Throughout
Swamp and alluvial forests, brooks, seeps	S of Potomac R.
Great variety of habitats; woodlands, grasslands, wetlands, mesic sites; leaps to catch flying insects	Throughout
Habitats not limited; similar in lifestyle and distribution to American toad; often breeds with *B. Americanus*	Throughout
Stays hidden in shallows of permanent bodies of water; does not climb; voices rapid clicks in chorus	Throughout
Occupies forested sandy loose soils, mainly bottomlands	Eastern Piedmont
Woodlands with streams and pools; in spring, calls from ground at edges of standing water, single ascending *peep*	Throughout

Amphibians of the Piedmont

Name	Food
Eastern gray treefrog *Hyla veriscolor*	Insects
Upland chorus frog *Pseudacris triseriata feriarum*	Insects
Eastern narrow-mouthed toad *Gastrophryne carolinensis*	Insects, insect eggs and larvae
Bullfrog *Rana catesbeiana*	Flying insects, occasionally small vertebrates
Wood frog *Rana sylvatica*	Insects
Green frog *Rana clamitans melanota*	Flying insects
Southern leopard frog *Rana cphenocephala*	Flying insects
Pickerel frog *Rana palustris*	Insects

continued

Habitat and Habits	Range in Piedmont
Forages in trees, shrubs near or in standing water; sings from ground in spring	Throughout
Near temporary pools, back-waters in bottomlands, open fields, moist meadows; voice an extended *crreeek*	Throughout
Usually found in decaying wood, under logs, stumps; squeezes into narrow openings	S of James R.
Larger lakes, rivers, marshes, where it hunts at water's edge, dives to avoid danger; voice a deep sonorous honk	Throughout
Distinguished by a dark mask through the eye; occupies moist woodlands; ducklike in voice	N of Potomac R.
The banks of streams and lakes, wooded and open; leaps into water when alarmed; voice an explosive croak	Throughout
Basically aquatic but found far from water on mesic and moist fields; woodland streams	S of James R.
Aquatic but wanders onto mesic sites; protected from snakes by glands which produce distasteful secretions	Throughout

Table A–4

Name	Food
Pied-billed grebe *Podilymbus podiceps*	Small fish
Mallard *Anas platyrhynchos*	Aquatic vegetation, insects, mussels
Black duck *Anas rubripes*	Aquatic vegetation, insects, mussels
Wood duck *Aix sponsa*	Acorns, seeds, aquatic plants
Ring-necked duck *Aythya collaris*	Aquatic animals, vegetation
Common merganser *Mergus merganser*	Small fish
Turkey vulture *Cathartes aura*	Carrion
Black vulture *Coragyps atratus*	Carrion, insects, mammal dung

Selected Birds of the Piedmont

Habitat and Habits	Range in Piedmont
Dives for its catch in rivers and farm ponds; must run on surface to fly; loses ring around bill in winter	Nests and winters throughout
Visits lakes and larger rivers in winter; tips up on surface to feed; jumps to flight	Winters throughout; nests are uncommon
Casual on major rivers; dark underparts of both sexes distinguish from mallard	Winters throughout
Closely identified with woodland streams, rivers; partially arboreal; nests are excavations of pileated woodpecker; tail long	Nests throughout; winters south of Potomac R.
Common on farm ponds and secluded inland waters; dives to feed; needs take-off run	Winters S of James R.
Wary; winters on large creeks and rivers; male not crested	Winters N of Roanoke R.
Soars over open country, wings held in positive dyhedral; rarely flaps	Nests and winters throughout
Inelegant of diet but graceful in the air; holds wings flat; able to descend into woods for food; soars, but flaps as needed. Most hawklike in flight of North American vultures	Winters throughout; nests as far N as Big Roundtop at Gettysburg

Selected Birds of the Piedmont

Name	Food
Cooper's hawk *Accipiter cooperii*	Birds
Sharp-shinned hawk *Accipiter striatus*	Small birds
Marsh hawk *Circus cyaneus*	Rodents
Red-tailed hawk *Buteo jamaicensis*	Rodents, snakes
Red-shouldered hawk *Buteo lineatus*	Reptiles, amphibians, rodents
Broad-winged hawk *Buteo platypterus*	Insects, rodents, reptiles
Bald eagle *Haliaeetus leucocephalus*	Fish
American kestrel *Falco sparverius*	Insects, small birds, rodents
Turkey *Meleagris gallopavo*	Acorns, fruits, seeds

continued

Habitat and Habits	Range in Piedmont
An uncommon woodland hawk; aggressive in pursuit of prey and in defense of territory	Nests and winters throughout—occasionally
Darts into hedgerows, woods' edges, abandoned fields after passerine birds	Nests and winters throughout; common
Quarters low over fields, mainly open bottomlands; perches on ground	Winters S of Potomac R.
Soars over open country or hunts from an exposed perch; very common	Nests and winters throughout
Limited to the bottomland forests in the Piedmont; endangered through loss of habitat	Nests and winters throughout
Hunts ecotones from woods' edge perch; Piedmont's most numerous nesting raptor	Nests throughout; absent in winter
Seen occasionally along the Piedmont's major rivers; rare, endangered	
Hunts from open perches and hovers above prey; smallest North American falcon and most colorful raptor	Nests N of Roanoke R.; winters throughout
A large scratching bird and a strong flier; mature woodlands; becoming more common	Locally common throughout; doesn't migrate

Selected Birds of the Piedmont

Name	Food
Bobwhite *Colinus virginianus*	Seeds, insects
Ringnecked pheasant *Phasianus colchicus*	Seeds, insects
Great blue heron *Ardea herodias*	Fish, amphibians, insects
Green heron *Butorides virescens*	Fish, amphibians, reptiles, insects
American bittern *Botaurus lentiginosus*	Amphibians, fish, insects
Black-crowned night heron *Nycticorax nycticorax*	Amphibians, fish, rodents
American coot *Fulica americana*	Aquatic vegetation
Spotted sandpiper *Actitis maculdria*	Lotic invertebrates
American woodcock *Philohela minor*	Earthworms

continued

Habitat and Habits	Range in Piedmont
Fields, hedgerows, thick cover; coveys burst into explosive flight, roost in tight circle on ground	Abundant throughout
Farmland, open fields, hedgerows; introduced from Asia and now well established N of the James R.	Locally common in N half of Piedmont
Common along streams and rivers, at pond sides; 6-ft span; only wintering heron N of Savannah	Present on flowing water all seasons
Prefers woodland streams; hunts from rocks, fallen logs; semiarboreal	Common throughout, nesting season only
Secretive and crepuscular in marshy areas, open pond sides	Nests N of James R.; winters throughout
Nocturnal and uncommonly seen; roosts by day; farm ponds, lakes, marshy areas; guttural croak	Nests locally throughout; winters S of Roanoke R.
Arrives in fall to dabble in vegetation of major rivers, farm ponds, lakes; dives to escape	Winters S of Roanoke R.
Prefers creeks, rivers, where it scours shoals, rocks for insects, crustaceans; flies with wings held stiffly downward	Nests N of Savannah R.
Common in moist lowland woods; nocturnal; late winter nuptial display is spectacular	Nests and winters effectively throughout

Selected Birds of the Piedmont

Name	Food
Killdeer *Charadrius vociferus*	Terrestrial insects
Rock dove (domestic pigeon) *Columba livia*	Seeds, insects
Mourning dove *Zenaidura macroura*	Grains, insects
Yellow-billed cuckoo *Coccyzus americanus*	Caterpillars, other insects
Screech owl *Otus asio*	Rodents, insects
Great horned owl *Bubo virginianus*	Birds, mammals to the size of house cats
Short-eared owl *Asio flammeus*	Rodents
Barn owl *Tyto alba*	Rodents

continued

Habitat and Habits	Range in Piedmont
A bird of lowland pastures and short-grass habitats; likes cattle, horses; only plover in the Piedmont	Nests and winters throughout
Introduced from Europe and generally limited to cities but becoming established in rock outcrop habitats similar to its pre-domestication habitats in the Old World	All seasons, throughout
Open fields, grain fields, woods, towns; raises 5 broods, Feb.–Oct.; migrates	Nests and winters throughout
Growing and mature forests; repetitive, gulping call identifies this reclusive controller of lepidopterans	Nests throughout; absent in winter
Woodlands; roosts by day in cavities, occasionally wood-duck boxes; vocal in summer	Nests and winters throughout
Hunts forests, open country, keeps to woods by day; nests Dec.–Mar.; Piedmont's largest owl; common	Nests and winters throughout
Hunts open country, mostly bottomlands	Winters throughout; absent in summer
Not vocal except for a prolonged hiss, rarely heard away from nest; may be more common than paucity of sightings suggests	Nests and winters throughout

Selected Birds of the Piedmont

Name	Food
Barred owl *Stryx varia*	Rodents, aquatic invertebrates
Chuck-will's-widow *Caprimulgus carolinensis*	Flying insects
Whippoorwill *Caprimulgus vociferus*	Flying insects
Common nighthawk *Chordeiles minor*	Flying insects
Chimney swift *Chaetura pelagica*	Flying insects
Ruby-throated hummingbird *Archilochus colubris*	Nectar, very small insects
Belted kingfisher *Megaceryle alcyon*	Small fish

continued

Habitat and Habits	Range in Piedmont
A woodland owl; bottomlands in S Piedmont, mesic woods in N; very vocal, particularly in late winter	Nests and winters throughout
Associated with pine woods; common S of Haw R. in N.C.; rarely seen; identified by call; nocturnal	Nests S of Roanoke R.; absent in winter
Nocturnal; smaller than *C. carolinensis*, more vocal; roosts in woods, on ground and limbs	Nests throughout; absent in winter
Crepuscular and nocturnal; call, a single rasping buzz; nests on rooftops in cities	Nests throughout; migrates in flocks; absent in winter
Cigar-shaped body, swept-back wings; on wing all day, never perches except to roost at night; gregarious	Nests throughout (in chimneys, hollow trees); absent in winter
Common, the only humming-bird in Piedmont; hovers to feed; very territorial; migrates to South and Central America over open sea	Nests throughout; absent in winter
Dives headfirst from perch or hovering flight; prefers streams, rivers; irregular wingbeat; protest call a raucous rattle	Nests and winters throughout

Selected Birds of the Piedmont

Name	Food
Common flicker *Colaptes auratus*	Wood-boring and other insects
Pileated woodpecker *Dryocopus pileatus*	Wood-boring insects
Red-bellied woodpecker *Centurus carolinus*	Wood-boring insects
Red-headed woodpecker *Melanerpes erythrocephalus*	Wood-boring insects
Yellow-bellied sapsucker *Sphrypicus varius*	Tree sap, insects
Hairy woodpecker *Dendrocopus villosus*	Wood-boring insects

continued

Habitat and Habits	Range in Piedmont
Most terrestrial North American woodpecker; prefers mature woods; white patch above rump and yellow primaries are field marks	Nests and winters throughout
Largest woodpecker in Piedmont (crow-size); mature forests, all drainages; becoming more common as forests mature	Nests and winters throughout
Common in growing and mature woods; only cap and cape are red; compare red-headed woodpecker	Nests and winters throughout
Uncommon; prefers mature deciduous woods; slightly predaceous (rodents, small birds); entire head and neck red	Nests and winters throughout
Common in deciduous woods; bores shallow wells to get sap and thereby makes this food source available to other birds, animals, insects	Winters throughout; absent in summer
Common in mature woods; similar to but substantially larger than downy woodpecker	Nests and winters throughout

Selected Birds of the Piedmont

Name	Food
Downy woodpecker *Dendrocopos pubescens*	Wood-boring insects
Eastern kingbird *Tyrannus tyrannus*	Flying insects
Great crested flycatcher *Myarchus crinitus*	Insects, mostly flying species
Eastern phoebe *Sayornis phoebe*	Flying insects
Acadian flycatcher *Empidonax virescens*	Flying insects
Eastern wood pewee *Contopus virens*	Flying insects
Horned lark *Eremophila alpestris*	Seeds, insects
Barn swallow *Hirundo rustica*	Flying insects
Rough-winged swallow *Stelgidopteryx ruficolis*	Flying insects

continued

Habitat and Habits	Range in Piedmont
Woods of all maturities, drainages; Piedmont's smallest woodpecker	Nests and winters throughout
Hawks insects from exposed perch; aggressive with larger birds; look for white terminal band in tail	Nests throughout; absent in winter
Common in growing and mature forests; nests in abandoned woodpecker cavities	Nests throughout; absent in winter
Piedmont's only wintering flycatcher; hunts from exposed perch, flips tail at rest	Nests N of Roanoke R.; winters S of Roanoke R.
Prefers wooded streamsides at maturity; Piedmont's only nesting *Empidonax*	Nests throughout; absent in winter
Common in growing and mature forests; call and wing bars distinguish from phoebe	Nests throughout; absent in winter
Gregarious; flocks common in short grass and plowed and freshly manured fields	Nests and winters throughout
A swift and tireless flier in open spaces; tail deeply forked; seen most often near barns	Nests throughout; absent in winter
Locally common near water; nests in burrows in steep banks; brown band across throat	Nests throughout

Selected Birds of the Piedmont

Name	Food
Purple martin *Progne subis*	Flying insects
Blue jay *Cyanocitta cristata*	Omnivorous
Common raven *Corvus corax*	Omnivorous
Common crow *Corvus brachyrhynchos*	Omnivorous
Fish crow *Corvus ossifragus*	Fish, carrion
Black-capped chickadee *Parus atricapillus*	Seeds, insects
Carolina chickadee *Parus carolinensis*	Seeds, insects
Tufted titmouse *Parus bicolor*	Seeds, insects

continued

Habitat and Habits	Range in Piedmont
A communal nester adapting readily to gourds and martin houses; urban and rural; needs help competing with starlings for nesting holes	Nests throughout; absent in winter
Common in woods of all maturities and drainages; frequents open spaces; predaceous; very vocal	Nests and winters throughout
Rare in Piedmont at western monadnocks; soars; is predaceous and takes carrion; V-shaped tail	Nests and winters at selected, craggy localities
Common throughout in all habitats; Gregarious in fall and winter; takes mice, carrion; vocal	Nests and winters throughout
Locally common along rivers, in cities; mixes with common crows; scavenges	Nests and winters in outer Piedmont
Common in woodlands; song, 2 notes; white edging on primaries	Nests and winters N of Roanoke R.
Common in all types of woodlands; larger than black-capped, lacks edging	Nests and winters S of Roanoke R.
Common in growing and mature forests	Nests and winters throughout

Selected Birds of the Piedmont

Name	Food
White-breasted nuthatch *Sitta carolinensis*	Seeds, insects
Red-breasted nuthatch *Sitta canadensis*	Seeds, insects
Brown-headed nuthatch *Sitta pusilla*	Pine seeds, insects
House wren *Troglodytes aedon*	Insects
Winter wren *Troglodytes troglodytes*	Insects, fruits
Carolina wren *Thryothorus ludovicianus*	Insects, fruits
Mockingbird *Mimus polyglottos*	Fruits, insects
Catbird *Dumetella carolinensis*	Fruits, insects
Brown thrasher *Toxostoma rufum*	Fruits, insects

continued

Habitat and Habits	Range in Piedmont
Common in deciduous woods; largest nuthatch; like all nuthatches, can descend tree trunk headfirst	Nests and winters throughout
Prefers maturing pines; note dark eye stripe	Winters throughout; absent in summer
Strongly associated with pinelands; uncommon in Piedmont N of James R.	Nests and winters S of James R.
A summer bird, lavish of song; nests in small cavities; tail erect, as in all wrens	Nests throughout; absent in winter
Confined to bottomland forests; smallest Piedmont wren; vanishes in ground cover	Winters throughout; absent in summer
Common in hedgerows, woods, ecotones; song, rollicking, very loud; makes domed nests in buildings, farm machinery	Nests and winters throughout
Common in overgrown fields, hedgerows; sings almost constantly, mimics variety of sounds	Nests and winters throughout
Common in brushy habitats; rusty coverts under tail; mewing call, does not repeat	Nests throughout; absent in winter
Woods' edges, hedgerows; common; repeats song once	Nests and winters throughout

Selected Birds of the Piedmont

Name	Food
Robin *Turdus migratorius*	Insects, earthworms
Wood thrush *Hylocichla mustelina*	Grubs, insects
Hermit thrush *Hylocichla guttata*	Insects
Eastern bluebird *Sialia sialis*	Insects, spiders, fruits
Blue-gray gnatcatcher *Polioptila caerulea*	Small insects
Golden-crowned kinglet *Regulus satrapa*	Insects
Ruby-crowned kinglet *Regulus calendula*	Insects
Water pipit *Anthus spinoletta*	Insects
Cedar waxwing *Bombycilla cedrorum*	Berries

continued

Habitat and Habits	Range in Piedmont
Common on lawns, meadows, forests; gregarious in winter	Nests and winters throughout
Common in growing and mature deciduous forests; flutelike song	Nests throughout; absent in winter
Common in deciduous woods of all drainages	Winters throughout; absent in summer
Locally common at edges of pastures; feeds in family groups by flying from hunting perch to ground; house sparrow usurps nests	Nests and winters S of Potomac R.
Flits among deciduous branches; resembles miniature mockingbird	Nests throughout; absent in winter
Forages high in pines, often with creepers, brown-headed nuthatches, pine warblers; shy	Winters throughout; absent in summer
Favors brushy areas, woods and edges; erects red crest when excited; tame—may feed from human hand	Winters throughout; absent in summer
Gregarious in plowed and open fields in winter; flies in tight flocks	Winters throughout; absent in summer
Tame and gregarious; feeds in tight flocks on fruits of black tupelo, holly, cedar	Winters throughout; may nest in extreme N Piedmont

Selected Birds of the Piedmont

Name	Food
Loggerhead shrike *Lanius ludovicianus*	Rodents, birds, large insects
Starling *Sturnus vulgaris*	Omnivorous
White-eyed vireo *Vireo griseus*	Insects
Red-eyed vireo *Vireo olivaceus*	Insects, mostly caterpillars
Black and white warbler *Mniotilta varia*	Insects
Prothonotary warbler *Protonotaria citrea*	Insects
Parula warbler *Parula americana*	Insects
Yellow warbler *Dendroica petechia*	Insects
Myrtle warbler *Dendroica coronata*	Insects

continued

Habitat and Habits	Range in Piedmont
Uncommon; seen from highway on wires, exposed hunting perches; a predator; impales prey on thorns	Nests and winters throughout
Regrettably abundant; an immigrant which proliferates at expense of native birds; all habitats, particularly those near man	Nests and winters throughout
Occupies bottomland forests, mesic ecotones to a lesser extent; call suggestive of that of chuck-will's-widow	Nests throughout; absent in winter
Abundant in deciduous forests; called "preacher bird" for didactic, emphatic phrases of continuous song	Nests throughout; absent in winter
Nuthatch-like in movements; common in deciduous woods	Nests throughout; absent in winter
Common along rivers and streams flowing through mature forests	Nests throughout; absent in winter
Prefers bottomland hardwoods in Piedmont	Nests throughout; absent in winter
Nests in willows at early woody phase of hydrosere; parasitized by cowbird	Nests throughout; absent in winter
Forages at all levels in all woody habitats; one of Piedmont's most common wintering birds	Winters throughout; absent in summer

Selected Birds of the Piedmont

Name	Food
Yellow-throated warbler *Dendroica dominica*	Insects
Pine warbler *Dendroica pinus*	Insects
Ovenbird *Seiurus aurocapillus*	Insects
Northern waterthrush *Seiurus motacilla*	Aquatic invertebrates
Yellowthroat *Geothlypis trichos*	Insects, fruits
Yellow-breasted chat *Icteria virens*	Insects
Kentucky warbler *Oporornis formosus*	Insects
Hooded warbler *Wilsonia citrina*	Insects
American redstart *Setophaga ruticilla*	Insects

continued

Habitat and Habits	Range in Piedmont
Common high in ashes and sycamores at riverside; generally the earliest warbler to arrive in spring	Nests throughout; absent in winter
Common in tall pines, uncommon elsewhere; one of the few warblers to winter in Piedmont	Winters S of Roanoke R.; nests throughout
Mature deciduous forests; common; nests on or near ground	Nests throughout; absent in winter
Fairly common along wooded streams; forages in shoals, at water's edge; hindparts in continuous swaying dance	Nests throughout; absent in winter
Common in overgrown fields, hedgerows, particularly if moist	Nests and winters throughout
Fairly common in abandoned fields with saplings established; issues incongruous series of squeeks, grunts, whistles	Nests throughout; absent in winter
A ground-nesting warbler common in moist and mesic deciduous woods	Nests throughout; absent in winter
Fairly common in maturing deciduous woods; forages and nests close to ground	Nests throughout; absent in winter
Common in deciduous understory; favors bottomland forests	Nests throughout; absent in winter

Selected Birds of the Piedmont

Name	Food
House sparrow *Passer domesticus*	Seeds, insects
Eastern meadowlark *Sturnella magna*	Insects, worms, beetle larvae
Red-winged blackbird *Agelaius phoeniceus*	Insects, invertebrates
Rusty blackbird *Euphagus carolinus*	Insects, invertebrates
Common grackle *Quiscalus quiscala*	Insects, some vegetative material
Brown-headed cowbird *Molothrus ater*	Seeds, insects
Orchard oriole *Icterus spurius*	Fruit, insects
Great (formerly Baltimore) oriole *Icterus galbula*	Insects
Scarlet tanager *Piranga olivacea*	Insects

continued

Habitat and Habits	Range in Piedmont
Introduced from Europe early this century; parasitic on man; not found away from towns, buildings; usurps bluebirds' nests	Nests and winters throughout
Abundant in pastures, hayfields; nests on ground, sings from rocks, fence posts	Nests and winters throughout
Abundant at farm ponds, pastures; prefers marshes; polygamous; 1st-year males don't breed	Nests and winters throughout
Winters in mature bottomland woods	Winters throughout; absent in summer
Abundant in farmlands; in winter, flies to and from roosts in long streams	Nests and winters throughout
Common in farmland; mixes with redwings, grackles, starlings; parasitic nester	Nests and winters throughout
Locally common; favors orchards (unsprayed) or individual fruit trees; builds hanging nest	Nests throughout; absent in winter
Locally common north of James R., but present throughout; feeds in treetops; only orange oriole in Piedmont	Nests throughout; absent in winter
Common in growing and mature mesic woods	Nests N of Haw R.; absent in winter

Selected Birds of the Piedmont

Name	Food
Summer tanager *Piranga rubra*	Insects
Cardinal *Cardinalis cardinalis*	Seeds, insects
Evening grosbeak *Hesperiphona vespertina*	Seeds
Blue grosbeak *Guiraca caerulea*	Insects, seeds
Indigo bunting *Passerina cyanea*	Insects, seeds
Purple finch *Carpodacus purpureus*	Seeds
House finch *Carpodacus mexicanus*	Seeds

continued

Habitat and Habits	Range in Piedmont
Common in mesic growing woods S of the Potomac R.	Nests S of Potomac R.; absent in winter
Common in overgrown fields, hedgerows; stays low	Nests and winters throughout
Has expanded range to eastern North America in this century; an irregular, wandering migrant in Piedmont in winter	Winters throughout; absent in summer
A common nesting species in early woody mesosere; resembles indigo bunting, distinguished by size and rusty wing bars	Nests S of Potomac R.; absent in winter
Prefers open brushy country, wood margins; sings from high perches; feeds low	Nests throughout; absent in winter
Common in tulip and sweet gum trees eating seeds; easily attracted to feeders with sunflower seed, as are most of the Piedmont's wintering fringillids	Winters throughout; absent in summer
A native North American bird but new east of the Rockies; range is expanding southward; a strong migrant in eastern North America, sedentary in west	Winters S to Savannah R.; nests north of Potomac R.

Selected Birds of the Piedmont

Name	Food
Pine siskin *Spinus pinus*	Seeds
American goldfinch *Spinus tristes*	Seeds, particularly those of thistles
Rufous-sided towhee *Pipilo erythrophthalmus*	Seeds, insects
Savannah sparrow *Passerculus sandwichensis*	Seeds, insects
Grasshopper sparrow *Ammodramus savannarum*	Insects, seeds
Dark-eyed (formerly slate-colored) junco *Junco hyemalis*	Seeds
Tree sparrow *Spizella arborea*	Seeds
Chipping sparrow *Spizella passerina*	Seeds
Field sparrow *Spizella pusilla*	Seeds

continued

Habitat and Habits	Range in Piedmont
A brown version of the goldfinch; flies and feeds in tight flocks; yellow in wing diagonal	Winters throughout
Prefers open weedy fields, also tall sweet gums; nesting cycle apparently synchronized with thistles	Nests and winters throughout
A ground-feeding bird of wood margins, hedgerows; those S of Savannah R. may have white eye	Nests and winters throughout
Prefers pastures and other short-grass habitats	Winters throughout; absent in summer
Feeds on grasshoppers, crickets in hayfields; look for clear buffy breast	Nests N of Haw R.; winters S of James R.
Abundant in weedy fields at woods' edge; likes fencerows	Winters throughout; absent in summer but nests in Blue Ridge, W of Piedmont
Prefers ecotones, hedgerows, woods' edges; feeds near ground	Winters N of James R.; absent in summer
In summer, forages in lawns and open, grassy woods; prefers hedgerows in winter	Nests throughout; winters S of James R.
Common in overgrown fields with trees established; hedgerows	Nests and winters throughout

Selected Birds of the Piedmont

Name	Food
White-crowned sparrow *Zonotrichia leucophrys*	Seeds
White-throated sparrow *Zonotrichia albicolis*	Seeds
Fox sparrow *Passerella iliaca*	Seeds, insects
Swamp sparrow *Melospiza georgiana*	Seeds
Song sparrow *Melospiza melodia*	Seeds

continued

Habitat and Habits	Range in Piedmont
Common in hedgerows N of James R.; less common in S Piedmont; hatch-year birds lack white	Winters throughout; absent in summer
Abundant in overgrown fields of all drainages	Winters throughout; absent in summer
Scratches on ground like towhee; prefers growing deciduous forests, woods' edges	Winters throughout; absent in summer
Common in shrubby low ground; feeds in mesic lands on migration; stays near ground	Winters throughout; absent in summer
Favors moist areas but common on shrubby mesic sites; ubiquitous in winter	Winters throughout; nests N of James R.

Table A-5

Name	Food
Opossum *Didelphis marsupialis*	Small mammals, fruits, nuts, carrion, eggs, insects
Smoky shrew *Sorex fumeus*	Worms, insects
Southeastern shrew *Sorex longirostris*	Insects, worms, young of small mammals
Long-tailed shrew *Sorex dispar*	Centipedes, spiders, insects
Least shrew *Cryptotis parva*	Insects, small mammals
Short-tailed shrew *Blarina brevicauda*	Insects, worms, snails, possibly young mammals
Star-nosed mole *Condylura cristata*	Worms, insects, aquatic invertebrates
Eastern mole *Scalopus aquaticus*	Worms, insects, vegetable matter
Hairytail mole *Parascalops breweri*	Worms, insects
Little brown myotis *Myotis lucifera*	Flying insects
Keen myotis *Myotis keeni*	Flying insects
Silver-haired bat *Lesionycteris noctivigans*	Flying insects

Mammals of the Piedmont

Habitat and Habits	Range in Piedmont
Active farms, woodlands, along rivers and streams; only U.S. marsupial	Throughout
Birch and hemlock forests	N of Rappahannock
Open fields, woods, prefers moist places	N of Rappahannock
Deciduous and mixed deciduous–coniferous woodlands	Md. and Pa.
Grassy areas	Throughout
Virtually all Piedmont habitats; saliva poisonous	Throughout
Moist open areas, low ground near water; good swimmer; does not make ridge	N of James R.
Open, moist sandy loams; makes visible ridges	Throughout
Mesic loams; burrows to 18 in (45 cm) deep	N of Potomac R.
Roosts in caves, mines, tunnels; flies near water, forests	Throughout
Roosts in caves, mines, tunnels, culverts, hollow trees	N of Roanoke R.
Roosts in trees, flies in woods; flight straight	Throughout

Mammals of the Piedmont

Name	Food
Eastern pipistrel *Pipistrellus subflavus*	Flying insects
Red bat *Lasurius borealis*	Flying insects
Big brown bat *Eptesicus fuscus*	Flying insects, mostly beetles
Hoary bat *Lasiurus cinereus*	Flying insects
Seminole bat *Lasiurus seminolus*	Flying insects
Evening bat *Nycticeius humeralis*	Flying insects
Black bear *Ursus americanus*	Nuts, fruits, insects, honey, garbage, carrion, eggs
Long-tailed weasel *Mustella frenata*	Small mammals, birds
Short-tailed weasel *Mustella erminea*	Mice, voles, birds
River otter *Lutra canadensis*	Fish, frogs, crayfish
Striped skunk *Mephitis mephitis*	Small rodents, insects, grubs, fruits, carrion

continued

Habitat and Habits	Range in Piedmont
Flight slow and erratic; smallest Piedmont bat	Throughout
Roosts and flies in wooded areas	Throughout
Probably most common Piedmont bat; woods and open spaces	Throughout
Woodlands; large, high-flying bat	Throughout
Woodlands; roosts in trees	All except N.J.
Hollow trees, buildings; straight, steady flight	Throughout
Forests and swamps; rare in Piedmont; wary, not strongly predaceous	S.C. and W Piedmont
All Piedmont habitats; nocturnal; terrestrial and arboreal	Throughout
Wooded and brushy areas not far from water	N of Potomac R.
Streams and lakes; dens in bank with entrance hidden; makes slides into water	Throughout
Semi-open lands, mixed deciduous–coniferous woods; a nocturnal omnivore	Uncommon S of Roanoke R.

Mammals of the Piedmont

Name	Food
Spotted skunk *Spilogale putoris*	Rodents, birds, eggs, insects, carrion, fruits
Red fox *Vulpes fulva*	Small mammals, insects, fruits
Gray fox *Urocyon cinereoargenteus*	Small mammals, birds, fruits, insects, carrion, acorns, grapes
Bobcat *Lynx rufus*	Small mammals, birds, carrion
Raccoon *Procyon lotor*	Fruits, insects, aquatic vertebrates and invertebrates, corn, nuts
Woodchuck *Marmota monax*	Tender shoots, cultivated grasses, and legumes
Eastern chipmunk *Tamais striatus*	Seeds, bulbs, fruits, nuts, insects, eggs
Eastern gray squirrel *Sciurus carolinensis*	Nuts, seeds, fruits, fungi
Eastern fox squirrel *Sciurus niger*	Seeds, nuts, eggs
Southern flying squirrel *Glaucomys volans*	Seeds, nuts, eggs

continued

Habitat and Habits	Range in Piedmont
Open woodlands, brushy fields, streamsides; nocturnal	S of Yadkin R.
All Piedmont habitats; prefers a mix of open and wooded country	Throughout
Mixed woods and open country; good climber	Throughout
Prefers bottomlands, mature forests; probably more common than is generally known	Throughout
Streams, bottomlands, deciduous woods; nocturnal	Throughout
Pastures, mesic open places; prefers active farmland; wary, diurnal; hibernates	N of Roanoke R., SW Piedmont
Deciduous forests, rocky places; diurnal; hibernates	Throughout
Deciduous forests; forested river bottoms	Throughout
Deciduous forests; sometimes forages in open; uncommon to rare	Isolated locations throughout
Deciduous and mixed deciduous–coniferous woods; common but rarely seen; nocturnal	Throughout

Mammals of the Piedmont

Name	Food
Beaver *Castor canadensis*	Bark, cambium, twigs
Eastern harvest mouse *Reithrodontomys humulis*	Seeds, tender shoots
White-footed mouse *Peromyscus leucopus*	Nuts, seeds, insects
Deer mouse *Peromyscus maniculatus*	Seeds, fruits, nuts
Golden mouse *Peromyscus nutalli*	Seeds, fruits, nuts
Rice rat *Oryzomys palustris*	Green shoots, seeds
Hispid cotton rat *Sigmodon hispidus*	Green shoots, eggs of ground-nesting birds
Meadow vole *Microtus pennsylvanicus*	Grasses, sedges, seeds, bark, insects
Pine vole *Pitymys pinetorum*	Bark from roots, tubers, bulbs, seeds
Muskrat *Ondatra zibethica*	Aquatic vegetation, frogs, clams, fish
Norway rat *Rattus norvegicus*	Omnivorous, including human infants and invalids

continued

Habitat and Habits	Range in Piedmont
Wooded streams and lakes; uncommon but widely distributed	Throughout
Prefers wet meadows	Md. southward
Mesic wooded, brushy, and open lands; caches food	Throughout
Mesic mixed wooded and open country; caches food	N of James R.
Woodlands, vine thickets; semiarboreal; builds nests in vines 5–10 ft (1.5–3 m) up	S of James R.
Wet meadows, marshy places; semiaquatic; nocturnal	S of Potomac R.
Mesic grasslands and weedy places	S of Roanoke R.
Hydric and mesic grassy sites	N of Savannah R.
Deciduous woodlands; sometimes in pine woods in S Piedmont	Throughout
Ponds, marshes, streams; chiefly aquatic; diurnal; active and common all season	Throughout
Buildings, rubbish heaps; associated with man; length to 18 in (45 cm); alien	Throughout

Mammals of the Piedmont

Name	Food
Black rat *Rattus rattus*	Omnivorous
Meadow jumping mouse *Zapus hudsonius*	Seeds, fruits, insects
Eastern cottontail *Sylvilagus floridanus*	*Summer:* green shoots; *Winter:* bark and twigs
Whitetail deer *Odocoileus virginicus*	Twigs, shoots, acorns, grass, fungi, herbs

continued

Habitat and Habits	Range in Piedmont
Buildings, cities, farms; may never touch soil; alien	Throughout
Mesic and hydric open and wooded sites; nocturnal; hibernates	Throughout
Brushy places, fencerows, edges of marshes	Throughout
Deciduous woods, meadows, abandoned fields, bottomlands	Throughout

Glossary

ACHENE: A small dry, one-seeded fruit which does not burst when ripe (e.g., an acorn).

AESTIVAL: Pertaining to the hottest part of summer; i.e., following the VERNAL aspect.

ALBEDO: The fraction of incident light reflected by a planet or satellite (such as the moon) expressed as a percentage.

ALGAE: Rootless, stemless, leafless plants containing chlorophyll. They may be one-celled, colonial or filamentous.

ALLUVIAL SOIL: That which is transported by wind or water; compare RESIDUAL SOIL.

ANADROMOUS: Going upstream to spawn.

ANNUAL: A plant that completes its growth cycle in one year or growing season.

ARACHNID: An ARTHROPOD of the class Arachnida, including the spiders, ticks, and mites.

ARTHROPOD: Any member of the phylum of nearly one million species of invertebrates having jointed legs and segmented bodies. It includes insects, crustaceans, arachnids, and other classes.

AUSTRAL: Southern.

AUTOTROPH: A free-living (not parasitic or SAPROPHYTIC) green plant capable of manufacturing its own food through photosynthesis; a self-nourishing organism.

BASE: In botany, the end of a leaf nearest the stem.

BENTHIC: Living at or near the bottom of a body of water.

BERRY: A fleshy or pulpy fruit developed from a single ovary with one or many seeds (e.g., a grape).

BIOMASS: The total mass of living organisms in a given volume.

BOREAL: Northern.

BRACONID: A member of the wasp family Braconidae.

BRACT: A modified leaf which encloses or subtends another structure, often a flower.

CALYX: The portion of a flower composing the outer, protective whorl of BRACTS and SEPALS.

Canid: A member of the mammalian family Canidae: the dogs, foxes, wolves, etc.

Canopy: Any closed covering of foliage.

Carapace: The horny or bony protective covering over the back of an animal (e.g., a turtle's shell).

Catkin: A flexible, dangling INFLORESCENCE (as in tag alder).

Chitin: A horny polysaccharide secreted by the skin and forming the tough outer cover of the bodies of most arthropods.

Clastic: Composed of fragments of transported rock.

Cleistogamous: Pertaining to a self-pollinating flower which does not open.

Cleptoparasite: An organism which lives by stealing from another.

Climax: The final phase in the succession of plants (and their attendant fauna) in a particular circumstance of soil and climate. The climax community, unlike previous successional phases, is self-perpetuating.

Coeval: Existing simultaneously; of the same age.

Coleopteran: A member of the order Coleoptera, the beetles.

Composite: A member of the botanical family Asteraceae, formerly the Compositae.

Compound leaf: One composed of separate leaflets (e.g., ash).

Conifer: A member of the botanic order Coniferales, those trees which bear seeds in a cone and have evergreen foliage (e.g., pine, cedar, yew, spruce).

Coprophagy: The practice of feeding on dung.

Corolla: The inner floral whorl, usually colored, sometimes fused. Compare CALYX.

Corvid: A member of the avian family Corvidae: the crows, jays, ravens.

Costal: Of or pertaining to the ribs.

Culm: The flowering stems of GRASSES and sedges.

Cycad: A tropical GYMNOSPERM resembling a palm but reproducing like lower forms.

Cyme: A broad, flattened INFLORESCENCE, the central FLORETS maturing first.

DBH: Abbreviation for "diameter at breast height," the standard measure of the thickness of trees.

DECIDUOUS: Not persistent or evergreen. Usually applies to foliage, sometimes to fruit.

DENTATE: Coarsely toothed (of leaves) with the teeth perpendicular to the margin.

DIABASE: A finely crystalline basaltic rock which decomposes to form soils with basic pH.

DISTAL: In anatomy, farther or farthest from the center or point of attachment; of leaves, toward the end.

DOMINANT: Any plant which predominates in a community by height, size, and numbers.

DRUPE: A fleshy fruit with a single seed enclosed in a hard case (e.g., a cherry or peach).

ECOTONE: The zone where two habitats meet, enriching each other biologically.

ECOTYPE: A group within a species which shows distinctive adaptations to a particular environment.

EDAPHIC: Of or pertaining to soil.

ELYTRA: The front wings of a beetle. The structures are not used in flight but are hardened into horny covers protecting the hindwings and body at rest.

ENDEMIC: Limited in distribution. A plant or animal is said to be endemic to a certain region or habitat if it grows nowhere else (e.g., *Amphianthus pusillus* is endemic to the granitic outcrops of the southern Piedmont).

EPHEMERAL: A short-lived plant that may grow, flower, and die within a matter of days.

EXFOLIATION: The shedding of leaves or layers.

FENCEROW: *See* HEDGEROW.

FLORET: A single floral unit within a group of similar units called, collectively, an INFLORESCENCE.

FOLLICLE: A dried fruit from a single ovary, opening along a single suture to release seeds.

FORB: An herb not in the GRASS family.

FRINGILLID: A member of the avian family Fringillidae: the sparrows, finches, cardinal, etc.

FUNGUS: Any of a large group of parasitic or saprophytic plants, including the mushrooms, molds, mildews, rusts, and smuts, which lack chlorophyll and reproduce by means of spores.

GABBRO: A dark, heavy, igneous rock containing pyroxene and feldspar.

GLABROUS: Without TRICHOMES or hairs; smooth.

GLOCHIDIUM: The parasitic larval stage of a freshwater mussel.

GNEISS: A course-grained metamorphic rock with a banded appearance.

GRAMINOID: *N.*, a plant in the GRASS family; *adj.*, of or relating to grasses.

GRASS: A member of the botanic family Poaceae (formerly Gramineae) and distinct from such grasslike plants as rushes, sedges, and certain herbaceous LEGUMES.

GYMNOSPERM: A member of the botanic class Gymnyspermae bearing ovules on open scales and usually lacking true vessels in the wood. This class includes the CONIFERS, CYCADS, and the ginko, among others.

HALICTID: A member of the insect family Halictidae, a cosmopolitan group of very small bees.

HEDGEROW: A thin strip of woody plants along a fence line separating open fields; extremely valuable to animals as cover and nesting habitat.

HEMIPTERAN: A member of the insect order Hemiptera, the true bugs.

HETEROTROPH: A plant or animal obtaining its nourishment from other organisms. Carnivores, parasites, SAPROPHYTES, and grazers are all heterotrophs.

HIEMAL: That aspect of the biological year when cold prevents most growth and metabolism (synonymous with hibernal); midwinter.

HYDRIC: Wet.

HYDROPHYTE: A plant growing in water or requiring moist soil.

HYDROSERE: Plant succession in wet soils.

HYPHAE: The individual tubelike threads of a fungus which collectively form the MYCELIUM.

ICTERINE: Of or relating to a member of the avian family Icteridae: the blackbirds and orioles.

IGNEOUS: Pertaining to rock formed by the process of hot magma cooling below the surface without erupting.

INFLORESCENCE: The flowering portion of a plant; may be composed of more than one flower.

INSTAR: Any of the stages in the life of an ARTHROPOD between successive molts.

LAGOMORPH: A mammal in the order Lagomorpha: the rabbits, hares, and pikas.

LANCEOLATE: Lance-shaped.

LEGUME: A member of the botanic family Fabaceae, the "bean family," which bears fruits in a beanlike pod.

LENTIC: Of or pertaining to still water.

LEPIDOPTERAN: A member of the insect order Lepidoptera: the moths, skippers and butterflies.

LIANA: A climbing vine.

LICHEN: A symbiotic association between a FUNGUS and an ALGA.

LIGNIN: A hard, cellulose-like substance that binds together the cellulose fibers of wood.

LITTORAL: Of or pertaining to the region along a shore. In a lake or pond, the littoral zone is occupied by rooted plants.

LOTIC: Pertaining to flowing water.

MESIC: Moderately moist.

MESOPHYTE: A plant having moderate requirements for moisture.

MESOSERE: Plant succession in soils of moderate moistness.

METAMORPHOSIS: A change of form, shape or substance. In geology, metamorphic rock is rock of sedimentary or IGNEOUS origin which has been deformed by intense heat and pressure within the earth's crust. In entomology, it is the change from the larval stage to adulthood. It can be accomplished in INSTARS (gradual metamorphosis) or by radical rearrangement of the cells in a pupal case (complete metamorphosis).

MICROCOMMUNITY: A small, localized community of plants and animals.

MONADNOCK: An isolated hill rising above a PENEPLAIN.

MORAINE: A terrace of TILL deposited at the sides or end of a glacier.

MYCELIUM: The mass of threadlike tubes (HYPHAE) which form the vegetative part of a fungus. The "mushroom" is the fruiting part.

NAIADS: The aquatic larvae of certain insects (e.g., the stoneflies, dragonflies, and damselflies).

NOCTUID: A member of the lepidopterous family Noctuidae, a large group of medium-sized moths.

OBOVATE: Inversely ovate, as a leaf with the narrower end at the base.

OPERCULUM: A protective flap or lidlike structure on a plant or animal.

ORB-WEAVING SPIDER: A member of the ARACHNID family Araneidae (formerly Argiopidae), those which spin vertically oriented orb webs.

OROGENY: The raising of mountains through folding and faulting of the earth's crust.

OUTCROP: An exposure of bare rock; a place where underlying rock protrudes from the soil.

OVATE: Egg-shaped; when applied to leaves, ovate indicates the broader end is nearer the base. Compare OBOVATE.

PAPPUS: The wispy airfoil that carries seeds of some asteraceous plants on the wind.

PEDICEL: The stalk of a single flower in a cluster.

PEDUNCLE: The stalk of a flower cluster.

PENEPLAIN: A large expanse of gently rolling land. The Latin *pene* means 'almost'; hence, almost a flat plain, i.e., preceding a flat plain in the erosion cycle.

PERENNIAL: A plant living three or more years that dies back seasonally and produces new growth from a root system.

PERIANTH: The combined CALYX and COROLLA.

Petiole: A leaf stalk.

Petrosere: Plant succession on a rocky surface.

pH: The measure of acidity or alkalinity. Values between pH 0 and pH 7 (neutral) are *acidic*; those between pH 7 and pH 14 are *basic*. Values lower than pH 4.5 and higher than pH 9.0 are uncommon in soils. The range between pH 6.5 and pH 7.4 is said to be *circumneutral* and is hospitable to most plants.

Phylogeny: The evolutionary development of a genetically related group of organisms.

Pinnate: Having the leaflets (of a COMPOUND LEAF) placed at opposite points on the RACHIS.

Piscivorous: Fish-eating.

Pistil: The female sexual organ of a flower.

Plankton: The very small organisms adrift in fresh or marine waters. Plankton are incapable of locomotion.

Plastron: The bottom shell of a turtle or tortoise.

Pome: A fleshy fruit with several embedded seeds (e.g., an apple).

Prevernal: The aspect of the biological year during late winter and early spring when the plumbing of woody plants is reactivated, buds swell, and spring ephemeral wildflowers bloom.

Primary feathers: The flight feathers anchored to the last joint of a bird's wing and providing the majority of thrust.

Primary producer: *See* AUTOTROPH.

Profundal: Of or pertaining to that portion of a lake or pond outside the littoral zone, where depth is great enough to exclude rooted plants.

Raceme: An elongated INFLORESCENCE with stalked flowers.

Rachis: A stem bearing flowers or leaflets, as in a COMPOUND LEAF.

Reniform: Kidney-shaped.

Residual soil: That which is formed in place; compare ALLUVIAL SOIL.

Rhizome: An underground stem producing sprouts and roots at intervals.

Riparian: Of or relating to or living on the bank of a river.

Riverine: Synonymous with riparian.

Samara: A dry, winged fruit (as of ash or elm) that does not open to release its seed.

Saprolite: Rock which has decomposed in place.

Saprophyte: A plant that lives on dead or decaying matter.

Savannah: Grassland with scattered trees. Not found in the Piedmont except on certain serpentine barrens.

Scarify: To abrade or weaken chemically or mechanically.

Scats: Feces.

Scatology: To the naturalist, inquiry into the territorial habits and diets of animals through examination of the scats.

Scree: Rock fragments at the base of a cliff. Synonymous with *talus*.

Scutellum: A small, horny or bony plate. A triangular scutellum pointing rearward from the head identifies true bugs (hemipterans).

Secondary feathers: Flight feathers anchored to the next-to-last section of a bird's wing.

Sepal: One of the modified leaves making up a calyx.

Sere: The plant succession process at a specific site.

Serotinal: Late summer and early fall. The aspect of the biological year colloquially called Indian summer and stemming from the adjective *serotine*, meaning 'late blooming'.

Serpentine: Rock containing high concentrations of hydrated magnesium silicate.

Serrate: A leaf having sharp, forward-pointing teeth.

Shrub: A woody perennial usually having multiple main stems and being 30 feet (9 meters) or less tall. Compare Tree.

Sinus: A deep indentation in the margin of a leaf.

Spadix: A fleshy spike in which florets are embedded (as in golden club).

Spathe: A large bract around an inflorescence (as in skunk cabbage).

Spore: The reproductive body (asexual) of a primitive plant or animal.

Sporophyll: A spore-bearing leaf or other structure.

STAMEN: The male or pollen-bearing part of a flower.

STRIDULATION: A shrill grating or chirping sound made by certain insects by scraping together parts of the body.

SUBCANOPY: The vegetative layer immediately below the CANOPY.

SUBDOMINANT: Existing vertically below the layer of DOMINANT vegetation.

SUBSTRATE: A support or base; a layer beneath.

SUCCESSION: In botany, the natural process by which vegetative communities replace one another until a stable CLIMAX is reached.

TAIGA: The CONIFEROUS forests of the far north in North America and Eurasia.

TALUS: *See* SCREE.

TECTONICS: The study of the earth's crust and the forces that change it.

TELSON: The terminal portion of the segmented body of an ARTHROPOD.

THERMOCLINE: A layer separating the warm waters near the surface from the cold waters near the bottom of a deep body of water. Convection currents do not circulate across the thermocline.

TILL: Unstratified, unsorted rocks, gravel, and debris deposited by a glacier. *See* MORAINE.

TOMENTUM: Dense, wooly hair (TRICHOMES) on the surface of a plant.

TRANSGRESSIVE: A plant found in a SUBDOMINANT vegetative layer through which it is passing, generally on its way to the CANOPY.

TRAPROCK: A layer of rock trapped between two others; e.g., the Triassic rocks of the Watchungs and the Palisades on the Hudson River.

TREE: A tall woody plant with a single main stem. Compare SHRUB.

TRICHOME: A bristle.

TYMPANUM: A drumlike structure covered by a vibratory membrane and serving as the receptor for sound in the hearing mechanisms of animals. In frogs it is a large visible disc rearward of the eye.

TYPE LOCATION: The location where the *type specimen* (from which a particular species was originally described) was collected.

UMBEL: An INFLORESCENCE with the PEDICELS or PEDUNCLES or both arising from a common point; e.g., the inflorescence of Queen Anne's lace. A member of the botanic family Umbeliferae.

UNGULATE: A hoofed animal.

UROPOD: A leaflike flattened appendage of the last abdominal segment of various crustaceans.

VALVE: A shell (as in *bivalve*).

VASCULAR PLANTS: Generally, the higher plants; that is, those with differentiated tissues that conduct fluids and synthesize food. Specifically, the Embryophyta (one of two plant subkingdoms), including the mosses, ferns, club mosses, conifers and flowering plants, are vascular plants, while fungi and bacteria are not.

VEGETATIVE REPRODUCTION: Reproduction by asexual means, generally by sending up numerous stems from a spreading root system.

VENTER: The belly or abdomen.

VERNAL: The aspect of the year following the prevernal and characterized by the flowering and growth of most plants.

XERIC: Dry.

XEROPHYTE: A plant with minimal requirements for moisture.

XEROSERE: Plant succession on dry soils.

A Piedmont Bibliography

Borror, Donald J., and White, Richard E. *A Field Guide to the Insects of America North of Mexico.* Boston: Houghton Mifflin, 1970.

Braun, E. Lucy. *Deciduous Forests of Eastern North America.* New York: Hafner Press, 1950

Brown, Andrew. *Geology and the Gettysburg Campaign.* Pennsylvania Topographic and Geologic Service, Educational Series no. 5. Harrisburg: Pennsylvania Department of Environmental Resources, 1962.

Burt, W. H., and Grossenheider, R. P. *A Field Guide to the Mammals.* Boston: Houghton Mifflin, 1952.

Butler, J. R.; Custer, Edward S.; and White, W. A. *Potential Geological Natural Landmarks: Piedmont Region, Eastern United States.* Chapel Hill: University of North Carolina, Department of Geology, 1975.

Chu, H. F. *How to Know the Immature Insects.* Dubuque, Iowa: Wm. C. Brown, 1949.

Claibourne, Robert. *Climate, Man and History.* New York: W. W. Norton, 1970.

Cobb, Broughton. *A Field Guide to the Ferns.* Boston: Houghton Mifflin, 1956.

Coker, R. E. *Streams, Lakes and Ponds.* Chapel Hill: University of North Carolina Press, 1954.

Comstock, Anne Botsford. *Handbook of Nature Study.* Ithaca, N.Y.: Cornell University Press, 1939.

Conant, Roger. *A Field Guide to the Reptiles and Amphibians.* Boston: Houghton Mifflin, 1958.

Deininger, R. W., et al. *Alabama Piedmont Geology.* Alabama Geological Society, 1964.

Egler, Frank E. *The Plight of the Right-of-way Domain: Victim of Vandalism.* Mt. Kisco, N.Y.: Futura Media Service, 1975.

Evans, Howard Ensign. *Wasp Farm.* New York: Doubleday, 1973.

Farb, Peter. *The Face of North America.* New York: Harper & Row, 1963.

Fenneman, N. M. *Physiography of the Eastern United States.* New York: McGraw-Hill, 1938.

Godfrey, Michael A. *Winter Birds of the Carolinas and Nearby States.* Winston-Salem, N.C.: John Blair, Publisher, 1977.

Hitchcock, A. S. *Manual of the Grasses of the United States.* Reprint of 2d (1950) ed., rev. by Agnes Chase. New York: Dover Publications, 1971.

Hunt, C. B. *Natural Regions of the U.S. and Canada.* San Francisco: W. H. Freeman, 1967.

Keever, Catherine. "Causes of Succession on Old Fields of the Piedmont, North Carolina." In Ecological Society of America, *Ecological Monographs* 20 (1950): 229–50.

Klots, Alexander B. *A Field Guide to the Butterflies.* Boston: Houghton Mifflin, 1951.

Lapham, D. M. "Geology of the Cedar Hill Serpentine Quarry." In *Guidebook,* 25th Field Conference of Pennsylvania Geologists, pp. 35–38. Harrisburg: Pennsylvania Geological Society.

Levi, H. W., and Levi, L. R. *Spiders and Their Kin.* New York: Golden Press, 1968.

Oosting, Henry J. "An Ecological Analysis of the Plant Communities of Piedmont, North Carolina." In *American Midland Naturalist* 28 (1942): 1–26.

Peet, Robert K. *A Bibliography of the Vegetation of the Carolinas.* Chapel Hill: University of North Carolina, Department of Botany, 1979.

Radford, A. E. *Natural Areas of the Southeastern United States: Field Data and Information.* Chapel Hill: University of North Carolina, Department of Botany, 1976.

Radford, A. E.; Ahles, H. E.; and Bell, C. R. *Manual of the Vascular Flora of the Carolinas.* Chapel Hill: University of North Carolina Press, 1964.

Radford, A. E., and Martin, David L. *Potential Ecological Natural Landmarks: Piedmont Region, Eastern United States.* Chapel Hill: University of North Carolina, Department of Botany, 1975.

Reid, George K. *Pond Life.* New York: Golden Press, 1967.

Robbins, Chandler S.; Brunn, Bertel; Zim, Herbert; and Singer, Arthur. *Birds of North America*. New York: Golden Press, 1966.

Robichaud, Beryl, and Buell, Murray F. *Vegetation of New Jersey: A Study of Landscape Diversity*. New Brunswick, N.J.: Rutgers University Press, 1973.

Ruffner, James A. *Climates of the United States*. Detroit, Mich.: Gale Research Co., 1978.

Shelton, John S. *Geology Illustrated*. San Francisco: W. H. Freeman, 1966.

Shepps, Vincent C. *Pennsylvania and the Ice Age*. Pennsylvania Topographic and Geologic Service, Educational Series no. 6. Harrisburg: Pennsylvania Department of Environmental Resources, 1962.

Sheve, Forest, et al. *Plant Life of Maryland*. Baltimore, Md.: Johns Hopkins University Press, 1910.

Thomas, Bill. *The Swamp*. New York: W. H. Norton, 1976.

Usinger, Robert L. *The Life of Rivers and Streams*. New York: McGraw-Hill, 1967.

Van Dersal, William R. *Native Woody Plants of the United States*. U.S. Department of Agriculture, Miscellaneous Publication no. 303. Washington: U.S. Government Printing Office, 1938.

Wells, B. W. *The Natural Gardens of North Carolina*. Chapel Hill: University of North Carolina Press, 1932.

White, W. A. *The Blue Ridge Front: A Fault Scarp*. Chapel Hill: University of North Carolina, Department of Geology. "Origin of the Granite Domes in the Southeastern Piedmont." *Journal of Geology* 53 (1945): 276–82.

Willard, Bradford. *Pennsylvania Geology Summarized*. Pennsylvania Topographic and Geologic Service, Educational Series no. 4. Harrisburg: Pennsylvania Department of Environmental Resources, 1962.

Zim, H. S., and Shoemaker, H. H. *Fishes*. New York: Golden Press, 1955.

INDEX

E

H

Q

R

S

Z

MEASURING SCALE IN CENTIMETERS AND MILLIMETERS

1 2 3 4 5 6 7 8 9 10 11 12 13 14 15 16 17 18

1

2

3

4

5

6

7

MEASURING SCALE IN INCHES